Heavy Metals in Mushrooms

Heavy Metals in Mushrooms

Editors

Miha Humar
Ivan Širić

MDPI

Basel • Beijing • Wuhan • Barcelona • Belgrade • Novi Sad • Cluj • Manchester

Editors
Miha Humar
Department of Wood Science and Technology
University of Ljubljana - Biotehnical Faculty
Ljubljana
Slovenia

Ivan Širić
Department of Animal Science and Technology
University of Zagreb Faculty of Agriculture
Zagreb
Croatia

Editorial Office
MDPI
St. Alban-Anlage 66
4052 Basel, Switzerland

This is a reprint of articles from the Special Issue published online in the open access journal *Journal of Fungi* (ISSN 2309-608X) (available at: www.mdpi.com/journal/jof/special_issues/heavy_metals_mushrooms).

For citation purposes, cite each article independently as indicated on the article page online and as indicated below:

Lastname, A.A.; Lastname, B.B. Article Title. *Journal Name* **Year**, *Volume Number*, Page Range.

ISBN 978-3-0365-9813-0 (Hbk)
ISBN 978-3-0365-9812-3 (PDF)
doi.org/10.3390/books978-3-0365-9812-3

Contents

About the Editors

Miha Humar

Miha Humar graduated from Wood Technology and then joined the Biotechnical Faculty at the University of Ljubljana as a research assistant, where he completed his PhD studies as the best PhD student in his year and was awarded the Jesenko Prize of the Biotechnical Faculty. After completing his PhD, Miha Humar first gave tutorials on wood pests and wood protection, and, later, he also started lecturing on these topics. In 2015, he was elected full professor in the field of wood pathology and wood protection. From 2010 to 2022, he held management positions at the Department of Wood Science and the Biotechnical Faculty. In 2016 and 2018, as Dean of the Faculty of Forestry and Wood Technology, he was responsible for the work of one of Slovenia's largest and most distinguished faculties. In 2023, Miha Humar was elected into Slovenian Academy of Science and Art as an associate member. Under his mentorship, more than 220 students graduated and 8 obtained their PhDs. Miha Humar is also active in the international field as a member of scientific committees of conferences and a member of international editorial boards. Together with Hojka Kraigher, he co-organised the scientific meeting called Forest and Wood for twelve years, and, in 2022, he organised the World Conference on Wood Protection (IRG/WP) in Bled. In addition to teaching, he is also active in research. He has coordinated several applied and basic projects and a research programme. He has published his scientific findings in more than 220 scientific papers. He is also a co-author of two international patents. His current research work addresses problems of wood life cycle assessment, as well as the development of classical biocidal and non-biocidal solutions for wood protection. For these purposes, he has set up the first comprehensive field trials. He works in close collaboration with the industry.

Ivan Širić

Ivan Širić graduated from the Faculty of Agriculture at the University of Zagreb on March 25, 2009. Since February 15, 2011, he has been employed at the Faculty of Agriculture as an asssistant novice researcher, where he is enrolled in the postgraduate doctoral programme "Agricultural Sciences". On October 10, 2014, he wrote his doctoral thesis entitled "Heavy metals in edible wild mushrooms in the area of northern and coastal Croatia". He is actively involved in teaching at the University of Zagreb, Faculty of Agriculture, at undergraduate, graduate, and doctoral levels, and at the University of Zadar, Department of Ecology, Agronomy and Aquaculture, at the graduate level. So far, he has supervised 35 theses and 17 doctoral theses and is a mentor of one doctoral thesis. Since the beginning of his work at the faculty, he has been involved in numerous research projects in the fields of ecology and environmental protection, mushrooms and mushroom cultivation, bioindicator properties of mushrooms, food safety, and quality. He received scientific education at the Biotechnical Faculty of the University of Ljubljana in the field of ecology and environmental protection with a focus on the monitoring and distribution of heavy metals in the ecosystem. He leads several scientific, professional, and technological projects in the country and collaborates on projects and studies in the field of ecology and environmental protection. He has been Editor-in-Chief of the proceedings and abstracts of agronomist symposia on several occasions. He is Guest Editor of Special Issues of renowned journals: *Chemosphere* (Q1), *Journal of Fungi* (Q1), and *Forests* (Q1). He is a member of the international research group SAEIRG (Sustainable Agro-Environment International Research Group). He was awarded the State Prize for Science in the year 2022.

Preface

Environmental pollution is one of the biggest issues today. The great progress of human civilization, achieved mainly through the development of new technologies, has inevitably led to the production of waste and pollutants, especially heavy metals. Their production is now so great that the ecological balance of many ecosystems has been shaken. As many forms of pollution spread over large areas, it is not enough to tackle the problem at a local level alone, but its solution, or rather containment, requires a comprehensive global commitment at many levels of human planning and action. Previous studies of pollutant concentrations in the environment have shown that many saprophytic and ectomycorrhizal fungal species are very sensitive to contamination of the substrate and water with heavy metals and metalloids. As mushrooms are a source of nutrients but also a common source of toxic metals and metalloids, the health consequences of toxic heavy metals from the consumption of edible mushrooms cannot be thoroughly assessed as data on bioavailability in humans are very limited.

Therefore, it is necessary to constantly monitor and control the levels of metals and metalloids in the human diet, both in foods of plant and animal origin and in food supplements, or in this case wild mushrooms, whose general popularity and availability on the market are constantly increasing. This Special Issue contains contributions on various topics, including studies on the following aspects: molecular tolerance mechanisms of fungi to metals, ectomycorrhizal fungi in heavy metal accumulation processes, the mycoremediation ability of fungi, the bioindicator properties of fungi, the estimation of heavy metal concentration in fungi and the effects on human health, and heavy metal concentrations in cultivated fungi. This reprint also provides a good insight into the development of science and technology related to the mentioned topics.

Miha Humar and Ivan Širić
Editors

Journal of *Fungi*

MDPI

Editorial

Preface to the Special Issue 'Heavy Metals in Mushrooms'

Ivan Širić [1,*] and Miha Humar [2]

1 Faculty of Agriculture, University of Zagreb, Svetosimunska 25, 10000 Zagreb, Croatia
2 Biotechnical Faculty, University of Ljubljana, Jamnikarjeva 101, 1000 Ljubljana, Slovenia; miha.humar@bf.uni-lj.si
* Correspondence: isiric@agr.hr

Population growth, intensive industrialization and urbanization have led to environmental pollution, especially soil and water pollution. The increasing presence of environmental pollutants, particularly heavy metals, has become an alarming problem. Heavy metals in the environment are often carcinogenic, mutagenic, neurotoxic and immunogenic. Thus, they enter the food chain, where they bioaccumulate and remain in the human body for a long time. According to the World Health Organization, lead poisoning causes neurological damage in humans, which manifests itself as a decline in intelligence, short-term memory loss, and problems with learning and movement coordination. Arsenic poisoning causes cardiovascular problems associated with skin cancer and other skin diseases, peripheral neuropathies and kidney damage. Cadmium also accumulates in the kidneys and lungs, and mercury damages the nervous system, causes uncontrolled tremors, muscle damage, partial vision loss and deformities in children. Due to the increasing presence of heavy metals and toxic elements in the environment, numerous scientific studies are being conducted worldwide to preserve and protect the environment. One of the first steps in solving the problem of environmental pollution is to monitor the presence and concentration of pollutants, especially heavy metals. To date, the study of pollutant concentrations in the environment has revealed that saprotrophic and ectomycorrhizal fungal species have a high sensitivity to the contamination of substrates and water with heavy metals and metalloids. The accumulation of Fe, Zn, Cu, Ni, Cr, Pb, Cd and Hg in saprophytic and ectomycorrhizal fungi is a complex phenomenon, which depends on numerous external factors, mechanisms within the fungi, their interactions, and genetic characteristics of the species. Fungal mycelium can accumulate in all heavy metals in a much higher concentration than the substrate on which it develops and lives. The density and depth of mycelium, which lives in the soil for several months or years, influence the content of heavy metals in the fruiting bodies of fungi. In addition, various ecological factors and soil properties (pH and organic matter) can influence the concentration of heavy metals in fungi.

The first article of this Special Issue addresses the molecular tolerance mechanism of *Paraisaria dubia* to Zn^{2+} stress and the possible application of *Paraisaria dubia* in the bioremediation of heavy metal pollution [1]. The second article analyzes the ectomycorrhizal communities in a former uranium mine and communities under experimental conditions to improve plants' tolerance to certain abiotic conditions [2]. The third article deals with the possibility of using *Yarrowia lipolytica* to remove heavy metals from woods treated with chromated copper arsenate (CCA) [3]. The fourth article describes a risk assessment of total mercury content in edible wild mushrooms in Slovakia, while the fifth article evaluates the concentration of six heavy metals in two edible oyster mushrooms [4,5]. Furthermore, the sixth article examines the mycomedial ability of *Agaricus bisporus* to grow on compost mixed with flotation residues [6]. The seventh article highlights the resistance of white and brown rot fungi to copper in the microcosm of copper/azole-treated wood [7]. The eighth article deals with the health risk assessment of cadmium accumulation in three fungal species in central and coastal Croatia [8]. In contrast, the ninth article underscores

Citation: Širić, I.; Humar, M. Preface to the Special Issue 'Heavy Metals in Mushrooms'. *J. Fungi* **2023**, *9*, 1163. https://doi.org/10.3390/jof9121163

Received: 27 November 2023
Accepted: 29 November 2023
Published: 3 December 2023

the mechanisms of cadmium bioconcentration using *Boletus griseus* and potential bioremediation fungi for alleviating cadmium contamination [9]. The tenth article investigates the exposure to essential and toxic elements upon the consumption of five edible wild mushroom species collected in southern Spain and northern Morocco [10]. The eleventh article reports on a spatial assessment of the concentration of eight toxic elements in *Agaricus bisporus* collected from local markets in India [11]. The twelfth article describes the synergy between agriculture and waste management, using sewage sludge as casing material for the cultivation of mushrooms (*Agaricus bisporus*) [12].

In conclusion, a global heavy metal monitoring program should be established to determine heavy metal concentrations for certain and commonly consumed saprophytic and ectomycorrhizal fungal species from the genera Agaricus, Boletus, Leccinum, Lactarius, Macrolepiota, Suillus, Tricholoma, etc. In addition, such a program should be intensified near significant sources of heavy metal emissions, such as smelters, power plants and major roads. Furthermore, there is a need for constant monitoring and controlling the levels of metals and metalloids in human diet, both in foods of plant and animal origin and in food supplements; in this case, it is edible wild mushrooms, whose availability on the market and general popularity are constantly increasing.

Author Contributions: Conceptualization, M.H. and I.Š.; data curation, M.H. and I.Š.; resources, M.H. and I.Š.; supervision, M.H. and I.Š.; validation, M.H. and I.Š.; visualization, M.H. and I.Š.; writing—original draft, I.Š.; writing—review and editing, M.H. and I.Š. All authors have read and agreed to the published version of the manuscript.

Funding: This research received no external funding.

Conflicts of Interest: The authors declare no conflict of interest.

References

1. Wang, Y.; Tong, L.-L.; Yuan, L.; Liu, M.-Z.; Du, Y.-H.; Yang, L.-H.; Ren, B.; Guo, D.-S. Integration of Physiological, Transcriptomic and Metabolomic Reveals Molecular Mechanism of *Paraisaria dubia* Response to Zn^{2+} Stress. *J. Fungi* **2023**, *9*, 693. [CrossRef] [PubMed]
2. Bogdanova, O.; Kothe, E.; Krause, K. Ectomycorrhizal Community Shifts at a Former Uranium Mining Site. *J. Fungi* **2023**, *9*, 483. [CrossRef] [PubMed]
3. Xing, D.; Magdouli, S.; Zhang, J.; Bouafif, H.; Koubaa, A. A Comparative Study on Heavy Metal Removal from CCA-Treated Wood Waste by *Yarrowia lipolytica*: Effects of Metal Stress. *J. Fungi* **2023**, *9*, 469. [CrossRef] [PubMed]
4. Šnirc, M.; Jančo, I.; Hauptvogl, M.; Jakabová, S.; Demková, L.; Árvay, J. Risk Assessment of the Wild Edible *Leccinum* Mushrooms Consumption According to the Total Mercury Content. *J. Fungi* **2023**, *9*, 287. [CrossRef] [PubMed]
5. Širić, I.; Kumar, P.; Adelodun, B.; Abou Fayssal, S.; Bachheti, R.K.; Bachheti, A.; Ajibade, F.O.; Kumar, V.; Taher, M.A.; Eid, E.M. Risk Assessment of Heavy Metals Occurrence in Two Wild Edible Oyster Mushrooms (*Pleurotus* spp.) Collected from Rajaji National Park. *J. Fungi* **2022**, *8*, 1007. [CrossRef] [PubMed]
6. Budzyńska, S.; Siwulski, M.; Budka, A.; Kalač, P.; Niedzielski, P.; Gąsecka, M.; Mleczek, M. Mycoremediation of Flotation Tailings with *Agaricus bisporus*. *J. Fungi* **2022**, *8*, 883. [CrossRef] [PubMed]
7. Pandharikar, G.; Claudien, K.; Rose, C.; Billet, D.; Pollier, B.; Deveau, A.; Besserer, A.; Morel-Rouhier, M. Comparative Copper Resistance Strategies of *Rhodonia placenta* and *Phanerochaete chrysosporium* in a Copper/Azole-Treated Wood Microcosm. *J. Fungi* **2022**, *8*, 706. [CrossRef] [PubMed]
8. Širić, I.; Kumar, P.; Eid, E.M.; Bachheti, A.; Kos, I.; Bedeković, D.; Mioč, B.; Humar, M. Occurrence and Health Risk Assessment of Cadmium Accumulation in Three *Tricholoma* Mushroom Species Collected from Wild Habitats of Central and Coastal Croatia. *J. Fungi* **2022**, *8*, 685. [CrossRef] [PubMed]
9. Tian, Z.; Wang, Y.; Zhuang, Y.; Mao, C.; Shi, Y.; Sun, L. Fungus–Fungus Association of *Boletus griseus* and *Hypomyces chrysospermus* and Cadmium Resistance Characteristics of Symbiotic Fungus *Hypomyces chrysospermus*. *J. Fungi* **2022**, *8*, 578. [CrossRef] [PubMed]
10. Barea-Sepúlveda, M.; Espada-Bellido, E.; Ferreiro-González, M.; Bouziane, H.; López-Castillo, J.G.; Palma, M.; Barbero, G.F. Exposure to Essential and Toxic Elements via Consumption of *Agaricaceae*, *Amanitaceae*, *Boletaceae*, and *Russulaceae* Mushrooms from Southern Spain and Northern Morocco. *J. Fungi* **2022**, *8*, 545. [CrossRef] [PubMed]

11. Kumar, P.; Kumar, V.; Eid, E.M.; AL-Huqail, A.A.; Adelodun, B.; Abou Fayssal, S.; Goala, M.; Arya, A.K.; Bachheti, A.; Andabaka, Ž.; et al. Spatial Assessment of Potentially Toxic Elements (PTE) Concentration in *Agaricus bisporus* Mushroom Collected from Local Vegetable Markets of Uttarakhand State, India. *J. Fungi* **2022**, *8*, 452. [CrossRef] [PubMed]
12. Kumar, P.; Kumar, V.; Adelodun, B.; Bedeković, D.; Kos, I.; Širić, I.; Alamri, S.A.M.; Alrumman, S.A.; Eid, E.M.; Abou Fayssal, S.; et al. Sustainable Use of Sewage Sludge as a Casing Material for Button Mushroom (*Agaricus bisporus*) Cultivation: Experimental and Prediction Modeling Studies for Uptake of Metal Elements. *J. Fungi* **2022**, *8*, 112. [CrossRef] [PubMed]

Journal of *Fungi*

MDPI

Article

Integration of Physiological, Transcriptomic and Metabolomic Reveals Molecular Mechanism of *Paraisaria dubia* Response to Zn^{2+} Stress

Yue Wang, Ling-Ling Tong, Li Yuan, Meng-Zhen Liu, Yuan-Hang Du , Lin-Hui Yang, Bo Ren * and Dong-Sheng Guo *

School of Food Science and Pharmaceutical Engineering, Nanjing Normal University, Nanjing 210023, China
* Correspondence: boren@nnu.edu.cn (B.R.); guodongs@njnu.edu.cn (D.-S.G.)

Abstract: Utilizing mycoremediation is an important direction for managing heavy metal pollution. Zn^{2+} pollution has gradually become apparent, but there are few reports about its pollution remediation. Here, the Zn^{2+} remediation potential of *Paraisaria dubia*, an anamorph of the entomopathogenic fungus *Ophiocordyceps gracilis*, was explored. There was 60% Zn^{2+} removed by *Paraisaria dubia* mycelia from a Zn^{2+}-contaminated medium. To reveal the Zn^{2+} tolerance mechanism of *Paraisaria dubia*, transcriptomic and metabolomic were executed. Results showed that Zn^{2+} caused a series of stress responses, such as energy metabolism inhibition, oxidative stress, antioxidant defense system disruption, autophagy obstruction, and DNA damage. Moreover, metabolomic analyses showed that the biosynthesis of some metabolites was affected against Zn^{2+} stress. In order to improve the tolerance to Zn^{2+} stress, the metabolic mechanism of metal ion transport, extracellular polysaccharides (EPS) synthesis, and microcycle conidiation were activated in *P. dubia*. Remarkably, the formation of microcycle conidiation may be triggered by reactive oxygen species (ROS) and mitogen-activated protein kinase (MAPK) signaling pathways. This study supplemented the gap of the Zn^{2+} resistance mechanism of *Paraisaria dubia* and provided a reference for the application of *Paraisaria dubia* in the bioremediation of heavy metals pollution.

Keywords: *Paraisaria dubia*; Zn^{2+} stress response; transcriptomic; metabolomic; metal ion transport; microcycle conidiation

Citation: Wang, Y.; Tong, L.-L.; Yuan, L.; Liu, M.-Z.; Du, Y.-H.; Yang, L.-H.; Ren, B.; Guo, D.-S. Integration of Physiological, Transcriptomic and Metabolomic Reveals Molecular Mechanism of *Paraisaria dubia* Response to Zn^{2+} Stress. *J. Fungi* **2023**, *9*, 693. https://doi.org/10.3390/jof9070693

Academic Editors: Miha Humar and Ivan Širić

Received: 25 May 2023
Revised: 13 June 2023
Accepted: 14 June 2023
Published: 21 June 2023

1. Introduction

Human activities such as mining, smelting, urban waste treatment, and electroplating continuously release heavy metal pollutants into water and soil, damaging the ecological environment [1]. In recent years, the issue of heavy metal pollution has been constantly emerging, and the problem of Zn^{2+} pollution has gradually become apparent, arousing people's attention. A high concentration of Zn^{2+} (16 mg/g) was found in soil near the smelting site [2]. Total soil Zn^{2+} concentrations in excess of 3 mg/g dry soil have been reported in contaminated agricultural fields [3,4]. The guideline values of Zn^{2+} in agriculture soil range from 120 to 250 mg/kg reported by various environmental departments (CCME, USEPA, EPA, etc.) [5]. Zn^{2+} is an essential trace element for plant growth and development, and the optimal Zn^{2+} level for healthy plants is 20–60 mg/kg dry weight [6]. However, above supra-optimal concentrations, this essential metal can have lethal effects on plant biology. Excessive Zn^{2+} inhibits seed germination, plant growth, and root development, disturbs photosynthesis, reduces the relative water content, and competes with other ions to disturb the physiological balance [7–9]. In addition, at the cellular level, excessive Zn^{2+} affects membrane integrity and permeability, causing oxidative damage and can even result in cell death [10]. Zn^{2+} concentration within the range of 25–150 mg/kg in plants is sufficient/tolerable, but when Zn^{2+} concentration reaches 100–400 mg/kg dry weight, the yield of plants will be reduced [5]. Zn^{2+} pollution damages plant growth, destroys the ecological environment, and hinders agricultural production. Therefore, the management of Zn^{2+} pollution is of great significance. Mycoremediation is a promising and effective method to control metal pollution, which is based on the bioaccumulation and biodegradation of macrofungi to eliminate toxic pollutants from the environment. Macrofungi can survive and accumulate large amounts of heavy metals from polluted environments, such as lead, copper, cadmium, chromium, and hydrargyrum [11]. In addition, previous studies have found a significant increase in the frequency of entomogenous fungi in the soil microfungal community of heavy metal-contaminated soil [12], suggesting that entomogenous fungi may be an excellent potential biomaterial for cleaning up Zn^{2+} pollution.

Cordyceps is a unique group of macrofungi that parasitizes insect larvae and pupae. More than 750 species have been reported, belonging to the order Ascomycota and classified into three families: Cordycipitaceae, Ophiocordycipitaceae, and Clavicipitaceae [13–15]. *Ophiocordyceps* (Ophiocordycipitaceae) is one of the most populous genera of entomopathogenic fungi, of which *Ophiocordyceps gracilis* (*O. gracilis*) is well-known traditional Chinese medicine [16]. *O. gracilis* is mainly distributed in China, Europe, and the Americas, a development that is generating more and more attention due to its excellent medicinal value [16,17]. The mycelia of *Paraisaria dubia* (*P. dubia*) is the asexual stage of *O. gracilis*, which is artificially cultivated as a substitute for natural *O. gracilis* [18]. It has been reported that the content of heavy metals in wild *Cordyceps* exceeds the standard, such as plumbum, hydrargyrum, and chromium [19]. This phenomenon indicates that heavy metal pollution has extended to the uninhabited wild environment, and the environmental pollution problem is becoming increasingly serious. Further, it also indicates that *Cordyceps* has a high capacity for heavy metal enrichment and is an ideal biological material to remediate heavy metal pollution [20]. However, there are no reports on the use of *Cordyceps* for bioremediation of heavy metal pollution, as previous studies have primarily focused on bioactive substances (polysaccharides, active peptides, cordycepin, etc.), pharmacologic effects, and artificial cultivation of *Cordyceps* [21–24]. Furthermore, the understanding of the molecular mechanism underlying heavy metal tolerance in *Cordyceps* remains limited, hampering the application of *Cordyceps* in heavy metal pollution control. Transcriptomics and metabolomics technologies offer efficient means to uncover information regarding adverse substance-induced molecular perturbations in cells and tissues. Moreover, the integration of transcriptomics and metabolomics has been employed to elucidate the molecular mechanisms in various macrofungi, including *Lentinula edodes*, *Ganoderma lucidum*, and *Agrocybe aegerita* [25]. In this study, *P. dubia* was used to investigate its tolerance capacity to Zn^{2+} stress, thereby expanding the application of mycoremediation in heavy metal pollution. Furthermore, by utilizing transcriptomics and metabolomics, the molecular tolerance mechanism underlying Zn^{2+} stress was explored, aiming to fill the knowledge gap concerning the molecular level of *Cordyceps*' tolerance to Zn^{2+} stress.

In this study, a comprehensive analysis of physiological response, transcriptomics, and metabolomics was performed to gain insight into the tolerance of *P. dubia* to Zn^{2+} stress. The objects of this study were to (a) evaluate *P. dubia*'s physiological response under Zn^{2+} stress by measuring dry mycelia biomass, intracellular polysaccharides (IPS), extracellular polysaccharides (EPS), microcycle conidiation (MC) and Zn^{2+} removal efficiency; (b) identify pivotal differentially expressed genes (DEGs), differential metabolites and their involved essential pathways; (c) reveal the molecular mechanisms of tolerance of *P. dubia* response to Zn^{2+} stress. This study provided valuable tolerance molecular information of *P. dubia* under Zn^{2+} stress, instructing its potential application of heavy metals remediation.

2. Materials and Methods

2.1. Fungal Strain and Zn^{2+} Treatment

The strain of *P. dubia* (CGMCC No. 20731) used in this study is preserved in the China General Microbiological Culture Collection Center. *P. dubia* was cultured in different liquid mediums. The control group was cultured in ordinary medium (2% glucose, 1% peptone, 0.2% KH_2PO_4, 0.1% $MgSO_4{\cdot}7H_2O$) [16], while the treatment group was cultured in

Zn^{2+}-contaminated medium (2% glucose, 1% peptone, 0.2% KH_2PO_4, 0.1% $MgSO_4 \cdot 7H_2O$, 0.1% $ZnSO_4 \cdot 7H_2O$). A 250 mL shake flask with 100 mL medium was cultured at 20 °C, 120 rpm. Three biological replicates were conducted per treatment condition. After being cultured for 10 days, mycelia pellets were separated from the fermentation broth containing spores by 200 mesh press cloth. Then, the fermentation broth was centrifuged at 8000 rpm for 10 min to take the precipitate to obtain spore powder. The mycelia and spores were washed three times with deionized water. Then place them in a freeze-drying machine (SCIENTZ-10N, NingBo Scientz Biotechnology, Zhejiang, China) to dry at −25 °C, 12 Pa. The weight of mycelia was recorded as biomass. In addition, mycelia and spores were ground into powder for subsequent experiments.

2.2. Microscopic Morphological Analysis

Micromorphological characteristics of *P. dubia* were observed under a conventional optical microscope (Olympus BX43, Tokyo, Japan), scanning electron microscopy (SEM) (SU8010 Tokyo, Japan) and transmission electron microscopy (TEM) (H-7650, Tokyo, Japan). Samples for SEM and TEM analysis were prepared according to the method described by Bo [26].

2.3. Preparation of IPS and EPS

The mycelia powder (0.5 g) of Zn^{2+} stress and the control were used for the extraction of IPS, and their fermentation broth was used for the extraction of EPS. The extraction method was conducted as described before [16], with minor modifications, as follows. Mycelia powder was extracted three times with ultrapure water (1:20, *w*/*v*) at 90 °C for 4 h. The mycelia extracting solution and fermentation broth were precipitated with a four-fold volume of 95% ethanol at 4 °C for 16 h and centrifuged at 8000× *g* for 10 min to obtain IPS and EPS. Then, the total polysaccharide content was determined using the phenol-sulfuric acid method.

2.4. Zn^{2+} Concentration Determination and Removal Rate

The content of Zn^{2+} in the mycelia, EPS, and conidia was determined by inductively coupled plasma-optical emission spectrometry/mass spectrometer (ICP-OES/MS, Thermo ICAP PRO) [27]. The concentration of Zn^{2+} in the samples was obtained by measuring the standard solution of Zn^{2+} and comparing which standard curve. The removal rate of Zn^{2+} was calculated using the following equation:

$$Zn^{2+} \text{ removal rate } (\%) = (C_2 + C_3 + C_4)/C_1 \times 100\% \quad (1)$$

C_1 was the initial content in fermentation broth (22.8 mg/100 mL). C_2, C_3 and C_4 were the content of Zn^{2+} in mycelia, conidia, and EPS obtained from 100 mL fermentation broth, respectively.

2.5. Nontargeted Metabolomic Profiling Analysis

Freeze-dried mycelia of Zn^{2+} stress and control ($n = 5$) were ground into a fine powder in liquid nitrogen, after which 60 mg powder was weighed in a centrifuge tube. Afterward, 500 μL pre-cooled methanol: H_2O (1:1 = *v*:*v*) was added, followed by vortexing for 30 s at 4 °C. The homogenates were subsequently centrifuged at 14,000× *g* and 4 °C for 10 min. All the supernatant was then transferred into a fresh 1.5 mL centrifuge tube, frozen, and dried under a vacuum. The residue was dissolved in 300 μL of 2-chlorobenzalanine (4 ppm) methanol aqueous solution (1:1, 4 °C). Finally, the supernatant was purified by passing through a 0.22-μm membrane filter for LC-MS/MS analysis.

Samples were analyzed by Ultimate 3000 UHPLC System (Thermo Fisher Scientific, Waltham, MA, USA) coupled to Q Exactive high-resolution mass spectrometer (Thermo Fisher Scientific, Waltham, MA, USA). Chromatographic separations were accomplished using an ACQUITY UPLC ® HSS T3 (150 × 2.1 mm, 1.8 μm) (Waters, Milford, MA, USA). The column was maintained at 40 °C. The flow rate and injection volume were set as

0.25 mL/min and 2 μL, respectively. For LC-ESI (+)-MS analysis, the mobile phases consisted of (C) 0.1% formic acid in acetonitrile (v/v) and (D) 0.1% formic acid in water (v/v). Separation was conducted under the following gradient: 0~1 min, 2% C; 1~9 min, 2%~50% C; 9~12 min, 50%~98% C; 12~13.5 min, 98% C; 13.5~14 min, 98%~2% C; 14~20 min, 2% C. For LC-ESI (−)-MS analysis, the analytes were carried out with (A) acetonitrile and (B) ammonium formate (5 mM). Separation was conducted under the following gradient: 0~1 min, 2% A; 1~9 min, 2%~50% A; 9~12 min, 50%~98% A; 12~13.5 min, 98% A; 13.5~14 min, 98%~2% A; 14~17 min, 2% A. Mass spectrometer settings for full-MS were as follows: the sheath gas pressure was set as 30 arb, the aux gas flow was set as 10 arb, the capillary temperature was set as 325 °C, the spray voltage: 3.50 kV (positive ion mode)/−2.50 kV (negative ion mode), the m/z range of MS1 was set as 100–1000, the MS1 resolving power was set as 70,000 FWHM, the number of data-dependent scans per cycle was set as 10, the MS/MS resolving power was set as 17,500 FWHM, the normalized collision energy was set as 30%, and the dynamic exclusion time set as automatic.

After the detection and analysis of four *Cordyceps* products by LC-MS/MS, the total ion chromatograms of all the samples were extracted. The raw data were processed using the XCMS package in R (V3.3.2) for feature detection, and retention time correction and alignment. The metabolites were identified by accuracy mass (<30 ppm) and MS/MS data which were matched with databases (Metlin, massbank, LipidMaps, and mzcloud). Metabolites are relatively quantified by peak area normalization. After normalization, only ion peaks with relative standard deviations (RSDs) less than 30% in QC were kept to ensure proper metabolite identification. OPLS−DA was performed using SIMCAP13.0 (Umetrics AB, Umea, Sweden) to visualize the differential metabolites among the samples. The selection of differentially abundant metabolites in the four sample classes was based on the combination of a statistically significant threshold of variable influence on projection (VIP) values and p values obtained using a two-tailed Student's t-test and, based on the normalized peak areas of the metabolites with VIP ≥ 1 and $p \leq 0.05$ were considered as differential metabolites. R^2X, R^2Y and Q^2 were used to evaluate the quality of the OPLS−DA mode. Metabolite profiling and metabolomics data analyses were executed by BioNovoGene Co., Ltd. (Suzhou, China).

2.6. Transcriptomic Analysis

Total RNA was extracted from mycelia of Zn^{2+} stress and control (n = 3) by using TRIzol® Reagent (Plant RNA Purification Reagent for plant tissue). The integrity and purity of the total RNA were determined using a 2100 Bioanalyzer and quantified using the ND-2000. Only high-quality RNA samples (OD260/280 = 1.8–2.2, OD260/230 ≥ 2.0, RIN ≥ 8.0, 28S:18S ≥ 1.0, >1 μg) were used to construct the sequencing library. RNA purification, reverse transcription, library construction and sequencing were performed at the Shanghai Majorbio Bio-pharm Biotechnology Co., Ltd. (Shanghai, China). RNA-seq transcriptomic libraries were prepared using the Illumina TruSeqTM RNA sample preparation Kit. The BLAST2GO (http://www.blast2go.com/b2ghome, accessed on 11 September 2022) program was used to obtain GO annotations of unique assembled transcripts for biological processes, molecular function, and cellular components. Metabolic pathway analysis was performed using the Kyoto Encyclopedia of Genes and Genomes (KEGG, http://www.genome.jp/kegg/, accessed on 11 September 2022). In order to identify differentially expressed genes (DEGs) between two groups, the expression level of each transcript was calculated according to the transcripts per million reads (TPM) method. RSEM (http://deweylab.biostat.wisc.edu/rsem/, accessed on 11 September 2022) was used to quantify the abundance of target genes. Differential expression analysis was performed using DEGs with |log2(foldchange)| ≥ 1 and $p \leq 0.05$ as criteria for significantly differentially expressed genes [28]. In addition, the GO functional enrichment and KEGG pathway analyses were conducted (significantly enriched DEGs with Bonferroni-corrected $p \leq 0.05$) using Goatools (https://github.com/tanghaibao/Goatools, accessed on 11 September 2022) and KOBAS (http://kobas.cbi.pku.edu.cn/home.do, accessed on 11 September 2022).

2.7. Statistical Analysis

Each experiment was performed with three replicates. The results are presented as means ± standard deviation (SD). The GraphPad prism 8.0 software was used to conduct one-way ANOVA tests for inter-group comparison, and $p \leq 0.05$ was considered statistically significant.

3. Results

3.1. Morphological Alterations of Mycelia and Reproduction of *P. dubia* under Zn^{2+} Stress

Under Zn^{2+} stress, the surface of mycelia pellets showed protuberances compared with the control group (Figure 1). Interestingly, microcycle conidiation (MC) was induced under Zn^{2+} stress, which produced a large number of conidia. MC is the phenomenon of directly repeated sporulation of filamentous fungi following sexual or asexual spore germination with no growth of vegetative hyphae or very weak hyphal growth, which generally appears to serve as a fail-safe mechanism that allows fungi to efficiently transfer resources from cellular growth to conidia under adverse conditions [29,30]. As shown in Figure 1A, the conidia began to appear in large numbers on the 4th day and reached their maximum abundance (1.3×10^7/mL) on the 12th day under Zn^{2+} stress. On the contrary, there were almost no conidia in the fermentation broth of the control group at the later stage. At the initial stage of the control group, the mycelia produced branched, after which the top of the hyphal branches produced conidia. The mature conidia disassociated from the top of the hyphae. As shown in Figure 1B, conidia produce new conidia by budding or on their germ tube. By utilizing TEM technology, yeast-like budding processes were captured. A conidium is about to be shed from the mother, and the cracks are clearly visible. Another mature conidium is producing a pro-conidium, and the channel between the pro-conidia and the mother conidium has not yet closed. Overall, the conidia formed by MC were rod-shaped (6.6–9.4 × 2.7–5.2 μm) or globular (2.6–3.7 μm), with smooth cell walls, large nuclei, and large lipid droplets. These conidia had a complete cellular structure and were capable of germination and reproduction.

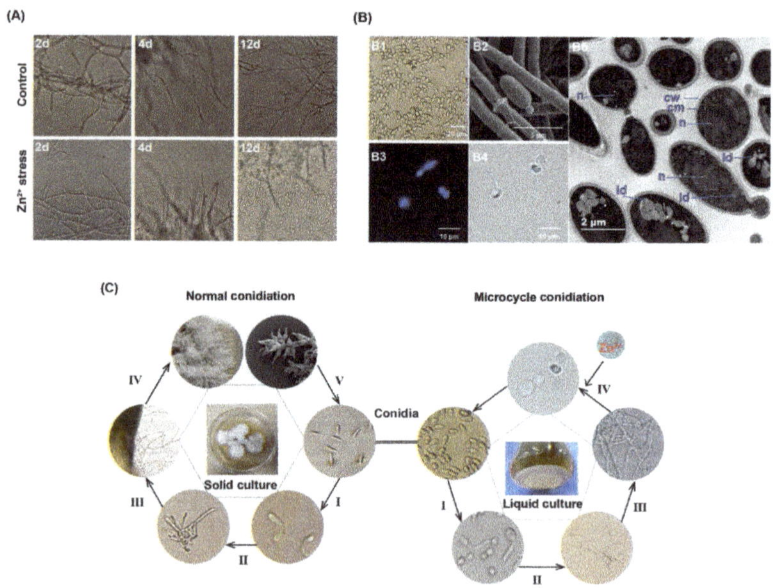

Figure 1. MC condition of *P. dubia* induced by Zn^{2+} stress. (**A**) The process of conidiation in control and Zn^{2+} stress. (**B**) Micrograph of the conidia produced by MC. (B1), (B2) brightfield micrograph; (B3) fluorescence micrograph; (B4) scanning electron micrograph; (B5) transmission electron micrograph showing the ultrastructure of microconidia of *P. dubia*. Abbreviations: n, nucleus; ld, lipid droplet; cw, cell wall; cm, cell membrane. (**C**) Asexual reproduction cycle of *P. dubia*.

Here, we documented the entire MC cycle of *P. dubia*, along with normal conidia production, through microscopic imaging. MC, as a subsidiary cycle of *O. gracilis's* complete life, was described for the first time (Figure 1C). The MC process consisted of four phases. In phase I, conidia expand and germinate into a mycelium. In phases II and III, the mycelia grow further and branch out, after which the primary conidia are produced at the tips of mycelial branches. In phase IV, the conidia produce a large number of secondary conidia through budding if exposed to Zn^{2+} stress. Finally, these conidia produced by MC germinate under suitable conditions and enter the next MC cycle. Alternatively, they undergo normal conidiation when cultured on a solid medium.

3.2. Physiological Characteristic and Zn^{2+} Removal Capacity of *P. dubia* under Zn^{2+} Stress

After 12 days of cultivation under Zn^{2+} stress, dry mycelia biomass, intracellular polysaccharides (IPS), extracellular polysaccharides (EPS), and conidia yield as well as Zn^{2+} removal capacity was measured to evaluate Zn^{2+} tolerance of *P. dubia* (Table 1). The mycelial biomass (6 ± 1.4 g/L) of *P. dubia* decreased significantly under Zn^{2+} stress, which was consistent with previous reports that mushroom biomass decreased dramatically under high-level heavy metal [25]. The yield of IPS (135.5 ± 5.3 mg/g) was similar to the control group (119.3 ± 6.6 mg/g). However, the yield of EPS (233 ± 14.8 mg/L) was increased and 1.28 times that of the control group (181 ± 13.5 mg/L). Additionally, Zn^{2+} levels in mycelia and EPS were 119 and 139 times (11.9 and 13.2 mg/g) higher than those in control, respectively. Furthermore, the conidia also contained higher Zn^{2+}, with a concentration of 6.2 mg/g. Based on the Zn^{2+} content of the above three substances, we calculated that the removal rate of Zn^{2+} was about 60%. These results indicate that *P. dubia* has great potential to effectively remove Zn^{2+} from high concentrations in the Zn^{2+} environment.

Table 1. The response and Zn^{2+} accumulate ability of *P. dubia* under Zn^{2+} stress.

	Mycelia Biomass (g/L)	IPS (mg/g)	EPS (mg/L)	Conidia (g/L)	Zn^{2+} Concentration (mg/g)		
					Mycelia	EPS	Conidia
Zn^{2+} stress	6 ± 1.4 ^{a}	135 ± 5.3 ^{a}	233 ± 14.8 ^{a}	0.162 ± 0.0062	11.9 ± 1.2 ^{a}	13.2 ± 1.6 ^{a}	6.2 ± 0.9
Control	13 ± 1.8 ^{b}	119 ± 6.6 ^{a}	181 ± 13.5 ^{b}	ND	0.1 ± 0.004 ^{b}	0.095 ± 0.003 ^{b}	ND

"ND" means conidia not be detected. Means with different letters are significantly different at $p \leq 0.05$.

3.3. Transcriptomic Sequencing and DEGs Identification

In order to explore the molecular response mechanism of *P. dubia* to Zn^{2+} stress, the total RNA was extracted from Zn^{2+} stress and control samples for transcriptomic analysis. A total of 54,816,258 clean reads were obtained from the six samples by transcriptomic sequencing, and the average Q20 and Q30 value, as well as GC contents in clean data, were 97.96%, 94.03%, and 58.35%, respectively. All clean reads were assembled into 22,663 transcripts with an average length of 3912.65 bp and an N50 of 5914 bp. These transcripts were further assembled into 8075 genes with an average length of 2514.75 bp and an N50 of 4396 bp (Table S1). In order to obtain comprehensive information on gene functions, six databases were used to annotate these genes. A total of 5924 (73.59%) genes were annotated. Among the annotated genes, NR (5720, 70.84%) and eggNOG (5357, 66.34%) had the largest match, followed by the GO, Pfam, Swissprot, and KEGG databases (Table S1). Through the DEG screening ($|\log2(\text{foldchange})| > 1$ and $p < 0.05$), 1533 DEGs (702 upregulated and 895 downregulated) in the mycelia of *P. dubia* were identified to be responsive to Zn stress (Figure 2A).

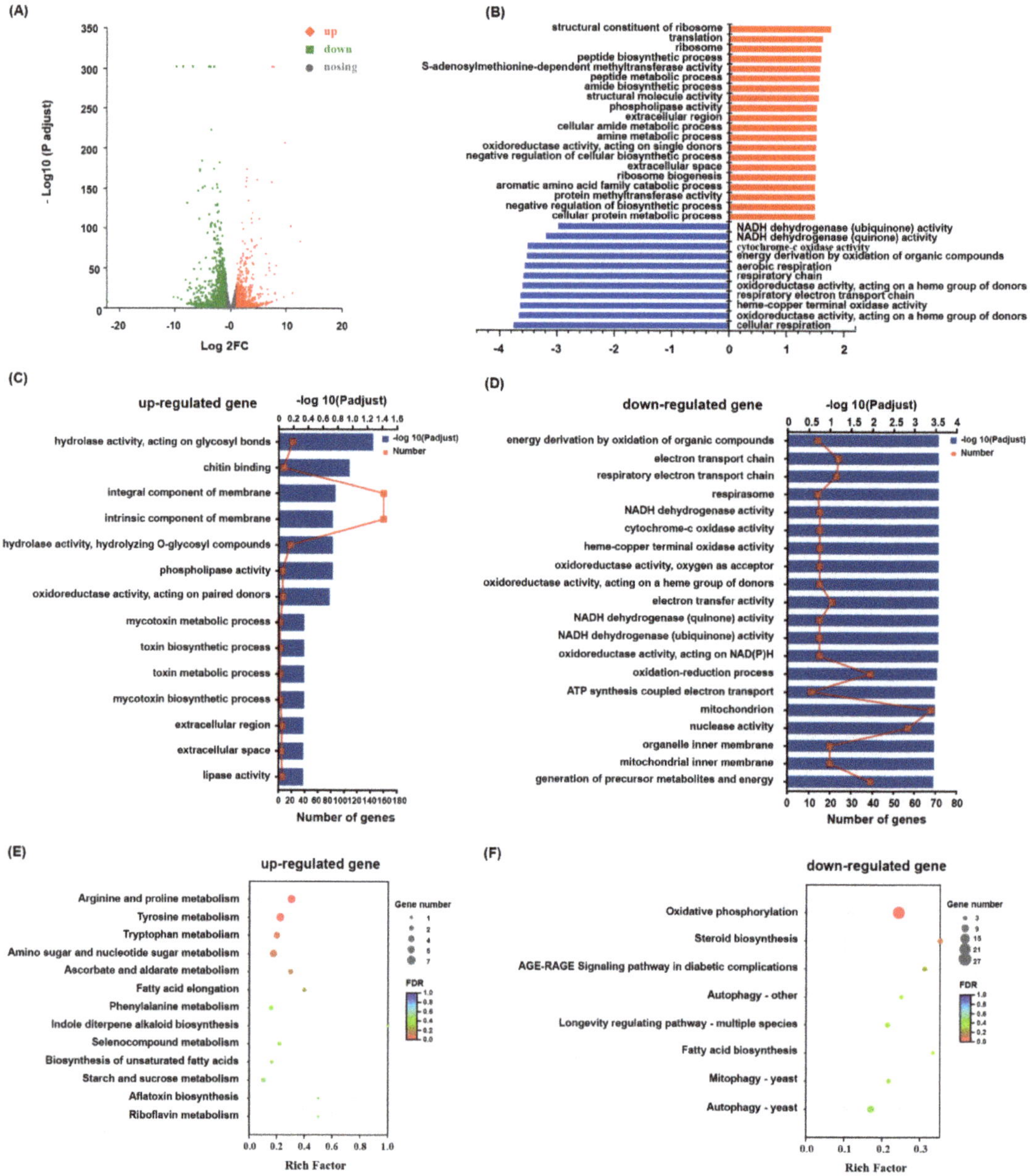

Figure 2. Distribution and function of DEGs related to Zn^{2+} stress. (**A**): Venn diagram and Volcano plot of DEGs; (**B**): GSEA analysis of Zn^2 treatment; (**C**,**D**): Scatter plot of GO enrichment of up- and downregulated genes. (**E**,**F**): Scatter plot of enriched KEGG pathways of upregulated and downregulated genes. The Rich factor is the ratio of the number of genes that were differentially expressed to the total number of genes in a certain pathway. The figure shows the top 20 enriched pathways.

3.4. Functional Annotation and Enrichment of DEGs

GO annotation analysis was performed to classify the DEGs functions. The results showed that the metabolic process and cellular process were the most enriched terms in the biological process ontology, which indicated that the cellular metabolism of *P. dubia* was affected under the condition of Zn^{2+} stress. Membrane part, cell part, and organelle were the most enriched terms in the cellular component ontology, which indicated that the membrane structure and mycelia growth of *P. dubia* were affected by Zn^{2+} stress. Catalytic activity and binding were the most enriched terms in the molecular function, which indicated that *P. dubia* may be resistant to Zn^{2+} stress by regulating, catalyzing, and activating various enzymes (Figure S1). The GO enrichment analysis was carried out to elucidate the specific biological functions of the DEGs. The upregulated DEGs were mainly enriched in the integral component membrane, intrinsic component membrane, hydrolase activity, acting on glycosyl bonds, and chitin-binding. Nevertheless, the downregulated DEGs were mainly enriched in the mitochondrion, nuclease activity, oxidation-reduction process, and generation of precursor metabolites and energy (Figure 2C,D). In order to confirm the biological pathways induced by Zn^{2+} stress, all DEGs were assigned to the KEGG database for KEGG pathway enrichment analysis. The upregulated DEGs were enriched in arginine and proline metabolism, tyrosine metabolism, tryptophan metabolism, and amino sugars and nucleotide sugars metabolism, whereas the downregulated DEGs were mainly enriched in oxidative phosphorylation, autophagy, steroid synthesis, longevity regulating pathway, and mitophagy (Figure 2E,F). Moreover, in order to reflect gene regulation more comprehensively, gene set enrichment analysis (GSEA) analysis was conducted. The results from it were similar to those in GO and KEGG enrichment analysis. The DEGs related to cellular respiration, respiratory electron transport chain, energy derivation by oxidation of organic compounds, cytochrome-c oxidase activity, and NADH dehydrogenase (quinone) activity were down-regulated (Figure 2B). Therefore, these pathways, glycolysis, tricarboxylic acid cycle (TCA), oxidative phosphorylation, antioxidant defense system, autophagy, metal ion transport, EPS synthesis, and amino acids metabolism, which should be connected by Zn^{2+} stress.

3.5. Critical Functional Genes in Response to Zn^{2+} Stress

According to the above enrichment analysis, we screened the critical genes related to oxidative phosphorylation, glycolysis, tricarboxylic acid cycle (TCA), antioxidant defense system, autophagy, metal ion transport, EPS synthesis, and amino acids metabolism (Tables S2–S5). In detail, 31, 29, 12, and 12 DEGs were identified, which participate in the oxidative phosphorylation pathway, glycolysis and TCA cycle pathway, enzyme antioxidant system, and autophagy, respectively. Further, almost of those DEGs were downregulated under the condition of Zn^{2+} stress (Table S2). Moreover, 14 genes associated with DNA repair and replication were identified, half of which were downregulated. (Table S2). These results indicated that the cells were damaged in response to Zn^{2+} stress. A series of genes responsible for metal ion transport were expressed differently in mycelia between the Zn^{2+} stress and control groups. A total of 37 genes belonging to the ZRT/IRT-like protein transporter (ZIP), ATP-binding cassette transporter (ABC), heavy metal ATPases (HMAs), major facilitator superfamily general substrate transporter (MFS), and efflux pumping were significantly upregulated (Table S3). A total of 51 genes were identified as participating in EPS synthesis, of which 34 genes were upregulated (Table S4). It is noteworthy that 29 genes regulating MC formation (e.g., *brlA*, *abaA*, *ste12*, *vosA*, *PpoA*, *ppoC*, *atg3*, etc.) were induced to express under Zn^{2+} stress (Table S5), which may be the important mechanism for *P. dubia* to tolerance Zn^{2+} stress.

3.6. Metabolic Changes of P. dubia under Zn^{2+} Stress

To elucidate the response of *P. dubia* under Zn^{2+} stress, we conducted a comparative analysis of the metabolites between the Zn^{2+}-treated and control groups by untargeted metabolomics (LC−MS/MS), aiming to confirm differential metabolites associated with Zn^{2+} stress. The score plots of the OPLS−DA model (a multivariate analysis) showed considerable separation between the Zn^{2+} treated and control groups. The model parameters R^2 and Q^2 values were 0.91 and 0.06, respectively, indicating that the OPLS−DA model had an excellent predictive ability. The results indicated that the metabolites composition of *P. dubia* had significantly varied under Zn^{2+} stress (Figure 3A,B). A total of 612 metabolites were identified. Metabolites with VIP ≥ 1 and $p \leq 0.05$ were selected as differential metabolites, and 207 were identified as differential metabolites (Tables S6 and S7). These differential metabolites were divided into seven categories, including amino acids and derivatives (21.26%), carbohydrates and derivatives (6.76%), nucleosides and derivatives (11.59%), fatty acids (4.35%), organic acids (3.86%), vitamins (1.93%), and others (39.61%). The category of others mainly includes alcohols, alkaloids, phenols, and some secondary metabolites, such as mannitol, ribitol, D-xylitol, choline, betaine, protocatechuic acid, etc. (Figure 3C). A large number of carbohydrates and amino acids were downregulated under Zn^{2+} stress. However, a few amino acids were shown to be upregulated under Zn^{2+} stress. Most of them have physiological functions of regulating osmotic pressure and antioxidants, such as arginine, taurine, tryptophan, and tyrosine (Figure 3C, Table S7). In addition, more than 50% of differential metabolites were upregulated in the categories of nucleosides, vitamins, and others under Zn^{2+} stress, including adenosine, 2-deoxyadenosine, inosine, biotin, mannitol, D-xylitol, ribitol, D-arabitol, tyrosol, betaine, choline, acetylcholine, etc. (Table S7). These upregulation metabolites may contribute to resistance to the Zn^{2+} stress of *P. dubia*.

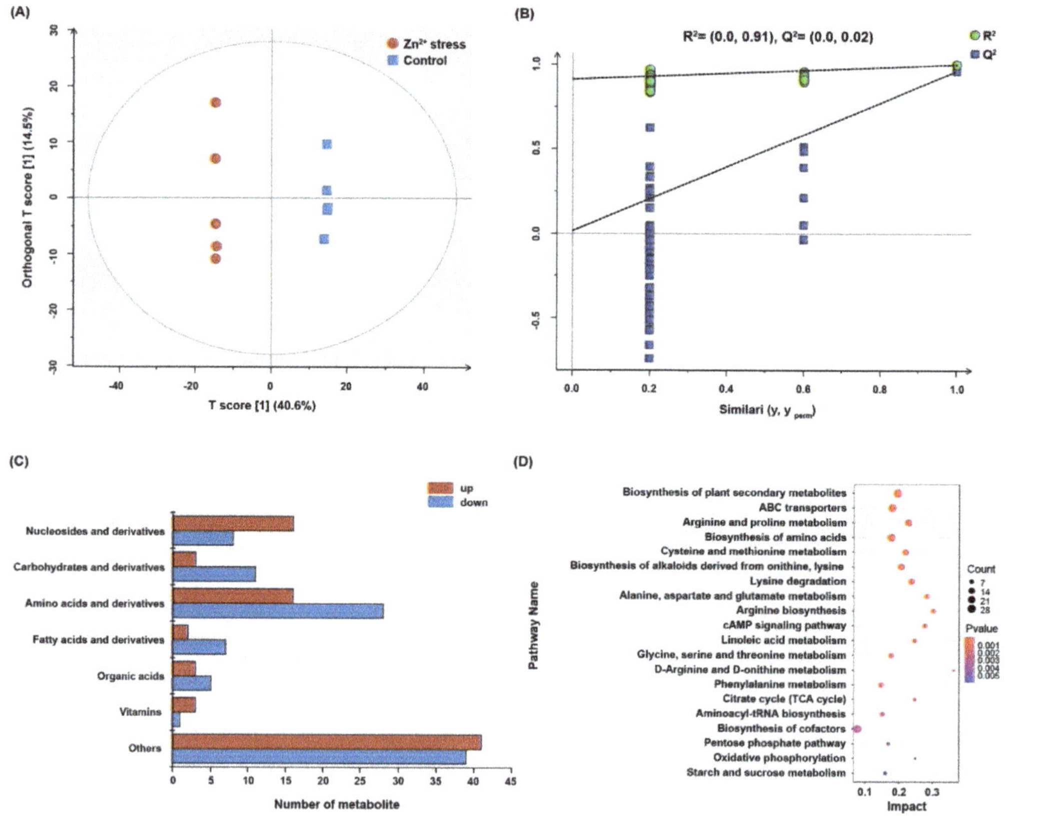

Figure 3. Metabolomic analysis of Zn^{2+} stress and normal cultivation of *P. dubia*. (**A**) OPLS−DA score plot. (**B**) Statistical validation of the OPLS−DA model with permutation. (**C**) The numbers of seven categories of differential metabolites. (**D**) KEGG pathway analysis of differential metabolites.

Furthermore, the KEGG enrichment pathways analysis was conducted on all differential metabolites to confirm the pivotal metabolic pathways of *P. dubia's* response to Zn^{2+}

stress. As shown in Figure 3D, differential metabolites were significantly enrichment in the biosynthesis of secondary plant metabolites, ABC transporters, biosynthesis of alkaloids, cAMP signaling pathway, linoleic acid metabolism, biosynthesis of cofactors, TCA cycle, and amino acids metabolism (such as arginine and proline metabolism, protein digestion and absorption, biosynthesis of amino acids).

4. Discussion

4.1. Zn^{2+} Stress Affected in Energy Metabolism, Oxidative Stress, and Autophagy

When metal ions are transported into the cell and heavily accumulated, the basic metabolism of the cell will be hindered, including the critical basic metabolism pathways for obtaining energy in microorganisms, such as glycolysis, tricarboxylic acid (TCA) cycle, and oxidative phosphorylation [31,32]. Here, genes involved in glycolysis and the TCA cycle were mostly downregulated, which may lead to an interruption in energy supply, thus explaining the reduction in mycelia biomass under Zn^{2+} stress (Figure 4A). Nearly all genes that participated in oxidative phosphorylation were downregulated (Figure 4B), which may block electrons transmission in the electron transport chain, and these excess electrons will directly or indirectly react with molecular oxygen to generate reactive oxygen species (ROS), causing an acceleration of cell damage [33]. In addition, almost all genes related to antioxidant enzymes (*SOD*, *CAT*, *POD*, and *GSH-Px*) were significantly downregulated. Simultaneously, the amount of glutathione (GSH) and its precursor glutamic acid decreased (Figure 4D), which suggested that *P. dubia* was likely to suffer from oxidative stress. GSH has antioxidant effects and can easily combine with heavy metals such as zinc, lead, and arsenic to eliminate the toxicity of heavy metals [34]. Interestingly, two genes involved in GSH metabolism were upregulated (Table S2), which may be a stress response to the depletion of GSH to scavenge ROS and Zn^{2+}. This phenomenon is consistent with previous research that GSH was clearly downregulated when yeast was exposed to a high concentration of Cu^{2+} [35]. ROS disrupts cell membranes and organelles, thereby affecting the growth of mycelia. Previous studies have shown that ROS can cause severe damage to DNA, including disrupting DNA repair and replication systems, which hinder gene expression and lead to the death of organisms [36]. In this study, the expression levels of genes related to DNA processes were significantly altered in response to Zn^{2+} stress. Nearly half of the DEGs involved in DNA repair and replication were downregulated (Table S2). The DNA replication licensing factor MCM, a component of the replication helicase necessary for the initiation and extension of DNA replication in eukaryotic cells, was significantly downregulated [37]. The results indicated that Zn^{2+} induced DNA damage by inhibiting the DNA repair and replication enzyme activity and disturbing the DNA-related protein activity. However, other upregulated DEGs were associated with DNA replication and repair, which played a crucial role in repairing DNA damage (Table S2). For example, a gene associated with the DNA damage checkpoint was upregulated, and related research revealed that once the DNA damage checkpoint is activated, it hinders the progression of the cell cycle until the DNA repair system resolves the damage [38]. Autophagy is activated in cells to promptly remove damaged cell membranes, organelles, and abnormal proteins to maintain cellular homeostasis and recycle metabolic components [39]. The *Atg1/13* kinase complex was identified as a key positive regulator of autophagy, which interacts with *Atg* protein to regulate autophagy, such as *Atg17*. When *Atg17* is absent, the autophagy that is formed becomes smaller. At the same time, PKA inhibits autophagy formation by phosphorylating *Atg1* [40]. In this study, the expression of *Atg1*, *Atg13*, and *Atg 17* was decreased, while *PKA* was increased, indicating that autophagy was inhibited (Figure 4C). Autophagy is a mechanism for cellular self-protection, and inhibition of autophagy may disturb cellular homeostasis and lead to apoptosis. The upregulation of the apoptosis-inducing factor (Table S2) and the downregulation of autophagy in this study are consistent.

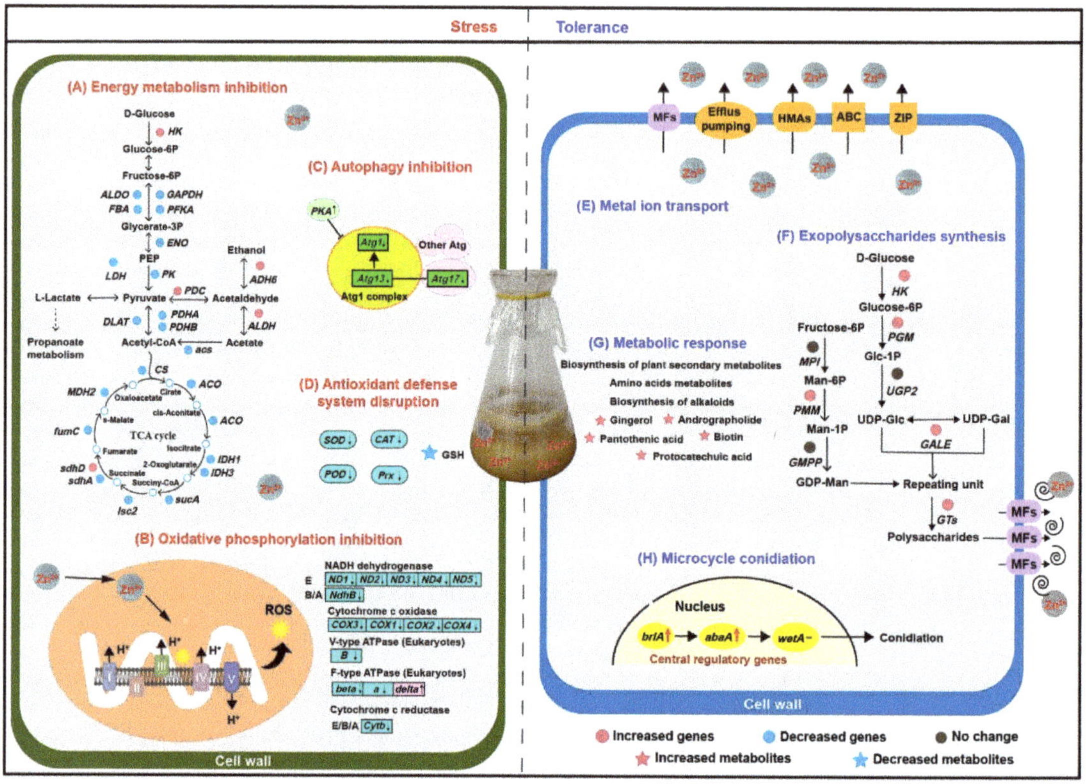

Figure 4. Zn^{2+} stress and tolerance mechanism in mycelia of *P. dubia*. (**A**) Energy metabolism inhibition. (**B**) Oxidative phosphorylation inhibition. (**C**) Autophagy inhibition. (**D**) Antioxidant defense system disruption. (**E**) Metal ion transport. (**F**) Exopolysaccharides synthesis. (**G**) Metabolic response. (**H**) Microcycle conidiation. ROS—reactive oxygen species; SOD—superoxide dismutase; CAT—catalase; Prx—peroxidase.

4.2. Lead-Induced Detoxification and Defense Mechanisms

In this study, the Zn^{2+} stress brings a series of harm to the mycelia of *P. dubia*, which disrupts energy metabolism, blocks electron transfer, damages the antioxidant system and DNA, obstructs autophagy, and even induces apoptosis. To enhance the tolerance of *P. dubia* capacity to Zn^{2+}, metal ion transport, EPS synthesis, and microcycle conidiation were activated. In addition, the biosynthesis of some metabolites was upregulated to resistant Zn^{2+} stress (Figure 4E–H). These coping strategies allow *P. dubia* successfully grow in the high Zn^{2+} environment.

4.2.1. Metal Ion Transport Response to Zn^{2+} Stress

The basic strategy to resist metal ion stress is to reduce the accumulation of metal ions in cells, such as active efflux, which is a typical mechanism for biomining bacteria to resist high concentrations of copper [41]. In this study, the expression of genes associated with ZRT/IRT-like protein transporter (ZIP), ATP-binding cassette transporter (ABC), heavy metal ATPases (HMAs), major facilitator superfamily general substrate transporter (MFS), and efflux pumping was affected (Figure 4E, Table S3). ZIP is known to mediate cellular uptake, intracellular trafficking, and detoxification of heavy metal cations such as zinc, iron, cadmium, copper, manganese, cobalt, and nickel [42]. In the present study, two ZIP transporters were downregulated, which may decrease the cellular uptake of

Zn^{2+}. However, *ZIP9* and a ZIP transporter were upregulated, and previous research found that under Zn^{2+} deficiency, the expression of *ZIP9* was upregulated in the shoot and root of *Thlaspi goesingense* to promote Zn^{2+} accumulation [43]. These indicate that the ion transport mechanism of ZIP in response to Zn^{2+} stress is complex and requires further investigation. ABC participates in various cellular processes, such as mitochondrial function, peroxisome biogenesis, and the export of Fe/S clusters, particularly in detoxifying heavy metals and removing toxic catabolic compounds [44]. Here, five ABC transporters were upregulated, with *ABCG* and *ABCB* showing upregulation. It has been reported that they can chelate heavy metals such as Cd^{2+} [25], which may contribute to enhancing tolerance to Zn^{2+}. Additionally, HMAs, MFS transporters, and efflux pumping are also responsible for maintaining proper metal homeostasis in cells, and some genes related to them were upregulated in this research. HMAs consume ATP to pump metals such as copper, zinc, cadmium, and lead across the membrane and are responsible for metal transport and detoxification [45]. Studies have shown that HMAs are involved in the active efflux of Pb^{2+} or Pb^{2+} containing toxic molecules across the plasma membrane, followed by sequestration into inactive organelles to avoid Pb^{2+}-induced toxicity [46]. MFS transporters can secrete a variety of toxic compounds, such as organic and inorganic ions, to enhance tolerance to heavy metals by enhanced efflux [27]. The efflux pumping serves as the foundation for resistance against heavy metal ions, such as P-type pumps, which play a critical role in mobilizing trace metals across the plasma membrane. The increased efflux of Zn^{2+} is one of the survival mechanisms of *P. dubia* mycelia under Zn^{2+} stress.

4.2.2. Exopolysaccharides Synthesis Promotion Response to Zn^{2+} Stress

Secreting EPS to scavenge Zn^{2+} in the environment is another mechanism for *P. dubia* to respond to Zn^{2+} stress. Previous studies have shown that EPS is an important biosorbent for removing heavy metals, and some microorganisms chelate heavy metal ions by synthesizing EPS to grow under high-concentration metal ions stress [47]. Our research data showed that EPS yield was enhanced, simultaneously genes involved in EPS synthesis showed varying degrees of upregulation under Zn^{2+} stress, such as *HK*, *PGM*, *GALE*, and *PMM*, which are key enzymes associated with EPS synthesis and conversion steps (Figure 4F). Hexokinase (*HK*) catalyzed glucose to produce glucose 6-phosphate, glucose 6-phosphate was further converted to glucose 1-phosphate by phosphoglucomutase (*PGM*), then the glucose-1-phosphate generates UDP-Glc under the action of UTP-glucose-1-phosphate uridylyltransferase (*UGP2*) [48]. Furthermore, UDP-glucose 4-epimerase (*GALE*) catalyzed the conversion of UDP-Glc to UDP-Gal, the upregulated of *GALE* indicating that Zn^{2+} stress may have altered the monosaccharide composition of polysaccharides [49]. In addition, mannose 6-phosphate isomerase (*MPI*), phosphomannomutase (*PMM*), and mannose 1-phosphate guanylyltransferase (*GMPP*) gradually catalyzed Fructose-6P conversion to GDP-Man [50]. Hence, the upregulation of PMM may promote the generation of GDP-Man. Except for the upregulation of glycosyl donor synthetase, a large number of glycosyltransferases (GTs) and MFS were upregulated, which contribute to the polymerization of polysaccharides and secretion of polysaccharides out of the cell, respectively. Herein, the expression levels of these genes may determine the accumulation of polysaccharides under Zn^{2+} stress. This provides a reference for future research on the response mechanism of polysaccharide biosynthesis under heavy metal stress.

4.2.3. Metabolic Response to Zn^{2+} Stress

KEGG analysis showed that differential metabolites were significantly enriched in plant secondary metabolite biosynthesis and alkaloid biosynthesis. Abiotic stress has been reported to induce secondary metabolites to form a mechanical barrier involved in osmotic stress resistance and ROS [45]. KEGG analysis showed that differential metabolites were significantly enriched in the biosynthesis of secondary plant metabolites and biosynthesis of alkaloids (Figures 3D and 4G). Studies have shown that fungi protect themselves from abiotic stresses such as high temperature, drought, and salinization by accumulating compatible solutes, which are mostly sugars, sugar alcohols (polyols), betaine, and amino acids [51]. Several fungal spores are rich in mannitol, exhibiting excellent stress resistance. For example, the conidia of *Aspergillus niger* contain a large amount of mannitol, accounting for 10.9% of the dry weight [30]. In this study, we have found that mannitol, betaine, and five kinds of alcohol (D-Xylitol, ribitol, D-Arabitol, myo-Inositol, and tyrosol) were upregulated, which may contribute to against Zn^{2+} stress by regulating osmotic. In addition, pantothenic acid, biotin, niacinamide, nicotinic acid, andrographolide, and protocatechuic acid were upregulated, which may serve as compensatory mechanisms for the deficiency of antioxidant enzyme and GSH, given their excellent antioxidant capacity [45]. In addition, the level of aromatic amino acids (tryptophan and tyrosine), arginine, and taurine were increased. Furthermore, amino acid metabolism (arginine and proline metabolism, biosynthesis of amino acids, phenylalanine metabolism, and so on) was enriched. This result is consistent with the KEGG enrichment analysis in transcriptomics. Amino acids metabolism is closely related to secondary metabolism, which may provide precursors for the formation of secondary metabolites [52].

4.2.4. Reproduction Response Based on MC in Zn^{2+} Stress

Apart from these aspects discussed above, the mycelia of *P. dubia* produced a large number of conidia through MC (Figure 4H). Conidia are more tolerant of adverse environmental conditions, and MC is a crucial survival tactic for *P. dubia* under Zn^{2+} stress. There are many factors that can induce the MC of fungi, such as nutrient deficiency, pH, light, and temperature [53–55]. While metal ions induce MC in fungi was found for the first time. In contrast to normal conidiation with a definite formation mechanism, the molecular mechanism of MC is still unclear, even though it was first observed in 1890 and has been reported in more than 100 fungal species. Only some research revealed the mechanism of MC preliminarily in some lower fungi which can't form fruiting bodies, such as *Metarhizium anisopliae* and *Fusarium graminearum* [56,57]. Although the MC phenomenon has been reported in natural *C. militaris*, its mechanism has not been investigated [58]. Therefore, it is necessary to understand the formation mechanisms of MC. The underlying mechanism of MC of *P. dubia* was preliminarily analyzed, as shown in Figure 5.

Signal-transduction of MC. Conidiation is controlled by upstream regulatory genes which respond to the external environment. Previous studies have revealed some upstream regulatory genes, such as *fluG*, *flbA*, *flbB*, *flbC*, *flbD*, and *flbE*, which participate in regulating the core regulatory pathway of conidiation (*brlA* → *abaA* → *wetA*) in the model fungus *Aspergillus* [59]. In the current study, only *flbA* was identified and had no significant difference, suggesting that the upstream regulatory factor of MC of *P. dubia* is special, which is different from some filamentous fungi that have been reported. The mitogen-activated protein kinase (MAPK) signaling pathway plays important roles in fungal growth and development, such as cell cycle, morphogenesis, stress resistance, virulence, cell wall rigidity and integrity, intercellular signal transduction, etc. In addition, some studies have also shown that the MAPK signaling pathway regulates development and is required for conidiation [60]. Based on the MAPK signaling pathway (map04011) of yeast, we suggest a signal transduction model of MC, in which *abaA* was upregulated by downregulated *ste12* in the MAPK pathway to produce conidia, which consistent with other research conclusions that deletion of the *ste12* gene could promote conidia production [61]. Based on previous studies, ROS is a critical cell differentiation signal promoting conidiation and sexual development, which accumulation can promote conidiation in *N. crassa* [62]. In this study, Zn^{2+} damages the oxidation-reduction system and leads to the accumulation of ROS, which may be an inducer of MC. In addition, a total of 16 transcription factors (TF) were screened out, and a majority of TFs belong to the Zn cluster family (Zcf) (Table S5). Zcf is a fungal-specific family of TF and is the largest family of TFs known in eukaryotes [63]. In taking these facts together, these data indicated that ROS may be an elicitor of MC, MAPK, and some factors which contain the zinc finger domain play an important role in regulating upstream regulation of MC.

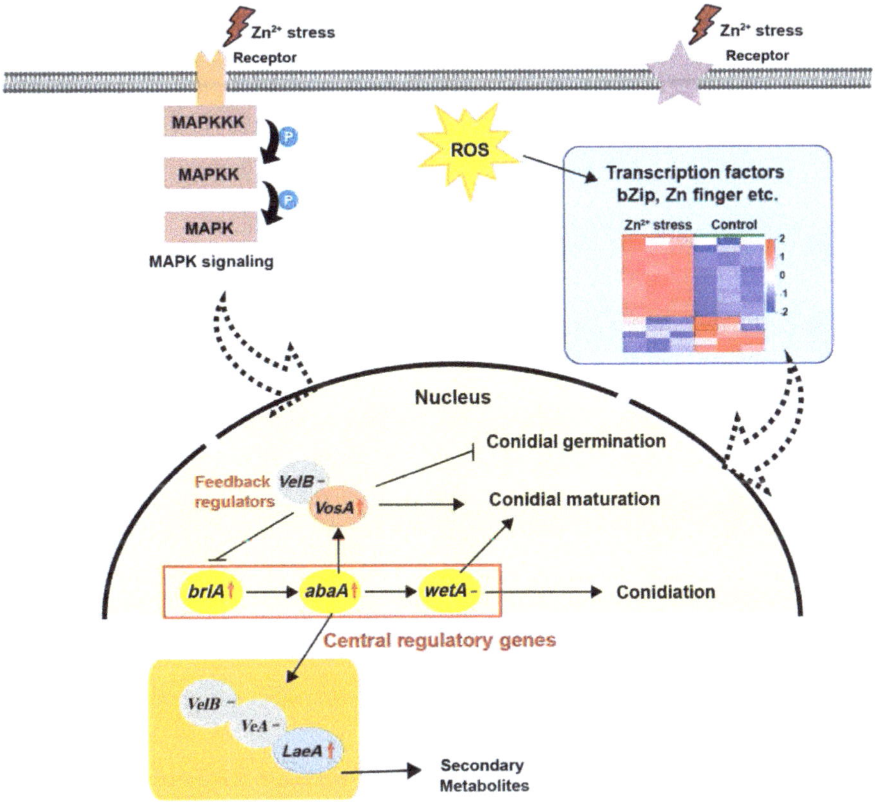

Figure 5. The predicted microcycle conidiation pathways induced by Zn^{2+} stress in *P. dubia*.

Genes involved in MC of *P. dubia* under Zn^{2+} stress. The expression of the central regulators *brlA* and *abaA* were upregulated under Zn^{2+} stress, which contribute to MC formation. *BrlA* is a key TF whose overexpression in vegetative cells leads to growth cessation and the formation of viable conidia directly from the hyphal tips [59]. In addition, *brlA* contains a C_2H_2 zinc finger DNA-binding domain that recognizes *brlA* response elements in the promoter regions of certain genes, such as *abaA*. AbaA is a transcriptional enhancer factor 1 family TF governing the expression of *wetA*, *vosA*, and *velB* [64]. Previous research reported that *abaA* also affects secondary metabolism by regulating the expression of *veA*, *velB*, and *laeA* [65]. These were consistent with the metabolomic analysis that the biosynthesis of plant secondary metabolites and biosynthesis of alkaloids was significant enrichment. Therefore, this connection may link secondary metabolite biosynthesis with conidiation through a shared signal transduction pathway. The *wetA* gene activated by *abaA* completes the developmental role in the late stage of conidiation and plays an important role in conidia maturity [59], while *wetA* has no difference in this study. It has been reported that velvet regulator *vosA* couples with *velB* as a functional unit conferring conidia maturation and attenuating conidial germination in *A. nidulans* [66]. In this study, *vosA* significantly upregulated, which may represent a compensatory mechanism contributing to the conidia maturation. Except for the above core genes regulating MC, *PpoA*, *ppoC*, and *atg3* may also participate in regulating MC formation. *PpoA* and *ppoC* are dioxygenases that jointly regulate the balance between conidia and ascospore formation. When *ppoA* is deficient, the proportion of conidia significantly increases, while when *ppoC* is deficient, the proportion of ascospores significantly increases [67]. Here, the downregulation degree of *ppoA* was much higher than that of *ppoC*, which was conducive to the formation of conidia. Autophagy has been reported to regulate conidiation. For example, when deletion *atg3* (E2-like enzyme) and *atg7* (E1-like enzyme), the conidia and mycelia were reduced considerably [68,69]. *Atg3* was significantly downregulated in this study, which may inhibit conidia production. However, experimental results showed that plenty of conidia were produced, this may be due to the presence of multiple pathways activating MC formation, and when one of the pathways was repressed, the repressive effect was offset by an increase in the expression of another pathway. In addition, cell walls play important roles in providing mechanical support and participating in conidiation [70]. Relevant studies have shown chitinases I and II contribute to the development of conidia, and a GPI-deficient strain of *Fusarium verticillioides* was not competent in conidiation and cell wall formation [71,72]. In this study, the expression level of chitinase I, GPI-anchored endo-1, and an uncharacterized GPI-anchored protein was significantly upregulated, which may facilitate the formation of MC.

5. Conclusions

Based on this study, we are convinced that *P. dubia* is a potential strain for bioremediation of Zn^{2+} pollution. There were 207 differential metabolites, and 1533 DEGs were identified from *P. dubia* under the condition of Zn^{2+} stress by transcriptomic and metabolomic analysis, which mainly enriched in TCA cycle, oxidative phosphorylation, autography, cell membrane and organelle, secondary metabolites biosynthesis, and amino acids metabolism. From the overall analysis, the mycelia of *P. dubia* enhances its tolerance to Zn^{2+} stress through the following strategies, such as restricting Zn^{2+} entering the cell by downregulating ion transport, reducing bioavailable Zn^{2+} concentration through EPS chelation, regulating metabolites to ensure cell viability, cell osmotic balance, and ROS scavenging, and inducing the production of a large number of conidia to maintain population continuity. Moreover, the phenomenon of MC was first observed in the liquid fermentation of *Cordyceps*. The MC may be an intelligent strategy employed by entomopathogenic fungi to resist adverse environments, which appear to be activated by increasing ROS levels. This study will provide a new strategy for the bioremediation of Zn^{2+} pollution and serve as a reference for the investigation of the heavy metal tolerance mechanism in *Cordyceps*.

Supplementary Materials: The following supporting information can be downloaded at: https://www.mdpi.com/article/10.3390/jof9070693/s1. Figure S1: Histogram of GO classification of *P. dubia* differential expression genes, Table S1: Summary of the de novo assembly of *P. dubia* transcriptomics, Table S2: List of the DEGs involved in oxidative phosphorylation, glycolysis and tricarboxylic acid (TCA) cycle pathway, enzyme antioxidant system, autophagy and DNA repair, Table S3: List of the DEGs involved in ion transport, Table S4: List of the DEGs involved in EPS synthesis, Table S5: List of the DEGs involved in microcycle conidiation, Table S6: The information of all identified metabolites, Table S7: List of differential metabolites identified by LC-MS/MS analysis platform.

Author Contributions: Y.W.: Investigation, Formal analysis, Writing—original draft. L.-L.T., L.Y., and M.-Z.L.: Formal analysis, Data curation, Software. Y.-H.D. and L.-H.Y.: Investigation, Data curation. B.R.: Formal analysis, Methodology, Software, Supervision. D.-S.G.: Conceptualization, Validation, Writing, review and editing, Supervision. All authors have read and agreed to the published version of the manuscript.

Funding: This work was financially supported by the National Key Research and Development Program of China (No. 2021YFC2103200), the Natural Science Foundation of Jiangsu Province of China (No. BK20220372), and the Natural Science Foundation of the Jiangsu Higher Education Institutions of China (No. 22KJB550011).

Institutional Review Board Statement: Not applicable.

Informed Consent Statement: Not applicable.

Data Availability Statement: The RNA-seq data reported in the current study are available in the NCBI SRA database. BioProject number is PRJNA763311.

Conflicts of Interest: The authors declare no conflict of interest.

References

1. Jiang, D.; Tan, M.T.; Zheng, L.; Wu, H.F.; Li, Y.N.; Yan, S.C. Cd exposure-triggered metabolic disruption increases the susceptibility of Lymantria dispar (Lepidoptera: Erebidae) larvae to Mamestra brassicae nuclear polyhedrosis virus: A multi-omits study. *Ecotoxicol. Environ. Saf.* **2022**, *232*, 113280. [CrossRef]
2. Bi, X.; Feng, X.; Yang, Y.; Qiu, G.; Li, G.; Li, F.; Liu, T.; Fu, Z.; Jin, Z. Environmental contamination of heavy metals from zinc smelting areas in Hezhang County, western Guizhou, China. *Environ. Int.* **2006**, *32*, 883–890. [CrossRef]
3. Long, X.X.; Yang, X.E.; Ni, W.Z.; Ye, Z.Q.; He, Z.L.; Calvert, D.V.; Stoffella, J.P. Assessing zinc thresholds for phytotoxicity and potential dietary toxicity in selected vegetable crops. *Commun. Soil Sci. Plan.* **2003**, *34*, 1421–1434. [CrossRef]
4. Arrivault, S.; Senger, T.; Kramer, U. The Arabidopsis metal tolerance protein AtMTP3 maintains metal homeostasis by mediating Zn exclusion from the shoot under Fe deficiency and Zn oversupply. *Plant J.* **2006**, *46*, 861–879. [CrossRef]
5. Natasha, N.; Shahid, M.; Bibi, I.; Iqbal, J.; Khalid, S.; Murtaza, B.; Bakhat, H.F.; Farooq, A.U.; Amjad, M.; Hammad, H.M.; et al. Zinc in soil-plant-human system: A data-analysis review. *Sci. Total Environ.* **2022**, *20*, 152024. [CrossRef] [PubMed]
6. De Almeida, H.; Carmona, V.M.V.; Inocencio, M.F.; Neto, A.E.F.; Cecilio, A.B.; Mauad, M. Soil Type and Zinc Doses in Agronomic Biofortification of Lettuce Genotypes. *Agronomy* **2020**, *10*, 124. [CrossRef]
7. Aydin, S.S.; Gokce, E.; Buyuk, I.; Aras, S. Characterization of stress induced by copper and zinc on cucumber (*Cucumis sativus* L.) seedlings by means of molecular and population parameters. *Mutat. Res.* **2012**, *746*, 49–55. [CrossRef]
8. Garg, N.; Singh, S. Arbuscular Mycorrhiza Rhizophagus irregularis and Silicon Modulate Growth, Proline Biosynthesis and Yield in *Cajanus cajan* L. Millsp (pigeonpea) Genotypes Under Cadmium and Zinc Stress. *J. Plant Growth Regul.* **2018**, *37*, 46–63. [CrossRef]
9. Tewari, R.K.; Kumar, P.; Sharma, P.N. Morphology and physiology of zinc-stressed mulberry plants. *J. Plant Nutr. Soil Sci.* **2008**, *171*, 286–294. [CrossRef]
10. Michael, P.I.; Krishnaswamy, M. The effect of zinc stress combined with high irradiance stress on membrane damage and antioxidative response in bean seedlings. *Environ. Exp. Bot.* **2011**, *74*, 171–177. [CrossRef]
11. Ab Khaman, S.M.S.; Naher, L.; Siddiquee, S. Mushroom Quality Related with Various Substrates' Bioaccumulation and Translocation of Heavy Metals. *J. Fungi* **2022**, *8*, 42. [CrossRef]
12. Nordgren, A.; Baath, E.; Soderstrom, B. Microfungi and microbial activity along a heavy metal gradient. *Appl. Environ. Microbiol.* **1983**, *45*, 1829–1837. [CrossRef]
13. Sung, G.H.; Hywel-Jones, N.L.; Sung, J.M.; Luangsa-Ard, J.J.; Shrestha, B.; Spatafora, J.W. Phylogenetic classification of *Cordyceps* and the clavicipitaceous fungi. *Stud. Mycol.* **2007**, *57*, 5–59. [CrossRef]
14. Kepler, R.M.; Sung, G.H.; Harada, Y.; Tanaka, K.; Tanaka, E.; Hosoya, T.; Bischoff, J.F.; Spatafora, J.W. Host jumping onto close relatives and across kingdoms by *Tyrannicordyceps* (Clavicipitaceae) gen. nov. and *Ustilaginoidea*_(Clavicipitaceae). *Am. J. Bot.* **2012**, *99*, 552–561. [CrossRef]
15. Kepler, R.; Ban, S.; Nakagiri, A.; Bischoff, J.; Hywel-Jones, N.; Owensby, C.A.; Spatafora, J.W. The phylogenetic placement of hypocrealean insect pathogens in the genus *Polycephalomyces*: An application of One Fungus One Name. *Fungal. Biol.* **2013**, *117*, 611–622. [CrossRef]
16. Tong, L.L.; Wang, Y.; Yuan, L.; Liu, M.Z.; Du, Y.H.; Mu, X.Y.; Yang, Q.H.; Wei, S.X.; Li, J.Y.; Wang, M.A.; et al. Enhancement of polysaccharides production using microparticle enhanced technology by *Paraisaria dubia*. *Microb. Cell Fact.* **2022**, *21*, 12. [CrossRef]
17. Perez-Villamares, J.C.; Burrola-Aguilar, C.; Aguilar-Miguel, X.; Sanjuan, T.; Jimenez-Sanchez, E. New records of entomopathogenous fungi of the genus *Cordyceps* s. l. (Ascomycota: Hypocreales) from the State of Mexico. *Rev. Mex. Biodivers.* **2017**, *88*, 773–783. [CrossRef]
18. Tong, L.L.; Wang, Y.; Du, Y.H.; Yuan, L.; Liu, M.Z.; Mu, X.Y.; Chen, Z.L.; Zhang, Y.D.; He, S.J.; Li, X.J.; et al. Transcriptomic Analysis of Morphology Regulatory Mechanisms of Microparticles to *Paraisaria dubia* in Submerged Fermentation. *Appl Biochem Biotechnol.* **2022**, *194*, 4333–4347. [CrossRef] [PubMed]
19. Zhou, L.; Wang, S.; Hao, Q.X.; Kang, L.P.; Kang, C.Z.; Yang, J.; Yang, W.Z.; Jiang, J.Y.; Huang, L.Q.; Guo, L.P. Bioaccessibility and risk assessment of heavy metals, and analysis of arsenic speciation in *Cordyceps sinensis*. *Chin. Med.* **2018**, *13*, 40. [CrossRef] [PubMed]
20. Song, Q.Y.; Zhu, Z.Y. Chemical structure and mechanism of polysaccharide on Pb^{2+} tolerance of *Cordyceps militaris* after Pb^{2+} domestication. *Int. J. Biol. Macromol.* **2020**, *165*, 958–969. [CrossRef]
21. Qian, G.M.; Pan, G.F.; Guo, J.Y. Anti-inflammatory and antinociceptive effects of cordymin, a peptide purified from the medicinal mushroom *Cordyceps sinensis*. *Nat. Prod. Res.* **2012**, *26*, 2358–2362. [CrossRef]
22. Wu, D.T.; Meng, L.Z.; Wang, L.Y.; Lv, G.P.; Cheong, K.L.; Hu, D.J.; Guan, J.; Zhao, J.; Li, S.P. Chain conformation and immunomodulatory activity of a hyperbranched polysaccharide from *Cordyceps sinensis*. *Carbohydr. Polym.* **2014**, *110*, 405–414. [CrossRef]
23. Wang, J.Q.; Kan, L.J.; Nie, S.P.; Chen, H.H.; Cui, S.W.; Phillips, A.O.; Phillips, G.O.; Li, Y.J.; Xie, M.Y. A comparison of chemical composition, bioactive components and antioxidant activity of natural and cultured *Cordyceps sinensis*. *LWT-Food Sci. Technol.* **2015**, *63*, 2–7. [CrossRef]
24. Ozenver, N.; Boulos, J.C.; Efferth, T. Activity of Cordycepin from *Cordyceps sinensis* Against Drug-Resistant Tumor Cells as Determined by Gene Expression and Drug Sensitivity Profiling. *Nat. Prod. Commun.* **2021**, *16*, 1934578X21993350. [CrossRef]

25. Xu, F.; Chen, P.; Li, H.; Qiao, S.Y.; Wang, J.X.; Wang, Y.; Wang, X.T.; Wu, B.H.; Liu, H.K.; Wang, C.; et al. Comparative transcriptome analysis reveals the differential response to cadmium stress of two Pleurotus fungi: *Pleurotus cornucopiae* and *Pleurotus ostreatus*. *J. Hazard. Mater.* **2021**, *416*, 125814. [CrossRef]
26. Bo, T.; Miao, L.; Cheng, Z.; Qian, Z.; Jia, S.R. Metabolomic Analysis of Antimicrobial Mechanisms of ε-Poly-l-lysine on Saccharomyces cerevisiae. *J. Agric. Food Chem.* **2014**, *62*, 4454–4465. [CrossRef] [PubMed]
27. Traxler, L.; Shrestha, J.; Richter, M.; Krause, K.; Schafer, T.; Kothe, E. Metal adaptation and transport in hyphae of the wood-rot fungus *Schizophyllum* commune. *J. Hazard. Mater.* **2022**, *425*, 127978. [CrossRef] [PubMed]
28. Xie, Y.; Ying, J.L.; Xu, L.; Wang, Y.; Dong, J.H.; Chen, Y.L.; Tang, M.J.; Li, C.; Muleke, E.M.; Liu, L.W. Genome-wide sRNA and mRNA transcriptomic profiling insights into dynamic regulation of taproot thickening in radish (*Raphanus sativus* L.). *BMC Plant Biol.* **2020**, *20*, 373. [CrossRef] [PubMed]
29. Sekiguchi, J.; Gaucher, G.M.; Costerton, J.W. Microcycle conidiation in *Penicillium urticae*: An ultrastructural investigation of conidiogenesis. *Can. J. Microbiol.* **1975**, *21*, 2069–2083. [CrossRef]
30. Anderson, J.G.; Smith, J.E. The Production of Conidiophores and Conidia by Newly Germinated Conidia of *Aspergillus niger* (*Microcycle conidiation*). *J. Gen. Microbiol.* **1971**, *69*, 185–197. [CrossRef]
31. Hu, T.; Sun, X.Y.; Zhao, Z.J.; Amombo, E.; Fu, J.M. High temperature damage to fatty acids and carbohydrate metabolism in tall fescue by coupling deep transcriptome and metabolome analysis. *Ecotoxicol. Environ. Saf.* **2020**, *203*, 110943. [CrossRef] [PubMed]
32. Huang, H.Y.; Ren, Q.Q.; Lai, Y.H.; Peng, M.Y.; Zhang, J.; Yang, L.T.; Huang, Z.R.; Chen, L.S. Metabolomics combined with physiology and transcriptomics reveals how Citrus grandis leaves cope with copper-toxicity. *Ecotoxicol. Environ. Saf.* **2021**, *223*, 112579. [CrossRef] [PubMed]
33. Zheng, X.W.; Liu, X.L.; Zhang, L.L.; Wang, Z.M.; Yuan, Y.; Li, J.; Li, Y.Y.; Huang, H.H.; Cao, X.; Fan, Z.Q. Toxicity mechanism of *Nylon microplastics* on *Microcystis aeruginosa* through three pathways: Photosynthesis, oxidative stress and energy metabolism. *J. Hazard. Mater.* **2022**, *426*, 128094. [CrossRef] [PubMed]
34. Mandal, P.K.; Roy, R.G.; Samkaria, A. Oxidative Stress: Glutathione and Its Potential to Protect Methionine-35 of Aβ Peptide from Oxidation. *Acs Omega* **2022**, *7*, 27052–27061. [CrossRef]
35. Farres, M.; Pina, B.; Tauler, R. LC-MS based metabolomics and chemometrics study of the toxic effects of copper on *Saccharomyces cerevisiae*. *Metallomics* **2016**, *8*, 790–798. [CrossRef]
36. He, R.J.; Wang, Z.; Cui, M.; Liu, S.; Wu, W.; Chen, M.; Wu, Y.C.; Qu, Y.J.; Lin, H.; Chen, S.; et al. HIF1A Alleviates compression-induced apoptosis of nucleus pulposus derived stem cells via upregulating autophagy. *Autophagy* **2021**, *17*, 3338–3360. [CrossRef]
37. Remus, D.; Beuron, F.; Tolun, G.; Griffith, J.D.; Morris, E.P.; Diffley, J.F.X. Concerted Loading of Mcm2-7 Double Hexamers around DNA during DNA Replication Origin Licensing. *Cell* **2009**, *139*, 719–730. [CrossRef]
38. Jiang, M.R.; Yang, Y.; Wu, J.R. Activation of DNA damage checkpoints in CHO cells requires a certain level of DNA damage. *Biochem. Biophys. Res. Commun.* **2001**, *287*, 775–780. [CrossRef]
39. Zhu, S.; Li, X.; Dang, B.R.; Wu, F.; Wang, C.M.; Lin, C.J. Lycium Barbarum polysaccharide protects HaCaT cells from PM2.5-induced apoptosis via inhibiting oxidative stress, ER stress and autophagy. *Redox Rep.* **2022**, *27*, 32–44. [CrossRef]
40. Mizushima, N. The role of the Atg1/ULK1 complex in autophagy regulation. *Curr Opin Cell Biol.* **2010**, *22*, 132–139. [CrossRef]
41. Narendrula-Kotha, R.; Theriault, G.; Mehes-Smith, M.; Kalubi, K.; Nkongolo, K. Metal Toxicity and Resistance in Plants and Microorganisms in Terrestrial Ecosystems. *Rev Environ Contam Toxicol.* **2020**, *249*, 1–27. [CrossRef] [PubMed]
42. Krishna, T.P.A.; Maharajan, T.; Roch, G.V.; Ignacimuthu, S.; Ceasar, S.A. Structure, Function, Regulation and Phylogenetic Relationship of ZIP Family Transporters of Plants. *Front. Plant Sci.* **2020**, *11*, 662. [CrossRef] [PubMed]
43. Gustin, J.L.; Loureiro, M.E.; Kim, D.; Na, G.; Tikhonova, M.; Salt, D.E. MTP1-dependent Zn sequestration into shoot vacuoles suggests dual roles in Zn tolerance and accumulation in Zn-hyperaccumulating plants. *Plant J.* **2009**, *57*, 1116–1127. [CrossRef] [PubMed]
44. Gonzalez-Guerrero, M.; Benabdellah, K.; Valderas, A.; Azcon-Aguilar, C.; Ferrol, N. GintABC1 encodes a putative ABC transporter of the MRP subfamily induced by Cu, Cd, and oxidative stress in *Glomus intraradices*. *Mycorrhiza* **2010**, *20*, 137–146. [CrossRef] [PubMed]
45. Leng, Y.; Li, Y.; Wen, Y.; Zhao, H.; Wang, Q.; Li, S.W. Transcriptome analysis provides molecular evidences for growth and adaptation of plant roots in cadimium-contaminated environments. *Ecotoxicol. Environ. Saf.* **2020**, *204*, 111098. [CrossRef] [PubMed]
46. Kumar, A.; Prasad, M.N.V. Plant-lead interactions: Transport, toxicity, tolerance, and detoxification mechanisms. *Ecotoxicol. Environ. Saf.* **2018**, *166*, 401–418. [CrossRef]
47. Lu, H.Y.; Liu, S.Y.; Zhang, S.L.; Chen, Q.H. Light Irradiation Coupled with Exogenous Metal Ions to Enhance Exopolysaccharide Synthesis from Agaricus sinodeliciosus ZJU-TP-08 in Liquid Fermentation. *J. Fungi.* **2021**, *7*, 992. [CrossRef]
48. Yang, S.L.; Yang, X.; Zhang, H. Extracellular polysaccharide biosynthesis in *Cordyceps*. *Crit. Rev. Microbiol.* **2020**, *46*, 359–380. [CrossRef]
49. Yuan, Y.; Imtiaz, M.; Rizwan, M.; Dai, Z.H.; Hossain, M.M.; Zhang, Y.H.; Huang, H.L.; Tu, S.X. The role and its transcriptome mechanisms of cell wall polysaccharides in vanadium detoxication of rice. *J. Hazard. Mater.* **2022**, *425*, 127966. [CrossRef]
50. Li, Q.Z.; Chen, J.; Liu, J.Y.; Yu, H.L.; Zhang, L.J.; Song, C.Y.; Li, Y.; Jiang, N.; Tan, Q.; Shang, X.D.; et al. De novo Sequencing and Comparative Transcriptome Analyses Provide First Insights into Polysaccharide Biosynthesis During Fruiting Body Development of Lentinula edodes. *Front. Microbiol.* **2021**, *12*, 627099. [CrossRef]

51. Xu, Y.; Du, H.M.; Huang, B.R. Identification of Metabolites Associated with Superior Heat Tolerance in Thermal Bentgrass through Metabolic Profiling. *Crop. Sci.* **2013**, *53*, 1626–1635. [CrossRef]
52. Hausler, R.E.; Ludewig, F.; Krueger, S. Amino acids—A life between metabolism and signaling. *Plant Sci.* **2014**, *229*, 225–237. [CrossRef]
53. Jung, B.; Kim, S.; Lee, J. Microcyle Conidiation in Filamentous Fungi. *Mycobiology* **2014**, *42*, 1–5. [CrossRef]
54. Hanlin, R.T.J.M. Microcycle conidiation—A review. *J. Gen. Microbiol.* **1994**, *35*, 113–123. [CrossRef]
55. Bosch, A.; Yantorno, O.J. Microcycle conidiation in the entomopathogenic fungus Beauveria bassiana bals. (vuill.). *Process Biochem.* **1999**, *34*, 707–716. [CrossRef]
56. Liu, J.; Cao, Y.; Xia, Y. Mmc, a gene involved in microcycle conidiation of the entomopathogenic fungus *Metarhizium anisopliae*. *J. Invertebr. Pathol.* **2010**, *105*, 132–138. [CrossRef] [PubMed]
57. Maheshwari, R. Microcycle conidiation and its genetic basis in *Neurospora crassa*. *J. Gen. Microbiol.* **1991**, *137*, 2103–2115. [CrossRef] [PubMed]
58. Shrestha, B.; Han, S.-K.; Yoon, K.-S.; Sung, J.-M. Morphological Characteristics of Conidiogenesis in *Cordyceps militaris*. *Mycobiology* **2005**, *33*, 69–76. [CrossRef]
59. Park, H.S.; Yu, J.H. Genetic control of asexual sporulation in filamentous fungi. *Curr. Opin. Microbiol.* **2012**, *15*, 669–677. [CrossRef]
60. de Vries, R.P.; Riley, R.; Wiebenga, A.; Aguilar-Osorio, G.; Amillis, S.; Uchima, C.A.; Anderluh, G.; Asadollahi, M.; Askin, M.; Barry, K.; et al. Comparative genomics reveals high biological diversity and specific adaptations in the industrially and medically important fungal genus *Aspergillus*. *Genome Biol.* **2017**, *18*, 28. [CrossRef]
61. Wendland, J.; Dunkler, A.; Walther, A. Characterization of alpha-factor pheromone and pheromone receptor genes of *Ashbya gossypii*. *FEMS Yeast Res.* **2011**, *11*, 418–429. [CrossRef]
62. Sun, X.Y.; Wang, F.; Lano, N.; Liu, B.; Hu, C.C.; Xue, W.; Zhang, Z.Y.; Li, S.J. The Zn(II)2Cys6-Type Transcription Factor ADA-6 Regulates Conidiation, Sexual Development, and Oxidative Stress Response in *Neurospora crassa*. *Front. Microbiol.* **2019**, *10*, 750. [CrossRef] [PubMed]
63. Son, Y.E.; Cho, H.J.; Lee, M.K.; Park, H.S. Characterizing the role of Zn cluster family transcription factor ZcfA in governing development in two Aspergillus species. *PLoS ONE* **2020**, *15*, e0228643. [CrossRef] [PubMed]
64. Wu, M.Y.; Mead, M.E.; Lee, M.K.; Neuhaus, G.F.; Adpressa, D.A.; Martien, J.I.; Son, Y.E.; Moon, H.; Amador-Noguez, D.; Han, K.H.; et al. Transcriptomic, Protein-DNA Interaction, and Metabolomic Studies of VosA, VelB, and WetA in Aspergillus nidulans Asexual Spores. *mBio* **2021**, *12*, e03128-20. [CrossRef]
65. Bayram, O.; Krappmann, S.; Ni, M.; Bok, J.W.; Helmstaedt, K.; Valerius, O.; Braus-Stromeyer, S.; Kwon, N.J.; Keller, N.P.; Yu, J.H.; et al. VelB/VeA/LaeA complex coordinates light signal with fungal development and secondary metabolism. *Science* **2008**, *320*, 1504–1506. [CrossRef] [PubMed]
66. Park, H.S.; Ni, M.; Jeong, K.C.; Kim, Y.H.; Yu, J.H. The Role, Interaction and Regulation of the Velvet Regulator VelB in *Aspergillus nidulans*. *PLoS ONE* **2012**, *7*, 45935. [CrossRef] [PubMed]
67. Tsitsigiannis, D.I.; Kowieski, T.M.; Zarnowski, R.; Keller, N.P. Endogenous lipogenic regulators of spore balance in *Aspergillus nidulans*. *Eukaryot Cell* **2004**, *3*, 1398–1411. [CrossRef] [PubMed]
68. Khalid, A.R.; Lv, X.; Naeem, M.; Mehmood, K.; Shaheen, H.; Dong, P.; Qiu, D.; Ren, M. Autophagy Related Gene (ATG3) is a Key Regulator for Cell Growth, Development, and Virulence of *Fusarium oxysporum*. *Genes* **2019**, *10*, 658. [CrossRef]
69. Lin, H.Y.; Wang, J.J.; Feng, M.G.; Ying, S.H. Autophagy-related gene ATG7 participates in the asexual development, stress response and virulence of filamentous insect pathogenic fungus *Beauveria bassiana*. *Curr. Genet.* **2019**, *65*, 1015–1024. [CrossRef]
70. Deniaud, E.; Le Gall, L.; Rusig, A.M.; Lahaye, M. Initial observations on glycoside deposition in cell walls of *Palmaria palmata* (L.) Kuntze (Rhodophyta) during spore germination. *Bot. Mar.* **2006**, *49*, 266–269. [CrossRef]
71. Zhang, S.Z.; Xia, Y.X.; Keyhani, N.O. Contribution of the gas1 Gene of the Entomopathogenic Fungus Beauveria bassiana, Encoding a Putative Glycosylphosphatidylinositol-Anchored beta-1,3-Glucanosyltransferase, to Conidial Thermotolerance and Virulence. *Appl. Environ. Microbiol.* **2011**, *77*, 2676–2684. [CrossRef] [PubMed]
72. Sheng, W.; Yamashita, S.; Ohta, A.; Horiuchi, H. Functional Differentiation of Chitin Synthases in *Yarrowia lipolytica*. *Biosci. Biotechnol. Biochem.* **2013**, *77*, 1275–1281. [CrossRef] [PubMed]

Journal of *Fungi*

MDPI

Article

Ectomycorrhizal Community Shifts at a Former Uranium Mining Site

Olga Bogdanova, Erika Kothe and Katrin Krause *

Microbial Communication, Institute of Microbiology, Friedrich Schiller University Jena, D-07743 Jena, Germany
* Correspondence: katrin.krause@uni-jena.de

Abstract: Ectomycorrhizal communities at young oak, pine, and birch stands in a former uranium mining site showed a low diversity of morphotypes with a preference for contact and short-distance exploration strategies formed by the fungi *Russulaceae, Inocybaceae, Cortinariaceae, Thelephoraceae, Rhizopogonaceae, Tricholomataceae*, as well as abundant *Meliniomyces bicolor*. In order to have better control over abiotic conditions, we established pot experiments with re-potted trees taken from the sites of direct investigation. This more standardized cultivation resulted in a lower diversity and decreased prominence of *M. bicolor*. In addition, the exploration strategies shifted to include long-distance exploration types. To mimic secondary succession with a high prevalence of fungal propagules present in the soil, inoculation of re-potted trees observed under standardized conditions for two years was used. The super-inoculation increased the effect of lower abundance and diversity of morphotypes. The contact morphotypes correlated with high Al, Cu, Fe, Sr, and U soil contents, the dark-colored short-distance exploration type did not show a specific preference for soil characteristics, and the medium fringe type with rhizomorphs on oaks correlated with total nitrogen. Thus, we could demonstrate that field trees, in a species-dependent manner, selected for ectomycorrhizal fungi with exploration types are likely to improve the plant's tolerance to specific abiotic conditions.

Keywords: ectomycorrhiza; morphotype; metal; plant; field; pot; inoculation

Citation: Bogdanova, O.; Kothe, E.; Krause, K. Ectomycorrhizal Community Shifts at a Former Uranium Mining Site. *J. Fungi* **2023**, *9*, 483. https://doi.org/10.3390/jof9040483

Academic Editors: Ivan Širić and Miha Humar

Received: 27 February 2023
Revised: 6 April 2023
Accepted: 8 April 2023
Published: 18 April 2023

1. Introduction

Post-mining areas constitute severe disturbances of the ecosystem with considerable soil damage. Waste rock heaps, their footprint, or open pit surfaces are generally low in nutrients, depleted of organic matter, and might contain high amounts of toxic metals due to acid mine drainage (AMD [1]). At the same time, the substrate is characterized by low porosity, decreased water-holding capacity, and the lack of water-stable aggregates due to compaction. As a result of the low content of clay-humic complexes and acidic pH, metals are almost not absorbed or sedimented, which facilitates their migration and transfer into food chains and, therefore, represents a threat to public health. Within the European Union, former mining areas are required to be rehabilitated to reduce severe effects on the environment and human health [2].

An example of such remediation activities is found at the former uranium mining site near Ronneburg, Germany. After the removal of the waste rock material, a topsoil cover was applied to stop further oxidative processes and to re-establish vegetation, including re-forestation in primary succession. The extreme fluctuations of abiotic conditions with the heterogeneous distribution of toxic metals and acidity, combined with low nutrient and soil moisture, resulted in locally successful tree growth, leaving large spots of topsoil unvegetated. Therefore, the sustainability and stability of land cover need improvement. Remediation methods developed for such sites might become especially helpful in the future when heavy rain events or drought spells become more prominent with climate change.

One measure that may reduce stress on planted trees is the mutually beneficial association with ectomycorrhizal fungi, mostly belonging to the *Basidiomycota*. *Betulaceae*, *Pinaceae*, and *Fagaceae* are prominent trees in temperate forest ecosystems and provide promising candidates for such approaches [3–5]. The mutually beneficial symbiosis, in addition to the general traits of nutrient exchange, alleviates abiotic stress like drought or can protect from bioavailable heavy metals, as well as increasing resistance against pathogens [5–7]. The positive effects on tree growth and metabolism can raise the survival of plants during afforestation [8–10].

The invasion of short roots by the ectomycorrhizal fungus leads to establishing the symbiotic morphotypes with a hyphal mantle surrounding the root and the Hartig' net providing an extended surface for the exchange of signal molecules as well as water and nutrients between the root cells of the tree, leaving the vascular tissue untouched [11]. Differentiation and distribution of extraradical mycelium reflect their functional traits and describe how fungi explore and interact with the environment defining fungal exploration types [12,13]. Here, different fungal species can be found to form specific exploration types, e.g., the short-contact type of *Inocybe* species [14]. While in nutrient-rich environments with amino acids, ammonium, and nitrate available, contact and short-distance exploration type fungi are generally found to be more prominent, scarce, and spatially dispersed and/or insoluble nutrient supplies lead to increased formation of medium and long-distance rhizomorphs that can transport nutrients towards the host tree over larger distances [15–17].

Like with plant communities, succession types for ectomycorrhiza can be defined, and these are distributed into "early-stage" or "late-stage" mycorrhizal interactions [18]. While *Inocybe*, *Laccaria*, and *Thelephora* are dominant at early successional stages, *Lactarius*, *Cortinarius*, *Russula*, or *Tricholoma* are typical for later stages of succession in different abiotic conditions of primary succession [19–21].

Fungi occurring in metal-contaminated soils are exposed to selective environmental pressure leading to adaption by developing particular metal resistance mechanisms [22,23]. Often these mechanisms are based on a modification of metal adaptation that causes decreased or increased metal mobility, e.g., by the release of organic acids [24]. It has been shown that some ectomycorrhizal fungal species demonstrate a high tolerance to environmental stress, including metal toxicity and low pH, and, therefore, can be applied in soil remediation programs [25,26]. Inoculation of plants with ectomycorrhizal fungi alone or together with associated bacteria might either intensify phytoextraction by plants and lead to the accumulation of heavy metals in aboveground biomass [27,28] or immobilize and sequester metal ions in fungal biomass by biosorption to chitin or melanin, and bioaccumulation with the transport of metals inside the cells by different compounds, e.g., metallothioneins, and surrounding soil limiting plant metal uptake [22,29–31]. While for metal removal in phytoextraction, often trees of the species *Salix* or *Populus* have been planted, a valid and sustainable land-use strategy might involve afforestation of the very large, heterogeneously and low-level contaminated lands with temperate forest climax species trees like oak, beech, birch, maple, or coniferous pine and spruce that can be harvested and at the same time provide recreational space without harmful dispersal of metals in the environment. In that case, phytostabilization is the process providing the highest potential. Since fungi can determine plant metal uptake, a stable mycorrhization should be reached, albeit with a high-stress impact, especially on young trees.

With this study, we combined direct observation of the roots of young trees in a former mining site with a set-up using re-potted trees from that same area in a pot experiment. This allowed us to create comparable abiotic conditions. In addition, the pot experiments could be inoculated in addition to the already established mycorrhiza to see whether the presence of additional, late-stage mycorrhiza propagules would improve tree health in a metal-contaminated substrate from a former mining heap site. Thus, this additional experiment was used to mimic secondary succession in a short time frame. The results may be useful to establish re-forestation programs after the closure of mine operations in AMD-impacted post-mining scenarios. Species and functional diversity of mycorrhiza were studied to gain insight into the establishment of mutual symbiosis under detrimental environmental conditions.

2. Materials and Methods

2.1. Experimental Settings

The field site, Kanigsberg, at the former uranium mining area near Ronneburg, Germany, is characterized by a high content of mobile metals, low pH, and high structural heterogeneity. Naturally occurring stands of young birch (*Betula* sp.), oak (*Quercus* sp.), and pine (*Pinus* sp.) trees, all approximately 2 to 3 years of age, were analyzed. Visually undamaged trees of approximately the same height were sampled: birches of 25–30 cm, oaks of 7–10 cm, and pines of 7–11 cm (Table 1). The topsoil layer was carefully removed to expose the root system, and several soil aggregates directly attached to the roots (along the whole root system when possible) were carefully sampled (rhizosphere soil; three replicates per tree species). Soil not directly connected to tree roots was taken for comparison (bulk soil; Supplementary Figure S1). Native roots were sampled for mycorrhiza characterization.

Table 1. Field sampling sites.

Sampling Site	Location
Birch and soil sampling site	50°49′35.35″ N, 12°9′11.68″ E
Oak sampling site	50°49′36.13″ N, 12°9′12.56″ E
Pine sampling site	50°49′39.65″ N, 12°9′17.87″ E

At the same site of the former uranium mining site, soil and plant material were retrieved to set up a pot experiment. This was established with soil sampled from Kanigsberg (50°49′35.27″ N, 12°9′12.13″ E) in a greenhouse experiment to control abiotic conditions. The pre-sieved control pot substrate was dried for three days, sieved (<2.8 mm), and filled into Mitscherlich pots (12 L; Supplementary Figure S2). To prevent direct contact with the soil substrate and pot metal walls, pots were inlaid with polyethylene film. Naturally occurring trees were sampled for the pot experiment, including 15 × 15 cm soil cores from the above-mentioned field site. Soil cores were carefully removed before planting with two trees of one species per pot with 6 kg of a substrate prepared as mentioned above. For additional mycorrhizal inoculation, a blend of mycorrhizal fungi in peat substrate containing *Amanita muscaria, Boletus edulis, Hebeloma crustuliniforme, Paxillus involutus, Pisolithus tinctorius, Cenococcum geophilum, Pisolithus arrhenius* and the endomycorrhizal fungus *Rhizophagus irregularis* was used (INOQ Forst, Schnega, Germany). The inoculation was performed according to the manufacturer's recommendations. All pots, including control variants, were watered with deionized water to approximately 15 volume % water content. Watering was performed weekly in the autumn-winter seasons and twice per week in the spring-summer seasons. To improve air exchange, the topsoil crust was carefully broken with non-metal equipment the day after watering. The experimental variants (Supplementary Table S1) contained at least three replicates of pots resulting in 10 to 14 plant replicates for each tree species. After two years, rhizosphere soil and roots were sampled.

2.2. Soil Chemical Analysis

To perform chemical analysis, the soil was air-dried at 40 °C, crushed, and sieved (<2.0 mm). Soil pH was measured with 5 g of pre-treated soil mixed with 25 mL of deionized water shaken for 1 h (pH electrode SenTix 81; Xylem Analytics, Weilheim, Germany) with three technical replicates. Elemental analysis was performed with the air-dried soil, finely ground in an agate mortar stored in 50 mL air-tight falcon tubes in darkness. Contents of total carbon (TC) and total nitrogen (TN) were measured with a VarioMAX CN Element Analyzer (Elementar Analysensysteme, Hanau, Germany). The contents were calculated corresponding to the absolute dry weight (measured at 105 °C). Total phosphorus (TP;

ICP-AES, PerkinElmer, Waltham, MA, USA) and the bioavailable fractions of toxic metals (ICP-OES, 725ES, Agilent, Waldbronn, Germany; Quadrupole-ICP-MS (XSeries II, Thermo Scientific, Bremen, Germany) were determined after sequential extraction [32]. The method used has been devised specifically for former mining site substrates. In our investigation, specifically, the two fractions containing bioavailable elements in the mobile fraction (F1), extracted with 1 M NH_4NO_3, and specifically adsorbed fraction (F2), extracted using 1 M NH_4OAc at pH 6,0, were considered. Standard reference material SPS-SW2 (Spectrapure Standards AS) and NIST 1643e (NIST) were analyzed, and the multielement standard solution (500 mg/L Ca, K, Mg, Bernd Kraft) was measured and compared to the certified values to prove the measurement accuracy. For every variant of the experiment, element analyses were performed in triplicates.

2.3. *Morphotyping*

To analyze ectomycorrhizal fungal communities, five trees of each species were sampled (20 × 20 cm soil cores, 10–15 cm depth, ca. 50–100 cm apart from each other). To perform morphotyping, the roots were soaked overnight in tap water at 4 °C, carefully washed on a sieve, and cut into 1–3 cm pieces before the fine roots were separated and observed with a dissecting microscope (Stemi, 2000-C, Zeiss, Jena, Germany).

To analyze the mycorrhizosphere of pot plants, the soil was carefully removed from pots in portions to prevent disturbance, and the tree was removed and immediately processed as described above. Root systems of three plants of each tree species, non-inoculated and inoculated separately, were observed.

The morphology was described according to Agerer [33], with color, type of ramification, the shape of unramified ends, mantle surface, characteristics of rhizomorph, and characteristics of emanating hyphae (compare Supplementary Table S2). The abundance of each morphotype per total root length was assessed. Mycorrhizal root tips of different morphology were collected for molecular identification.

2.4. *Molecular Identification*

The small sample size of individual mycorrhizal root tips combined with a substrate rich in heavy metals that may hinder polymerase amplification and the presence of iron and manganese hydroxides that unspecifically bind nucleic acids warranted special care to be taken for DNA sequencing. Root sections from selected short roots were placed in 1.5 mL microcentrifuge tubes containing 0.5 mL sterile distilled water and vortexed for 30 s to remove soil particles. Roots were then transferred to 1.5 mL microcentrifuge tubes with 200 µL 30% H_2O_2 and vortexed for 10 s for surface sterilizing, rinsed in sterile distilled water three times, and transferred into a sterile 1.5 mL microcentrifuge tube kept at −20 °C until further processing. Direct PCR [34] using primers specific for fungal internal transcribed spacer (ITS) was performed then on a small piece of the hyphal mantle taken under the binocular. This piece was placed in 20 µL of sterile distilled water and taken for PCR amplification. This led to the successful identification of some of the morphotypes.

When the direct PCR approach was not successful, DNA was extracted from sampled morphotypes. The different individual samples were first extracted using PowerSoil DNA Isolation Kit (MoBio, Carlsbad, CA, USA) after milling with sterile glass beads (Sigma Aldrich, Taufkirchen, Germany) using a plastic pestle, vortexing with 100 µL of sterile distilled water for 30 s and transfer of the supernatant. Bovine serum albumin (Carl Roth, Karlsruhe, Germany) was added at a final concentration of 0.4 µg/µL to reduce the inhibition of *Taq* polymerase. After mixing by pipetting, in a final 50 µL volume, the PCR was performed with 6.25 µL 10x Dream *Taq* Buffer, 1.25 µL 10 mM dNTPs, 5 µL 10 µM each forward (ITS1: 5′-TTCGTAGGTGAACCTGCGG-3′) and reverse (ITS4: 5′-TCCTCCGCTTATTGATATGC-3′) primers [35], and 1.25 U *Taq* polymerase. After preheating for 5 min at 95 °C, 35 cycles (95 °C 30 s, 56 °C 30 s, 72 °C 50 s) and final elongation at 72 °C for 10 min followed. Amplified DNA was visualized and documented after agarose gel electrophoresis using 1 µg/mL ethidium bromide under UV light.

The PCR products were purified (QIAquick, Qiagen, Germany) and sequenced (GATC, Konstanz, Germany) using primer ITS1. Sequences were compared to entries in NCBI (http://www.ncbi.nlm.nih.gov; accessed on 17 March 2023) and UNITE [36] databases.

2.5. Data Processing

Ectomycorrhizal (ECM) communities of field trees were characterized based on the relative abundance of morphotypes with community diversity indices (Shannon diversity index (H_{SH})), Gini-Simpson index (H_{GS}), Simpson dominance index (H_{SD}), Berger–Parker index (H_{BP})). Trees of one species were compared to each other within each variant of the experiment (field trees, non-inoculated pot trees, inoculated pot trees) to determine the similarity/dissimilarity of ECM communities with Sørensen, Jaccard, and Bray–Curtis indices and to estimate representability of replicates.

Canonical correspondence analysis (CCA) was performed to estimate the correlation of exploration types of ectomycorrhiza and field morphotypes with environmental variables. A p-value lower than 0.05 was considered significant. An open-source software, PAST 4.03, was used for all multivariate analyses. The significance of multivariate analysis results was checked with a one-way analysis of similarities (ANOSIM) and one-way permutational analysis of variance (PERMANOVA) and calculated by permutation of group membership (N = 9999). Bonferroni correction was applied.

Shapiro–Wilks's test of normality was performed with JASP 0.14.0.0 open-source software. If the data were normally distributed, one-way ANOVA analysis with posthoc Tukey test was applied to identify significant differences between groups of values (relative abundance of taxa, diversity indices, or environmental variables value). If Shapiro–Wilks's test failed, a non-parametric Kruskal–Wallis test with Bonferroni correction was performed.

Diversity indices were calculated with PAST 4.03. Platform SPADE R online (Species Prediction and Diversity Estimation [37]) was used to calculate community similarity indices (Sørensen, Jaccard, Bray-Curtis) for each variant. The number of bootstrap replications was 100.

Microsoft Excel was used to calculate mean values and standard deviation of the variables and calculate and depict the relative abundance of the most representative taxa as well as differences between groups of variants.

3. Results

3.1. Meliniomyces Dominates Ectomycorrhizae on Birch, Oak, and Pine in the Field

For the trees grown in the field, the naturally occurring mycorrhizae were determined. Eleven morphotypes could be distinguished, three on birches, four on oaks, and four on pines (Figures 1 and S3; Supplementary Tables S3 and S4). The most abundant morphotype for all field trees, irrespective of tree species, was formed with ECM fungus *Meliniomyces bicolor* (the morphotype description corresponding to the dark-colored type of fungus *Cenococcum geophilum*). The remaining morphotypes were specific to the three tree species. While birches were associated with the basidiomycete *Lactarius mammosus* in almost equal amounts as *Meliniomyces*, *Inocybe lacera* was less widely distributed. The oak trees, in addition to the most representative fungus *M. bicolor* (with the morphotype here previously described as *Pinirhiza bicolorata*), showed a subdominant morphotype with a smooth hyphal mantle of light brown to beige color without emanating hyphae that resisted molecular approaches for sequencing. The remaining morphotype was formed by *Cortinarius bivelus*. With the third tree species, pine from the field site, the most abundant morphotypes were formed again by *M. bicolor*, while other basidiomycetes present were *Tomentella ellisii* and minor representatives *Rhizopogon mohelensis* and *Tricholoma argyraceum*.

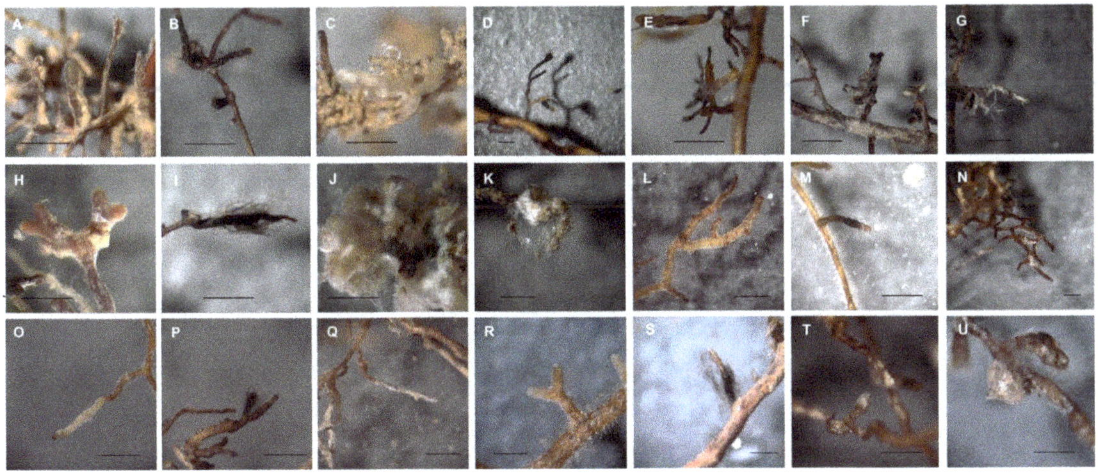

Figure 1. Morphotypes observed on field plants. Birch roots with (**A**) *Lactarius mammosus*, (**B**) *Meliniomyces bicolor* (*Pinirhiza bicolorata*-type with emanating hyphae), and (**C**) *Inocybe lacera*; oak roots with (**D**) *Meliniomyces bicolor* (*Cenococcum geophilum*-type), (**E**) non-identified O_F_MT2, (**F**) *Meliniomyces bicolor* (*Pinirhiza bicolorata*-type w/o emanating hyphae), and (**G**) *Cortinarius bivelus*; pine roots with (**H**) *Tomentella ellisii*, (**I**) *Meliniomyces bicolor* (*C. geophilum*-type), (**J**) *Rhizopogon mohelnensis*, and (**K**) *Tricholoma argyraceum*. With inoculated pot plants, birch roots formed mycorrhizae with (**L**) non-identified B_inoc_MT1, (**M**) non-identified B_inoc_MT2, and (**N**) non-identified B_inoc_MT3; oaks roots with (**O**) non-identified O_inoc_MT1, (**P**) non-identified O_inoc_MT2, (**Q**) *Pisolithus arhizus*; and pine roots with (**R**) *Inocybe lacera*, (**S**) *Meliniomyces* sp., (**T**) *Meliniomyces bicolor*, and (**U**) *Rhizopogon mohelnensis*. Bar always represents 1.0 mm.

3.2. Comparison to Morphotypes Present on the Trees Planted in Pots

An extensively enlarged root system by forming new roots was visible for pot birches and, to a lesser extent, pot oaks. At the same time, the original roots of all pot plants looked dried, and some were even non-viable. Mycorrhiza observed on pot plants was characterized by a wrinkled mantle surface and light brown to reddish color; the more stable abiotic stress factors in pots resulted in less dominance of the dark *Meliniomyces* ectomycorrhizae. Without artificial inoculation, 12 morphotypes were identified: five for birches, three for oaks, and four for pines (Figure 2).

Among them, a member of the genus *Meliniomyces* and subdominant *Inocybe lacera* occurred (Supplementary Table S3). These mycorrhiza types, therefore, had increased in abundance from the time of tree transplantation, and *Meliniomyces* had decreased in the abundance of mycorrhizae after the two years in pots.

To mimic a secondary succession where fungal propagules would already be present in sufficient amounts, potted trees were inoculated in addition to the naturally acquired mycorrhizae. Here, the diversity was slightly lower with 10 instead of 12 morphotypes on birches, while again, three morphotypes were observed on oaks and four on pines (Figure 1). *Pisolithus arhizus* was identified on inoculated pot oaks. However, the low numbers do not allow for direct comparison. Hence more statistical approaches were needed.

A higher diversity was confirmed with Shannon diversity (1.03) and Gini–Simpson indices (0.61) for non-inoculated pot oaks and the lowest for field oaks and inoculated pot birches. The Berger–Parker index had the highest values for inoculated pot birches (0.75) and field oaks, suggesting that common morphotypes mainly dominated ectomycorrhizal communities in these variants (Tables 2 and S5).

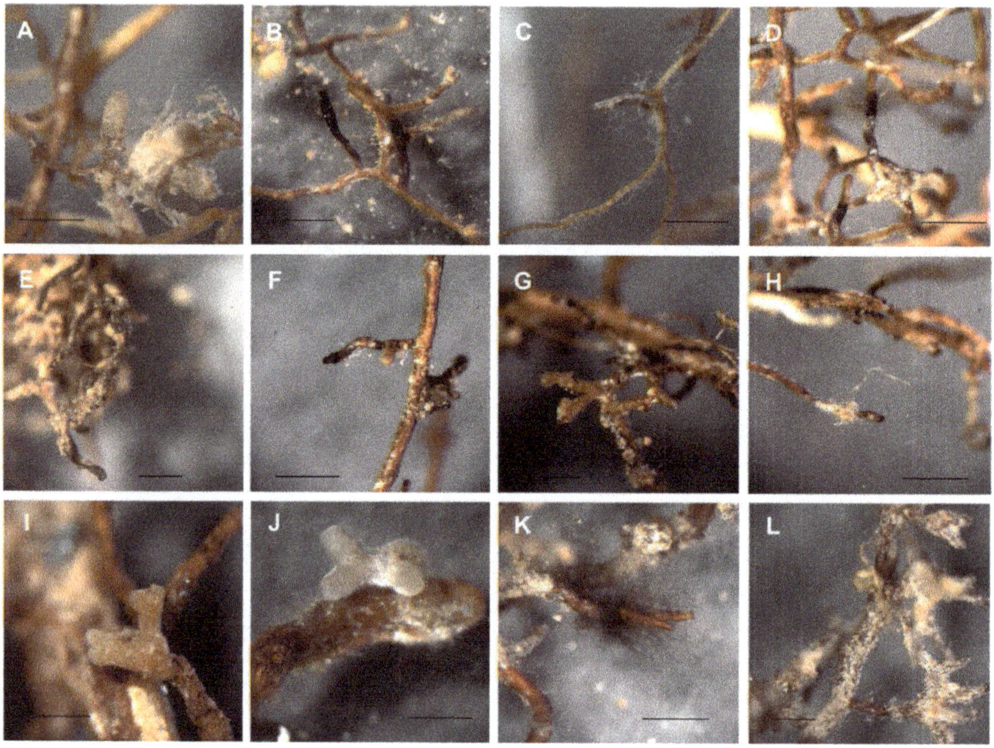

Figure 2. Morphotypes observed on non-inoculated pot plants. Birch roots with (**A**) non-identified B_non_inoc_MT1, (**B**) non-identified B_non_inoc_MT2, (**C**) non-identified B_non_inoc_MT3, (**D**) non-identified B_non_inoc_MT4, (**E**) non-identified B_non_inoc_MT5; oak roots with (**F**) non-identified O_non_inoc_MT1, (**G**) non-identified O_non_inoc_MT2, (**H**) *Meliniomyces*; pine roots with (**I**) *Inocybe lacera*, (**J**) non-identified P_non_inoc_MT2, (**K**) non-identified P_non_inoc_MT3, (**L**) non-identified P_non_inoc_MT4. Bar always represents 1.0 mm.

Table 2. Diversity indices calculated for ectomycorrhizal fungi communities in different variants of the experiment.

Variant	H_{SD}	H_{GS}	H_{SH}	H_{BP}
Field birches	0.60 ± 0.11	0.39 ± 0.11	0.64 ± 0.16	0.72 ± 0.13
Field oaks	0.66 ± 0.20	0.34 ± 0.20	0.55 ± 0.30	0.73 ± 0.21
Field pines	0.55 ± 0.14	0.45 ± 0.14	0.71 ± 0.24	0.66 ± 0.15
Non-inoculated pot birches	0.48 ± 0.05	0.52 ± 0.05	0.90 ± 0.13	0.60 ± 0.07
Non-inoculated pot oaks	0.39 ± 0.06	0.61 ± 0.06	1.03 ± 0.08	0.50 ± 0.13
Non-inoculated pot pines	0.57 ± 0.13	0.43 ± 0.13	0.77 ± 0.16	0.69 ± 0.16
Inoculated pot birches	0.68 ± 0.36	0.32 ± 0.36	0.56 ± 0.55	0.75 ± 0.31
Inoculated pot oaks	0.51 ± 0.11	0.49 ± 0.11	0.77 ± 0.18	0.59 ± 0.14
Inoculated pot pines	0.49 ± 0.13	0.51 ± 0.17	0.82 ± 0.21	0.57 ± 0.17

H_{SH}—Shannon diversity index, H_{GS}—Gini–Simpson index, H_{SD}—Simpson dominance index, H_{BP}—Berger–Parker index.

This was supported by a comparison of similarity indices which revealed that the ectomycorrhizal communities among the trees in one variant were similar (Supplementary Table S6), with a high Sørensen similarity index (0.833 for field oaks and field and inoculated pot pines) to 1.000 (determined for non-inoculated pot oaks). Similarly, the Bray–Curtis similarity index, which also includes species abundance, had high values from 0.852 for inoculated pot pines to 1.000 calculated for field and inoculated pot birches, pot oaks, and non-inoculated pot pines. Low values of the Jaccard coefficient were calculated for field oaks (0.500) and field pines (0.500), suggesting that these trees had only 50% of common species. For trees in other variants, the Jaccard coefficient ranged from 0.700 to 1.000 representing high similarity within one tree species.

3.3. Functional Diversity of Ectomycorrhizal Community

Among all field trees, the most abundant exploration types were contact and short-distance exploration types (Figure 3, Table 3). ECM fungi *Lactarius mammosus* and *Inocybe lacera* observed on field birches, unidentified field oak morphotype O_F_MT2 with smooth hyphal mantle and no rhizomorphs, dark-colored without emanating hyphae field oak morphotype of ECM fungus *Meliniomyces* and field pine morphotype formed by fungus *Tomentella ellisii* contributed to the group of contact exploration type.

Table 3. Summarized data of morphotypes, the identification of ECM fungus, and exploration type.

Host	Variant	Morphotype	Molecular Identification	Exploration Type
Birch	Field plant	B_F_MT1	*Lactarius mammosus*	contact
		B_F_MT2	*Meliniomyces bicolor*	short
		B_F_MT3	*Inocybe lacera*	contact
	Pot non-inoculated plant	B_non_inoc_MT1	non-identified	contact
		B_non_inoc_MT2	non-identified	short
		B_non_inoc_MT3	non-identified	short
		B_non_inoc_MT4	non-identified	short
		B_non_inoc_MT5	non-identified	short
	Pot inoculated plant	B_inoc_MT1	non-identified	contact
		B_inoc_MT2	non-identified	short
		B_inoc_MT3	non-identified	contact
Oak	Field plant	O_F_MT1	*Meliniomyces bicolor*	short
		O_F_MT2	non-identified	contact
		O_F_MT3	*Meliniomyces*	contact
		O_F_MT4	*Cortinarius bivelus*	medium, fringe
	Pot non-inoculated plant	O_non_inoc_MT1	non-identified	short
		O_non_inoc_MT2	non-identified	contact
		O_non_inoc_MT3	*Meliniomyces*	medium, smooth subtype
	Pot inoculated plant	O_inoc_MT1	non-identified	contact
		O_inoc_MT2	non-identified	short
		O_inoc_MT3	*Pisolithus arhizus*	short
Pine	Field plant	P_F_MT1	*Tomentella ellisii*	contact
		P_F_MT2	*Meliniomyces bicolor*	short
		P_F_MT3	*Rhizopogon mohelnensis*	medium, fringe
		P_F_MT4	*Tricholoma argyraceum*	medium, mat
	Pot non-inoculated plant	P_non_inoc_MT1	*Meliniomyces*	contact
		P_non_inoc_MT2	non-identified	contact
		P_non_inoc_MT3	non-identified	short
		P_non_inoc_MT4	non-identified	long
	Pot inoculated plant	P_inoc_MT1	*Inocybe lacera*	contact
		P_inoc_MT2	*Meliniomyces*	short
		P_inoc_MT3	*Meliniomyces bicolor*	short
		P_inoc_MT4	*Rhizopogon mohelnensis*	short

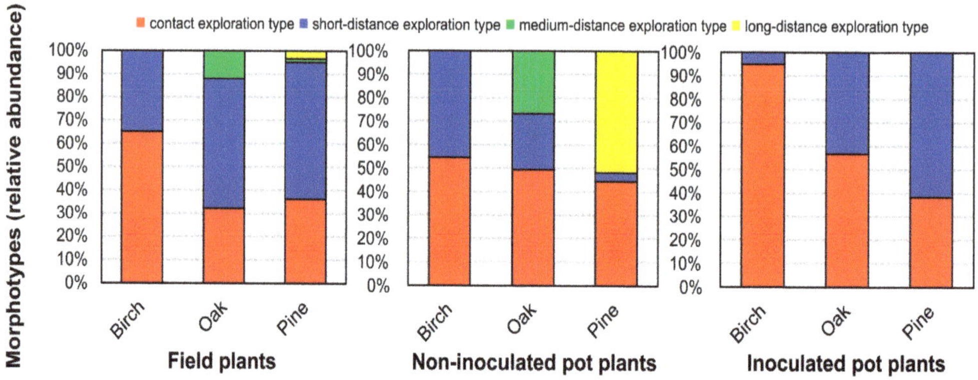

Figure 3. Relative abundance of exploration types of field plants, non-inoculated pot plants, and inoculated pot plants.

Dark-colored, with voluminous black emanating hyphae of *M. bicolor,* constituted the single morphotype with short-distance exploration type in trees from the field. *Cortinarius* sp., observed on field oaks, formed fans of ramified rhizomorphs and was attributed to medium-distance fringe subtype exploration type. On one field pine, *T. argyraceum* formed white-colored mycorrhiza with a hairy hyphal mantle, extensive emanating hyphae, and interconnected rhizomorphs that was placed in the medium-distance fringe subtype exploration type. Pine was the only tree species on which long-distance exploration type, formed by *R. mohelnensis*, was identified. It showed a coralloid ramification and formed moderately hairy rhizomorphs.

Pot birches had mycorrhiza grouped into contact and short-distance exploration types. The ratio of relative abundance of exploration types in pots differed from that observed from field trees. While for non-inoculated birches, the amount of short-distance exploration type significantly increased to 45.5%, for the inoculated variants, the contact exploration type constituted 95%.

Pot oaks were dominated by contact exploration type of mycorrhiza. The relative abundance of short-distance exploration type was lower for non-inoculated pot variants compared to the field; however, according to the Kruskal–Wallis test, this difference was not significant (Supplementary Table S7). The medium-distance exploration type observed in non-inoculated pots was formed by the fungus related to the genus *Meliniomyces.* The relative abundance of this exploration type was significantly higher than on field plants. Inoculation of oaks led to the loss of medium-distance exploration type mycorrhizae. The subdominant short-distance exploration type in inoculated pots was formed by the fungus *Pisolithus arhizus*, characterized by the woolly hyphal mantle with moderately present emanating hyphae.

Non-inoculated pines in pots formed mycorrhiza with grainy hyphal mantle and elongated well-differentiated rhizomorphs and were classified as a long-distance exploration type. Unidentified short-distance exploration type mycorrhiza with dark-colored, voluminous emanating hyphae was rare and observed only on one tree. Two morphotypes formed contact exploration types with a smooth hyphal mantle, mainly dichotomous branching and no emanating hyphae. One was identified to be formed by *Inocybe lacera.* Inoculation of pot pines resulted in the loss of long-distance exploration type mycorrhizae and led to the dominance of contact and short-distance exploration types. The fungal partner of the contact exploration type was represented by *Inocybe lacera.* Short-distance exploration types consisted of three morphotypes of different morphology. The most representative, with short-distance exploration type, was formed by *M. bicolor* and had a morphology different from morphotypes formed by this species on other trees in all variants; this new morphotype exhibited a brown color, woolly hyphal mantle with rare, white emanating hyphae. A morphotype with abundant dark-colored emanating hyphae was observed only on one inoculated tree with ECM fungus identified at the genus level as *Meliniomyces*. The least abundant morphotype was formed by *R. mohelnensis*, which had a woolly hyphal mantle and short emanating hyphae. Additional inoculation of pines did not significantly change the relative abundance of contact and short-distance exploration types compared to trees from the field or non-inoculated pot plants. Overall, additional inoculation of all plants in pots caused the loss of mycorrhiza with rhizomorphs.

3.4. Community Correlations to Soil Parameters

Different soil parameters were measured, such as the content of total carbon (TC), total nitrogen (TN), and total phosphorus (TP), as well as soil pH values (Table 4) and the content of metals in the soil (Figure 4).

Table 4. Selected soil chemical parameters.

Variant of Experiment	pH	TC (%)	TN (%)	TP (mg/kg)
Birch_mycorrhizosphere	4.44 ± 0.32	0.87 ± 0.24	0.11 ± 0.01	669 ± 173
Birch_bulk soil	3.75 ± 0.30	1.07 ± 0.05	0.13 ± 0.01	603 ± 198
Birch_pot substrate	6.03 ± 0.36	0.39 ± 0.05	0.07 ± 0.00	769 ± 285
Oak_mycorrhizosphere	3.54 ± 0.10	3.45 ± 0.35	0.31 ± 0.05	786 ± 127
Oak bulk soil	3.56 ± 0.09	2.60 ± 0.85	0.27 ± 0.07	802 ± 193
Oak pot substrate	5.32 ± 0.42	0.36 ± 0.01	0.07 ± 0.00	843 ± 100
Pine_mycorrhizosphere	5.20 ± 1.47	0.87 ± 0.13	0.09 ± 0.01	347 ± 61
Pine_bulk soil	6.21 ± 0.53	1.11 ± 0.44	0.10 ± 0.03	369 ± 37
Pine_pot substrate	5.52 ± 0.22	0.37 ± 0.01	0.07 ± 0.00	718 ± 178
Control pot substrate	3.47 ± 0.01	0.33 ± 0.01	0.07 ± 0.00	801 ± 121

TC, TN, TP—total carbon, nitrogen, phosphorus.

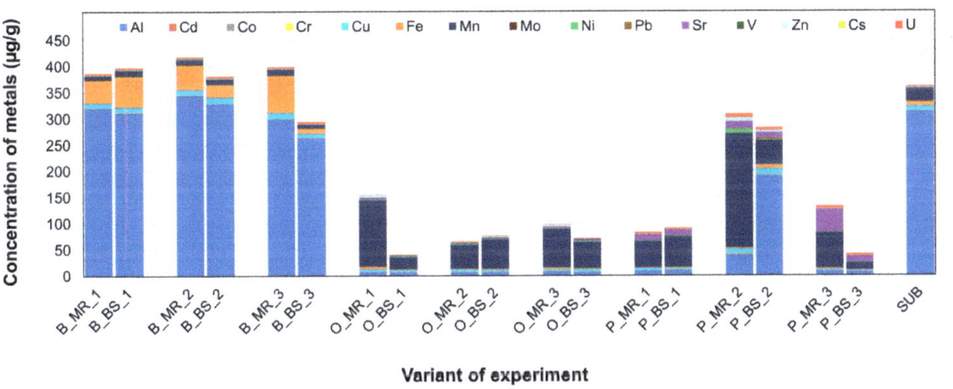

Figure 4. Sum content of toxic metals' bioavailable fractions determined in test field soil. B—birch, O—oak, P—pine; MR—mycorrhizosphere of field plant, BS—bulk soil, SUB—control pot substrate.

Canonical correspondence analysis shows a correlation between soil characteristics and the relative abundance of exploration types (Figure 5A). Contact exploration types correlated with the concentration of Al, Cu, Fe, Sr, and U. Higher concentrations of Pb, Sr, Fe, and U, as well as total phosphorus and C/N ratio, resulted in a higher abundance of short-distance exploration types. Medium-distance exploration types correlated with the concentration of Cs and Mn, while the long-distance exploration types were found to be associated with the concentrations of Zn, Ni, Co, Al, Cu, and soil pH. Although a permutation test (N = 999, $p = 0.444$) did not reveal an overall significant association between soil parameters and exploration types, correlation analysis demonstrated significant positive correlations between short-distance exploration types and TC, TN, and C/N ratio, between medium-distance exploration type and the concentration of Cs, and a negative correlation was found between contact exploration types and TN (Supplementary Table S8).

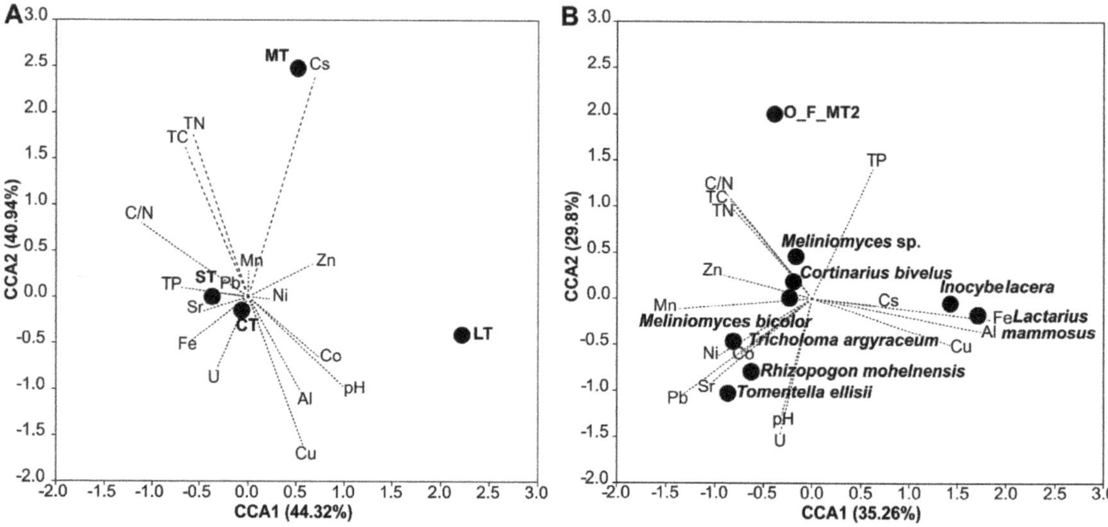

Figure 5. Canonical correspondence analysis biplot representing the correlation (**A**) between soil parameters and exploration types of mycorrhizae and (**B**) between soil parameters and mycorrhizal taxa described for field plants. CT—contact exploration type, ST—short-distance exploration type, MT—medium-distance exploration type, LT—long-distance exploration type.

Lactarius mammosus and *Inocybe lacera* (both described for birches and forming contact exploration type mycorrhiza) positively correlated in the CCA biplot with the concentrations of Fe, Al, and Cu, and negatively correlated with Mn, Ni, Pb, Sr, and TC and C/N ratios (compare Figure 5B). The non-identified oak morphotype O_F_MT2 (contact exploration type) revealed a positive correlation with nutrients and negatively correlated with the concentrations of Al, U, and soil pH value. A positive correlation was also recorded for the pine morphotype formed by *Tomentella ellisii* (contact exploration type) for Mn, Pb, Sr, and U concentrations as well as soil pH; a negative correlation was observed towards Fe and P. Correlation analysis revealed that associations of all field morphotypes with contact exploration and soil characteristics were significant (Table 5). The morphotype formed by *Cortinarius bivelus* (medium-distance exploration type) on oaks had a positive correlation with total nitrogen. *M. bicolor* and *Meliniomyces* sp. (short-distance exploration type) described for oaks did not correlate with soil characteristics. Pine morphotypes formed by *T. argyraceum* and *R. mohelnensis* correlated with the concentrations of Ni, Pb, Sr, Mn, U, and soil pH; however, correlation analysis did not confirm the significance of these correlations.

3.5. Plant Inoculation

Tree species showed different performances related to additional inoculation (Table 6, Supplementary Figure S4). Non-inoculated birches and pines showed higher survival (66.7 and 80%, respectively), whereas non-inoculated oaks did not perform successfully, and only two oaks out of 12 had leaves at the end of the experiment. Inoculation of the rhizosphere with the blend of mycorrhizal fungi resulted in a decrease in birches' survival rate (14.3%) and an increase in survival for oaks (69.2%). Although the number of inoculated pines at the end of the experiment was low, the overall survival rate after inoculation remained relatively high (60.0%).

Table 5. Coefficients of correlation between soil characteristics and ectomycorrhizal morphotypes and ECM fungi described for the field plants.

Soil Characteristics	*Lactarius mammosus*	*M. bicolor*	*Inocybe lacera*	O_F_MT2	*Meliniomyces*	*Cortinarius bivelus*	*Tomentella ellisii*	*R. mohelnensis*	*T. argyraceum*
Al	**0.76** *	−0.36	**0.79** *	**−0.69** *	−0.44	−0.35	0.11	0.17	0.10
Co	−0.44	0.02	−0.40	0.19	0.26	0.21	0.26	0.45	0.38
Cu	**0.75** *	−0.49	**0.75** *	−0.42	−0.37	−0.41	−0.13	0.24	0.10
Fe	**0.79** *	−0.37	**0.78** *	−0.03	−0.13	0.10	**−0.65** *	−0.38	−0.17
Mn	**−0.77** *	0.11	**−0.77** *	0.15	−0.04	0.20	**0.57** *	0.45	0.38
Ni	**−0.78** *	0.23	**−0.75** *	0.20	0.25	0.10	0.53	0.45	0.38
Pb	**−0.78** *	0.21	**−0.78** *	−0.07	−0.04	0.06	**0.79** *	0.45	0.28
Sr	**−0.76** *	0.17	**−0.76** *	−0.06	−0.09	0.00	**0.84** *	0.24	0.38
Zn	−0.48	−0.11	−0.49	0.36	0.14	0.31	0.12	0.45	0.31
Cs	0.44	−0.02	0.45	−0.04	0.09	0.43	−0.52	0.00	−0.14
U	−0.02	−0.01	−0.01	**−0.67** *	−0.36	−0.50	**0.82** *	0.45	0.38
TC	**−0.57** *	0.34	**−0.54** *	**0.61** *	0.52	0.47	−0.20	−0.28	−0.03
TN	**−0.54** *	0.33	−0.50	**0.54** *	0.42	**0.54** *	−0.22	0.24	0.07
C/N	**−0.61** *	0.35	**−0.61** *	**0.69** *	0.44	0.35	−0.24	−0.38	−0.17
TP	0.17	0.03	0.22	**0.58** *	0.28	0.14	**−0.84** *	−0.24	−0.38
pH	0.14	−0.18	0.14	**−0.64** *	−0.39	−0.27	**0.65** *	0.45	0.38

Asterisks and bold print represent a significant correlation ($p < 0.05$). TC, TN, TP—total carbon, nitrogen, phosphorus; O_F_MT2: non-identified.

Table 6. Pot plant's survival rates.

Tree Species		Total	Plant Survival Rate (%)
Birch	Number of plants non-inoculated	12	66.7
	Number of plants inoculated	14	14.3
Oak	Number of plants non-inoculated	12	16.7
	Number of plants inoculated	14	69.2
Pine	Number of plants non-inoculated	10	80.0
	Number of plants inoculated	10	60.0

All three species contained different concentrations of metals depending on the variant of the experiment. Overall, additional fungal inoculation did not affect metal content in pot plant biomass. Although significant differences between field variants and one or both pot variants were observed, no trend in metal accumulation in plant aboveground biomass was determined, and different environmental factors might have caused the changes (Supplementary Figure S5).

4. Discussion

4.1. The Former Mining Site Is Characterized by Low Ectomycorrhizal Diversity

In this study, field trees of three species of typical trees in temperate forests were investigated. Using young birch, oak, and pine stands that developed naturally at a former mining site, the native mycorrhizal community could be evaluated and compared to re-potted trees from that same site kept under greenhouse conditions for two years. The mycorrhizal assemblies and their environmental functions could be assessed with that approach under more standardized conditions. Nevertheless, mycorrhizal root tips were limited, and this limited taxon sampling might bias the similarity indices, which might be overcome with a probabilistic approach incorporating [37,38]. Our approach could confirm high values of similarity between ectomycorrhizal communities between trees of one species. The trees' age, succession stage, and contamination level have been described as influencing ectomycorrhizal associations, with metal-polluted soils sharing low diversity and a highly uneven taxa distribution [39–41]. Similar trends were observed in the current study. Experiments on the influence of applied heavy metals on mycorrhiza formation confirmed that the increase in metal content reduced, in general, the diversity of mycorrhiza [42,43]. The site's age seems to be crucial for fungal community diversity [1,44]. Since metal-polluted soils generally contain a lower number of vital fungal spores and a low density of fine roots is typical for younger trees, a decreased community diversity might be expected [45].

4.2. Potential for Afforestation Programs

The success of vegetation establishment depends on adaptation to the newly created conditions in post-mining landscapes [46]. The substrate generally is spacially heterogenous, contains low amounts of nutrients, and is prone to have a low pH, increasing the mobility of heavy metals with capillary transport of water. Even if (a limited layer of) allochthonous soil cover is applied, this will not provide the water holding capacity nor the abundance of ectomycorrhizal fungal propagules present in native sites. However, ectomycorrhiza can be beneficial for plant establishment, specifically under detrimental conditions. Fertilization will alleviate the scarcity of nutrients but, at the same time, is known to suppress the formation of sustainable mycorrhiza. Hence, we investigated the addition of fungal inoculum that might provide a measure easily applicable in such afforestation programs on former mining sites.

Trees to be planted should be able to compete and establish a viable forest ecosystem. Among the typical woody pioneer species which can naturally colonize post-mining areas are birches [47]. While they facilitate soil functioning and biodiversity, birches are known to be particularly sensitive to intraspecific competition [48]. In afforestation, plant biomass should be evaluated for the uptake of metals. Here again, mycorrhizae provide a barrier with the hyphal mantle protecting root tips from excess heavy metals, and metal accumulation at and inside the hyphae without transfer to the host tree can protect the trees from toxic element concentrations under field conditions. In our analysis, we, therefore, checked our pot birches and found that they accumulated only Mn and Cd, while in the field, Fe, Cu, Sr, Pb, and U uptake was higher. Still, the measured concentrations are below phytotoxicity levels. Thus, (mycorrhizal) birches can be considered phytostabilizing as the biomass does not contain levels exceeding threshold concentrations. Hence, in addition to providing a positive effect on stabilizing the substrate from erosion and allowing for recreational use, birch wood from mining site reforestation provides a potential crop for wood harvesting with minimal risk of soil re-contamination through leaf fall.

Oak, in contrast to birch, is attributed to climax or late successional-stage species in the European temperate forests [49]. At Kanigsberg, sites with better soil quality and nutrient contents would be specifically prone to harbor oaks known to accumulate heavy metals in their biomass [49]. After transfer in pots, oaks accumulated Mn, potentially related to low P contents, as plants are known to exude more organic acids to obtain P, which facilitates metal uptake, specifically under moist conditions [50]. This is accompanied by the observed high mortality in non-inoculated pots, while mycorrhiza could stabilize oak performance showing that the implementation of mycorrhizal blends might considerably improve the growth of oaks in post-mining areas.

Pines represent typical pioneer trees widely used in reclamation [46,51,52]. Fertilization improved their growth in polluted soils. However, that may come at the cost of phytoextraction, visible in both field and pot pines containing elevated Al [53]. Here, the pot experiment showed that pines were the least affected by disturbance, regardless if they were inoculated with mycorrhizal fungi. Transfer in pots led to the accumulation of Mn, Co, and Cu into the needles, while Cr, Fe, Ni, and Sr were excluded. Overall, as the turnover of aboveground biomass in coniferous tree stands is markedly slower than in deciduous tree stands, coniferous trees are a good alternative for the reclamation of areas contaminated with heavy metals.

4.3. Tree Transplanting Affected Functional Diversity

Transfer of the plants from the field to pots affected the tree species differently. The low survival rate of non-inoculated variants (lowest for oak with 16.7%) was correlated with an enlarged root system by forming new roots for birches, less for oaks, and remained unchanged for pines, consistent with the trees' ability to perform well in primary, early succession. In non-inoculated oaks and pines, an abundance of ectomycorrhizae with rhizomorphs increased significantly. The development of rhizomorphs could facilitate the extensive exploration of newly created niches.

Besides, the hydrophobicity of rhizomorphs can indirectly control the access of water into mycelium. Therefore, for oaks and pines, the development of hydrophobic rhizomorphs might be a strategy for improving effective nutrient transport, as hydrophobic properties will likely prevent the loss of solutes during transfer to the host plant [16]. The application of the ectomycorrhizal fungi by inoculation generally could contribute to the improvement of the water regime but might lead to the loss of rhizomorphs as more carbon-costly and unnecessary in pot conditions.

4.4. Relationships to Ecosystem Functioning and Succession

Among the field birches, *Lactarius* was abundant, potentially related to the high tolerance to unfavorable local conditions, particularly high aluminum contents [54,55]. For both *Cortinarius* and *Tricholoma*, extensive hydrophobic rhizomorphs have been reported at contaminated sites [16,41,56] that enhance the acquisition of nitrogen from organic sources [57]. *Rhizopogon mohelnensis* would be typically found in early, primary succession [58,59]. *Thelephoraceae* with *Tomentella ellisii* and *Meliniomyces* have been described commonly in coniferous forests and are often found in contaminated soils [60,61].

Field morphotypes formed mycorrhiza preferentially with contact and short-distance exploration types usually described for undisturbed forest stands [62]. However, the exploration types in metal-affected soils consisted of increasing numbers of morphotypes with short and medium-distance exploration types [40]. It has been discussed that abundant emanating hyphae and ramifying rhizomorphs typical for medium-distance exploration types can function as a filter hindering the entry of heavy metals into the host plant cells. At the same time, they can explore a larger soil volume for nutrient acquisition, which is even better for long-distance exploration types that are usually described for trees in metal-contaminated areas [56]. Here, permutation analysis did not reveal a significant correspondence between metals and exploration types, leading to the conclusion that abiotic parameters of the disturbed and coarse substrate might override the intricate relationships between exploration types and heterogeneous metal bioavailability. Only pines formed mycorrhiza with woolly silver mantle with infrequent emanating hyphae and no rhizomorphs with *Rhizopogon mohelnensis*. This species also was detected on field pine roots, where it formed distinctive rhizomorphs. Again, hydrophobic mycelium and rhizomorphs seem to indicate rather dry conditions [63,64].

With respect to succession, the spatial position of short roots might influence the choice of the mycorrhizal partner fungus. Bruns [58] argued that short roots close to the stem accumulate more carbohydrates, supporting infection by carbon-demanding, late-stage fungal symbionts. At the same time, these are prone to produce rhizomorphs, as the center of the plant root is deprived of mineral nutrients. Since we selected healthy short roots found more often at a distance from the stem, the study might carry an intrinsic bias towards early-stage mycorrhizal fungi.

Overall, it can be concluded that field plants promoted ecological filtering toward selecting specific symbiotic fungi with specific exploration types, which would contribute to plants' tolerance to abiotic conditions specific at each sampling site.

5. Conclusions

With our approach, we could show that even on substrates in post-mining landscapes, afforestation is possible, and the trees can establish a viable root system. This will be essential to stabilize the soil and protect it from water and wind erosion. At the same time, the presence of mycorrhizal fungi can positively affect water availability. While the young trees did not yet develop high amounts of long-distance exploration type mycorrhizae, still a succession from constant to medium-distance exploration types were found, showing that establishing sustainable ecosystems even under these detrimental conditions is possible. The addition of mycorrhizal fungal inoculum could help to facilitate this process of increasing mycorrhization that, in the end, helps the trees survive the planting. The use of a mixed forest consisting of deciduous birches and oaks, and coniferous pines seems well suited to provide not only a sustainable land use for society's recreational activities but the phytostabilization features observed will also allow for forest harvesting of these newly established forests in future generations.

Supplementary Materials: The following supporting information can be downloaded at: https://www.mdpi.com/article/10.3390/jof9040483/s1, Figure S1. Schemes of soil sampling. A, at the test field; B, from pots. MR, mycorrhizosphere; BS, bulk soil. Figure S2. Images of pot variants. A, pot birches; B, pot oaks; C, pot pines; D, the process of inoculation of plants with the ectomycorrhizal fungi. Table S1. Overview of pot experiment variants. Table S2. Characteristics of exploration types of mycorrhiza according to Agerer [10]. Table S3. Morphological description of morphotypes. Table S4. ECM fungus identification by sequencing. For morphotypes with grey cells, sequence analysis was not successful. Figure S3. ITS sequences for mycorrhiza fungus identification. Table S5. One-Way ANOVA, Kruskal–Wallis, and post hoc tests for significance of differences between variants of the experiment in diversity indices of ECM communities. Table S6. Indices of similarity between ECM communities determined for trees within variants of the experiment. Table S7. Kruskal–Wallis test and post hoc test output for significance of differences between variants of the experiment in relative abundance of exploration types of mycorrhiza. Table S8. Correlation coefficients between soil characteristics and exploration types of mycorrhizae. Figure S4. Plant growth dynamics during pot experiment. B—birch, O—oak, P—pine; non_inoc—non-inoculated variant of the experiment, inoc—inoculated variant of the experiment. Figure S5. Content of toxic metals measured in aboveground dried plant biomass in variants of the experiment. y-axis represents concentration in µg/g. Brackets represent significant differences ($p < 0.05$).

Author Contributions: Conceptualization, O.B., K.K. and E.K.; methodology, O.B.; software, O.B.; validation, K.K.; formal analysis, O.B.; investigation, O.B.; resources, E.K.; data curation, O.B.; writing—original draft preparation, O.B. and K.K.; writing—review and editing, K.K. and E.K.; visualization, O.B. and K.K.; supervision, K.K. and E.K.; project administration, E.K.; funding acquisition, E.K. All authors have read and agreed to the published version of the manuscript.

Funding: This research was funded by the International Max Planck Research School "Global Biogeochemical Cycles", Jena, Germany; German Science Foundation under Germany's Excellence Strategy, EXC 2051 project ID 390713860 and BMBF (UserII, project ID 15S9417).

Institutional Review Board Statement: Not applicable.

Informed Consent Statement: Not applicable.

Data Availability Statement: All data are provided with this submission.

Acknowledgments: We are grateful for assistance to Dirk Merten, Applied Geology, Friedrich Schiller University; Ines Hilke, Michael Rässler, Gerd Gleixner and Steffi Rothhardt, Max Planck Institute for Biogeochemistry. We wish to acknowledge the help of Sebastian Pietschmann and Markus Riefenstahl. Further, the pot plant experiments were possible with the help of Hubert Schröter at Thüringer Landesamt für Landwirtschaft und Ländlichen Raum.

Conflicts of Interest: The authors declare no conflict of interest.

References

1. Gebhardt, S.; Neubert, K.; Wöllecke, J.; Münzenberger, B.; Hüttl, R.F. Ectomycorrhiza communities of red oak (*Quercus robur* L.) of different age in the Lusatian lignite mining district, East Germany. *Mycorrhiza* **2007**, *17*, 279–290. [CrossRef] [PubMed]
2. Scannell, Y. The regulation of mining and mining waste in the European Union. *J. Energy Clim. Environ.* **2012**, *3*, 3.
3. van der Heijden, M.G.A.; Martin, F.M.; Selosse, M.-A.; Sanders, I.R. Mycorrhizal ecology and evolution: The past, the present, and the future. *New Phytol.* **2015**, *205*, 1406–1423. [CrossRef] [PubMed]
4. Dahlberg, A. Community ecology of ectomycorrhizal fungi: An advancing interdisciplinary field. *New Phytol.* **2001**, *150*, 555–562. [CrossRef]
5. Bruns, T.D.; Bidartondo, M.I.; Taylor, D.L. Host specificity in ectomycorrhizal communities: What do the exceptions tell us? *Int. Comp. Biol.* **2002**, *42*, 352–359. [CrossRef]
6. Kuo, A.; Kohler, A.; Martin, F.M.; Grigoriev, I.V. Expanding genomics of mycorrhizal symbiosis. *Front. Microbiol.* **2014**, *5*, 582. [CrossRef]
7. Plett, J.M.; Miyauchi, S.; Morin, E.; Plett, K.; Wong-Bajracharya, J.; de Freitas Pereira, M.; Kuo, A.; Henrissat, B.; Drula, E.; Wojtalewicz, D.; et al. Speciation underpinned by unexpected molecular diversity in the mycorrhizal fungal genus *Pisolithus*. *Mol. Biol. Evol.* **2023**, *40*, msad045. [CrossRef]
8. Menkis, A.; Vasiliauskas, R.; Taylor, A.F.S.; Stenlid, J.; Finlay, R. Afforestation of abandoned farmland with conifer seedlings inoculated with three ectomycorrhizal fungi—Impact on plant performance and ectomycorrhizal community. *Mycorrhiza* **2007**, *17*, 337–348. [CrossRef]
9. Oliveira, R.S.; Franco, A.R.; Castro, P.M.L. Combined use of *Pinus pinaster* plus and inoculation with selected ectomycorrhizal fungi as an ecotechnology to improve plant performance. *Ecol. Eng.* **2012**, *43*, 95–103. [CrossRef]
10. Luo, S.; Phillips, R.P.; Jo, I.; Fei, S.; Liang, J.; Schmid, B.; Eisenhauer, N. Higher productivity in forests with mixed mycorrhizal strategies. *Nat. Commun.* **2023**, *14*, 1377. [CrossRef]
11. Sanchez-Zabala, J.; Majada, J.; Martín-Rodrigues, N.; Gonzalez-Murua, G.; Ortega, U.; Alonso-Graña, M.; Arana, O.; Duñabeitia, M.K. Physiological aspects underlying the improved outplanting performance of *Pinus pinaster* Ait. seedlings associated with ectomycorrhizal inoculation. *Mycorrhiza* **2013**, *23*, 627–640. [CrossRef]
12. Morel, M.; Jacob, C.; Kohler, A.; Johansson, T.; Martin, F.; Chalot, M.; Brun, A. Identification of genes differentially expressed in extraradical mycelium and ectomycorrhizal roots during *Paxillus involutus-Betula pendula* ectomycorrhizal symbiosis. *Appl. Environ. Microbiol.* **2005**, *71*, 382–391. [CrossRef]
13. Agerer, R. Exploration types of ectomycorrhizae. *Mycorrhiza* **2001**, *11*, 107–114. [CrossRef]
14. Seress, D.; Dima, B.; Kovács, G.M. Characterisation of seven *Inocybe* ectomycorrhizal morphotypes from a semiarid woody steppe. *Mycorrhiza* **2016**, *26*, 215–225.
15. Hobbie, E.A.; Agerer, R. Nitrogen isotopes in ectomycorrhizal sporocarps correspond to belowground exploration types. *Plant Soil* **2010**, *327*, 71–83. [CrossRef]
16. Lilleskov, E.A.; Hobbie, E.A.; Horton, T.R. Conservation of ectomycorrhizal fungi: Exploring the linkages between functional and taxonomic responses to anthropogenic N deposition. *Fungal. Ecol.* **2011**, *4*, 174–183. [CrossRef]
17. Moeller, H.V.; Peay, K.G.; Fukami, T. Ectomycorrhizal fungal traits reflect environmental conditions along a coastal California edaphic gradient. *FEMS Microbiol. Ecol.* **2014**, *87*, 797–806. [CrossRef]
18. Mason, P.A.; Wilson, J.; Last, F.T.; Walker, C. The concept of succession in relation to the spread of sheathing mycorrhizal fungi on inoculated tree seedlings growing in unsterile soils. *Plant Soil* **1983**, *71*, 247–256. [CrossRef]
19. Newton, A.C. Towards a functional classification of ectomycorrhizal fungi. *Mycorrhiza* **1992**, *2*, 75–79. [CrossRef]
20. Iordache, V.; Gherghel, F.; Kothe, E. Assessing the effect of disturbances on ectomycorrhiza diversity. *Int. J. Environ. Res. Public. Health* **2009**, *6*, 414–432. [CrossRef]
21. Twieg, B.D.; Durall, D.M.; Simard, S.W. Ectomycorrhizal fungal succession in mixed temperate forests. *New Phytol.* **2007**, *176*, 437–447. [CrossRef] [PubMed]
22. Gadd, G.M. Microbial influence on metal mobility and application for bioremediation. *Geoderma* **2004**, *122*, 109–119. [CrossRef]
23. Epelde, L.; Lanzén, A.; Blanco, F.; Urich, T.; Garbisu, C. Adaptation of soil microbial community structure and function to chronic metal contamination at an abandoned Pb-Zn mine. *FEMS Microbiol. Ecol.* **2015**, *91*, 1–11. [CrossRef] [PubMed]
24. Gadd, G.M. Metals, minerals and microbes: Geomicrobiology and bioremediation. *Microbiology* **2010**, *156*, 609–643. [CrossRef]
25. Dodd, J.C.; Thomson, B.D. The screening and selection of inoculant arbuscular mycorrhizal and ectomycorrhizal fungi. *Plant Soil* **1994**, *159*, 149–158. [CrossRef]
26. Cullings, K.; Makhija, S. Ectomycorrhizal fungal associates of *Pinus contorta* in soils associated with a hot spring in Norris Geyser Basin, Yellowstone National Park, Wyoming. *Appl. Environ. Microbiol.* **2001**, *67*, 5538–5543. [CrossRef]
27. Baum, C.; Hrynkiewicz, K.; Leinweber, P.; Meißner, R. Heavy-metal mobilization and uptake by mycorrhizal and nonmycorrhizal willows (*Salix* × *dasyclados*). *J. Plant. Nutr. Soil Sci.* **2006**, *169*, 516–522. [CrossRef]
28. Zimmer, D.; Baum, C.; Leinweber, P.; Hrynkiewicz, K.; Meissner, R. Associated bacteria increase the phytoextraction of cadmium and zinc from a metal-contaminated soil by mycorrhizal willows. *Int. J. Phytorem.* **2009**, *11*, 200–213. [CrossRef]
29. Colpaert, J.V.; van Assche, J.A. Zinc toxicity in ectomycorrhizal *Pinus sylvestris*. *Plant Soil* **1992**, *143*, 201–211. [CrossRef]

30. Fernández-Fuego, D.; Keunen, E.; Cuypers, A.; Bertrand, A.; González, A. Mycorrhization protects *Betula pubescens* Ehr. from metal-induced oxidative stress increasing its tolerance to grow in an industrial polluted soil. *J. Haz. Mat.* **2017**, *336*, 119–127. [CrossRef]
31. Cejpková, J.; Gryndler, M.; Hršelová, H.; Kotrba, P.; Řanda, Z.; Synková, I.; Borovička, J. Bioaccumulation of heavy metals, metalloids, and chlorine in ectomycorrhizae from smelter-polluted area. *Env. Pollut.* **2016**, *218*, 176–185. [CrossRef]
32. Zeien, H.; Brümmer, G.W. Chemische Extraktion zur Bestimmung von Schwermetallbindungsformen in Böden. *Mitt. Dtsch. Bodenkdl. Ges.* **1989**, *59*, 505–510.
33. Agerer, R. *Colour Atlas of Ectomycorrhizae*; Einhorn-Verlag: Schwäbisch Gmünd, Germany, 1987–2006.
34. Lotti, M.; Zambonelli, A. A quick and precise technique for identifying ectomycorrhizas by PCR. *Mycol. Res.* **2006**, *110*, 60–65.
35. Gardes, M.; Bruns, T.D. ITS primers with enhanced specificity for basidiomycetes—Application to the identification of mycorrhizae and rusts. *Mol. Ecol.* **1993**, *2*, 113–118. [CrossRef]
36. Nilsson, R.H.; Larsson, K.H.; Taylor, A.F.S.; Bengtsson-Palme, J.; Jeppesen, T.S.; Schigel, D.; Kennedy, P.; Picard, K.; Glöckner, F.O.; Tedersoo, L.; et al. The UNITE database for molecular identification of fungi: Handling dark taxa and parallel taxonomic classifications. *Nucl. Acids. Res.* **2019**, *47*, D259–D264. [CrossRef]
37. Chao, A.; Chazdon, R.L.; Colwell, R.K.; Shen, T.-J. A new statistical approach for assessing similarity of species composition with incidence and abundance data. *Ecol. Lett.* **2005**, *8*, 148–159. [CrossRef]
38. Hao, M.; Corral-Rivas, J.J.; González-Elizondo, M.S.; Ganeshaiah, K.N.; Nava-Miranda, M.G.; Zhang, C.; Zhao, X.; von Gadow, K. Assessing biological dissimilarities between five forest communities. *For. Ecosys.* **2019**, *6*, 30. [CrossRef]
39. Hartley, J.; Cairney, J.W.G.; Meharg, A.A. Do ectomycorrhizal fungi exhibit adaptive tolerance to potentially toxic metals in the environment? *Plant Soil* **1997**, *189*, 303–319. [CrossRef]
40. Rudawska, M.; Leski, T.; Stasińska, M. Species and functional diversity of ectomycorrhizal fungal communities on Scots pine (*Pinus sylvestris* L.) trees on three different sites. *Ann. For. Sci.* **2011**, *68*, 5–15. [CrossRef]
41. Staudenrausch, S.; Kaldorf, M.; Renker, C.; Luis, P.; Buscot, F. Diversity of the ectomycorrhiza community at a uranium mining heap. *Biol. Fertil. Soils* **2005**, *41*, 439–446. [CrossRef]
42. Chappelka, A.H.; Kush, J.S.; Runion, G.B.; Meier, S.; Kelley, W.D. Effects of soil-applied lead on seedling growth and ectomycorrhizal colonization of loblolly pine. *Environ. Poll.* **1991**, *72*, 307–316. [CrossRef] [PubMed]
43. Duñabeitia, M.K.; Hormilla, S.; Garcia-Plazaola, J.I.; Txarterina, K.; Arteche, U.; Becerril, J.M. Differential responses of three fungal species to environmental factors and their role in the mycorrhization of *Pinus radiata* D. Don. *Mycorrhiza* **2004**, *14*, 11–18. [CrossRef] [PubMed]
44. Peter, M.; Ayer, F.; Egli, S.; Honegger, R. Above- and below-ground community structure of ectomycorrhizal fungi in three Norway spruce (*Picea abies*) stands in Switzerland. *Can. J. Bot.* **2001**, *79*, 1134–1151.
45. Leyval, C.; Turnau, K.; Haselwandter, K. Effect of heavy metal pollution on mycorrhizal colonization and function: Physiological, ecological and applied aspects. *Mycorrhiza* **1997**, *7*, 139–153. [CrossRef]
46. Pietrzykowski, M.; Daniels, W.L. Estimation of carbon sequestration by pine (*Pinus sylvestris* L.) ecosystems developed on reforested post-mining sites in Poland on differing mine soil substrates. *Ecol. Eng.* **2014**, *73*, 209–218. [CrossRef]
47. Clegg, S.; Gobran, G.R. Effects of aluminium on growth and root reactions of phosphorus stressed Betula pendula seedlings. *Plant Soil* **1995**, *168*, 173–178. [CrossRef]
48. Dubois, H.; Verkasalo, E.; Claessens, H. Potential of birch (*Betula pendula* Roth and *B. pubescens* Ehrh.) for forestry and forest-based industry sector within the changing climatic and socio-economic context of Western Europe. *Forests* **2020**, *11*, 336. [CrossRef]
49. Frouz, J.; Vobořilová, V.; Janoušová, I.; Kadochová, S.; Matějíček, L. Spontaneous establishment of late successional tree species English oak (*Quercus robur*) and European beech (*Fagus sylvatica*) at reclaimed alder plantation and unreclaimed post mining sites. *Ecol. Eng.* **2015**, *77*, 1–8. [CrossRef]
50. Hocking, P. Organic acids exuded from roots in phosphorus uptake and aluminum tolerance of plants in acid soils. *Adv. Agron.* **2001**, *74*, 63–97.
51. Baumann, K.; Rumpelt, A.; Schneider, B.U.; Marschner, P.; Hüttl, R.F. Seedling biomass and element content of *Pinus sylvestris* and *Pinus nigra* grown in sandy substrates with lignite. *Geoderma* **2006**, *136*, 573–578. [CrossRef]
52. Chodak, M.; Niklińska, N. The effect of different tree species on the chemical and microbial properties of reclaimed mine soils. *Biol. Fertil. Soils* **2010**, *46*, 555–566. [CrossRef]
53. Reimann, C.; Koller, F.; Kashulina, G.; Niskavaara, H.; Englmaier, P. Influence of extreme pollution on the inorganic chemical composition of some plant. *Environ. Poll.* **2001**, *115*, 239–252. [CrossRef]
54. Egerton-Warburton, L.M. Aluminum-tolerant *Pisolithus* ectomycorrhizas confer increased growth, mineral nutrition, and metal tolerance to *Eucalyptus* in acidic mine spoil. *Appl. Environ. Soil Sci.* **2015**, 803821. [CrossRef]
55. Moyer-Henry, K.; Silva, I.; Macfall, J.; Johannes, E.; Allen, N.; Goldfarb, B.; Rufty, T. Accumulation and localization of aluminium in root tips of loblolly pine seedlings and the associated ectomycorrhiza *Pisolithus Tinctorius*. *Plant. Cell Environ.* **2005**, *28*, 111–120. [CrossRef]
56. Bierza, W.; Bierza, K.; Trzebny, A.; Greń, I.; Dabert, M.; Ciepał, R.; Trocha, L.K. The communities of ectomycorrhizal fungal species associated with *Betula pendula* Roth and *Pinus sylvestris* L. growing in heavy-metal contaminated soils. *Plant Soil* **2020**, *457*, 321–328. [CrossRef]

57. Bödeker, I.T.M.; Clemmensen, K.E.; de Boer, W.; Martin, F.; Olson, A.; Lindahl, B.D. Ectomycorrhizal *Cortinarius* species participate in enzymatic oxidation of humus in northern forest ecosystems. *New Phytol.* **2014**, *203*, 245–256. [CrossRef]
58. Bruns, T.D. Thoughts on the processes that maintain local species diversity of ectomycorrhizal fungi. *Plant Soil* **1995**, *170*, 63–73. [CrossRef]
59. Horton, T.R.; Cázares, E.; Bruns, T.D. Ectomycorrhizal, vesicular-arbuscular and dark septate fungal colonization of bishop pine (*Pinus muricata*) seedlings in the first 5 months of growth after wildfire. *Mycorrhiza* **1998**, *8*, 11–18. [CrossRef]
60. Kõljalg, U.; Dahlberg, A.; Taylor, A.F.S.; Larsson, E.; Hallenberg, N.; Stenlid, J.; Larsson, K.H.; Fransson, P.M.; Kårén, O.; Jonsson, L. Diversity and abundance of resupinate thelephoroid fungi as ectomycorrhizal symbionts in Swedish boreal forests. *Mol. Ecol.* **2000**, *9*, 1985–1996. [CrossRef]
61. Vrålstad, T.; Myhre, E.; Schumacher, T. Molecular diversity and phylogenetic affinities of symbiotic root-associated ascomycetes of the *Helotiales* in burnt and metal polluted habitats. *New Phytol.* **2002**, *155*, 131–148. [CrossRef]
62. Rosinger, C.; Sandén, H.; Matthews, B.; Mayer, M.; Godbold, D.L. Patterns in ectomycorrhizal diversity, community composition, and exploration types in European beech, pine, and spruce forests. *Forests* **2018**, *9*, 445. [CrossRef]
63. Unestam, T.; Sun, Y.-P. Extramatrical structures of hydrophobic and hydrophilic ectomycorrhizal fungi. *Mycorrhiza* **1995**, *5*, 301–311. [CrossRef]
64. Bakker, M.R.; Augusto, L.; Achat, D.L. Fine root distribution of trees and understory in mature stands of maritime pine (*Pinus pinaster*) on dry and humid sites. *Plant Soil* **2006**, *286*, 37–51. [CrossRef]

Journal of
Fungi

MDPI

Article

A Comparative Study on Heavy Metal Removal from CCA-Treated Wood Waste by *Yarrowia lipolytica*: Effects of Metal Stress

Dan Xing [1], Sara Magdouli [2], Jingfa Zhang [3], Hassine Bouafif [4] and Ahmed Koubaa [1,*]

1 Institut de Recherche sur les Forêts, Université du Québec en Abitibi-Témiscamingue, Rouyn-Noranda, QC J9X 5E4, Canada; dan.xing@uqat.ca
2 Lassonde School of Engineering, York University, Toronto, ON M3J 1P3, Canada
3 State Key Laboratory of Biobased Material and Green Papermaking, Qilu University of Technology, Shandong Academy of Sciences, Jinan 250353, China
4 Centre Technologique des Résidus Industriels en Abitibi Témiscamingue, 433 Boulevard du Collège, Rouyn-Noranda, QC J9X 0E1, Canada
* Correspondence: ahmed.koubaa@uqat.ca

Abstract: Bioremediation is an effective way to remove heavy metals from pollutants. This study investigated the effects of *Yarrowia lipolytica* (*Y. lipolytica*) on the bioremediation of chromated copper arsenate (CCA)-treated wood wastes. Copper ions stressed the yeast strains to improve their bioremediation efficiency. A comparison of changes in morphology, chemical composition, and metal content of CCA wood before and after bioremediation was conducted. The amount of arsenic (As), chromium (Cr), and copper (Cu) was quantified by microwave plasma atomic emission spectrometer. The results showed that yeast strains remained on the surface of CCA-treated wood after bioremediation. The morphologies of the strains changed from net to spherical because of the Cu^{2+} stress. Fourier-transform infrared spectroscopy showed that carboxylic acid groups of wood were released after removing heavy metals. A large amount of oxalic acid was observed when the optical density (OD_{600nm}) was 0.05 on the 21st day. Meanwhile, the highest removal rate of Cu, As, and Cr was 82.8%, 68.3%, and 43.1%, respectively. Furthermore, the Cu removal from CCA-treated wood increased by about 20% after Cu^{2+} stress. This study showed that it is feasible to remove heavy metals from CCA-treated wood by *Y. lipolytica* without destroying the wood structure, especially by copper-induced *Y. lipolytica*.

Keywords: bioremediation; heavy metals; *Y. lipolytica*; CCA-treated wood; copper stress

Citation: Xing, D.; Magdouli, S.; Zhang, J.; Bouafif, H.; Koubaa, A. A Comparative Study on Heavy Metal Removal from CCA-Treated Wood Waste by *Yarrowia lipolytica*: Effects of Metal Stress. *J. Fungi* **2023**, *9*, 469. https://doi.org/10.3390/jof9040469

Academic Editors: Miha Humar and Ivan Širić

Received: 13 February 2023
Revised: 4 April 2023
Accepted: 11 April 2023
Published: 13 April 2023

1. Introduction

The quality of human life is highly dependent on the conditions of the terrestrial environment. The harmfulness of solid pollutants, such as plastic, batteries, and preservative-treated wood (PTW) wastes, attracts human attention due to their negative impacts on the environment and human health [1]. PTW has been widely used in construction, railway, boardwalk, and electricity poles by humans for many decades. However, heavy metals in wood preservatives are harmful to the environment. Previous reports show that inorganic metal salts leaching from solid pollutants are toxic and damage human health even at low concentrations [1,2]. Therefore, the decontamination of PTW is an important and urgent issue rather than being directly released into the environment.

CCA has historically been one of the most widely used waterborne preservatives for wood due to its efficiency and low cost [3]. Arsenic, copper, and chromium in the CCA preservative have a great risk to human health. They are leached into soil or water when the treated wood is exposed to the environment. Thus, the use of CCA-treated wood has been banned in Europe since 2004 [4]. Additionally, the EPA and the wood industry decided in 2003 to stop using CCA-treated wood in most residential construction because it contained arsenic. However, due to its low cost, much CCA-treated wood waste is still in use, especially in several countries (e.g., India, China, and Brazil). As a result of scientific concerns, finding an economical and feasible way to alleviate environmental contamination is significant when CCA-treated wood is at the end of its service life.

Various approaches, such as landfilling, incineration, pyrolysis, gasification, chemical extraction, electro-dialysis, and bioremediation, have been investigated to deal with CCA-treated wood wastes [5–7]. Among those methods, incineration is the most used method. During the incineration of CCA-treated wood, Cr and Cu compounds mostly remain in the ash, while As will be diffused in the air [8]. Therefore, the premise of incineration treatment should be equipped with an advanced filtration system. Pyrolysis is usually performed at 400–800 °C accompanied by the release of harmful traces of arsenic [9]. Worryingly, incineration and gasification will accelerate carbon emissions and are not conducive to realizing carbon neutrality goals. Landfilling may contaminate the groundwater because of the leaching potential of contaminated substances. Thus, landfilling is not a legal option in most countries. Chemical pretreatment may generate intermediate by-products and secondary pollutants even if it can efficiently remove heavy metals from CCA-treated wood. Bioremediation is considered a promising and safe method. It is also the least mature solution and requires further analysis and discussion.

Microorganisms such as fungi can secrete oxalic acids outside the cells, which is important in removing metal compounds in CCA wood. Many wood-rot fungi have been used to treat wood waste by removing toxic metals or bioconverting heavy metals into less toxic or completely harmless forms [10]. In addition to *Saccharomyces cerevisiae*, *Cryptococcus* sp., *Candida* sp. *PS33*, and *black mold* have also been studied to bioremediate CCA-treated wood wastes and pollutants [11–15]. Fungi absorb heavy metals by chemical transformation through redox reactions and bioaccumulation through passive diffusion, facilitated diffusion, or active transport [5]. In addition to the microbes' biofilm's metal tolerance, their secondary metabolites play key roles in decontamination [16]. Notably, *Y. lipolytica* and *Aspergillus niger* adsorb Pb(II), Cr(III), Cr(VI), Zn(II), Cu(II), As(V), and Ni(II) ions from aqueous solutions and successfully purify sewage [17,18]. *Y. lipolytica* is a model yeast for organic acid production, especially citric acids [19]. Moreover, the changes in different morphological forms of *Y. lipolytica* help it easily adapt to new environmental conditions [20]. Numerous studies have investigated the yeast's response to different domesticated conditions, such as heavy metal stress, temperature, pH, nutrient availability (carbon, nitrogen, trace elements, etc.), dissolved oxygen concentration, electrochemical stress, agitation, and aeration [20–22]. Previous literature also found that *Y. lipolytica* is easy to domesticate using oleic acid as the oxygen vector [23]. In summary, the yeasts were more resistant to heavy metal environments after domestication.

However, studies on removing metal ions from solid contamination (e.g., CCA-treated wood) by *Y. lipolytica* are rare. Although *Y. lipolytica* effectively removes heavy metal ions from aqueous solutions, its ability to remove metals from CCA-treated wood is uncertain. Removing metal ions from aqueous environments is much easier than from CCA wood because the metals in CCA wood are bound to the wood components. It is important to reveal the mechanism of the metal removal from CCA-treated wood by *Y. lipolytica* to provide more theoretical support for the biological decontamination of CCA wood wastes.

Thus, *Y. lipolytica* was used to remove heavy metals from CCA-treated wood to investigate its decontamination properties in this study. The change in the wood structure and surficial chemicals were explored. Yeast communities are shielded from environmental stress by secreting secondary metabolites and transforming their morphology. Cu^{2+} solutions with different concentrations were used to induce *Y. lipolytica* to increase heavy metals removal further. The effect of Cu^{2+} stress on the morphology of *Y. lipolytica* and the removal rate of heavy metals were also explored. This study provides a promising potential for further application of *Y. lipolytica* for toxic metals removal from solid wastes.

2. Materials and Methods

2.1. Materials and Microorganisms

Sigma-Aldrich Canada (Oakville, ON, Canada) supplied all reagent-grade chemicals used in this study. The Global Bioresource Center provided the *Yarrowia lipolytica* 20,460 strain. The cells of activated *Y. lipolytica* were preserved at −80 °C in a mixture of glycerol and yeast malt (YM) medium. The waste CCA-treated wood used in this study came from shredded poles supplied by Tred'si (North America). They were crushed into small pieces (0.3 mm). The CCA-treated wood waste contained Cu of 2.46 $mg \cdot g^{-1}$, Cr of 4.54 $mg \cdot g^{-1}$, and As of 1.57 $mg \cdot g^{-1}$ before bioremediation.

2.2. Culture Medium and Conditions

The *Y. lipolytica* strain was first cultivated in a YM medium for 48 h. The YM medium comprised 3 g of yeast extract, 3 g of malt extract, 10 g of glucose anhydrous, 5 g of tryptic soy broth, and 1 L of deionized (DI) water. The pH of the activated strain was 3.5. The activated strain was then transferred to a growth medium and cultivated in a shaker. The growth medium contained 50 g of glucose, 0.25 g of $(NH_4)_2SO_4$, 1.7 g of KH_2PO_4, 12 of NaH_2PO_4, 1.25 g of $MgSO_4 \cdot 7H_2O$, 0.5 g of yeast extract, and 1 L of DI water. The pH of the growth medium was 4.72.

2.3. Metal Stress of *Y. lipolytica*

Y. lipolytica was inoculated in the growth medium, and its optical density (OD) was adjusted to 0.6. The OD value was measured using a UV-vis photo spectrometer (GEN10S, Madison, WI, USA) at an excitation wavelength of 600 nm. Growth mediums with high copper ion concentrations of 0.5, 0.6, 0.7, 0.8, and 1.0 $g \cdot L^{-1}$ were prepared for metal stress using $CuCl_2 \cdot 2H_2O$. An inoculated growth medium without metal ions served as a control. The above medium of 100 μL was withdrawn and transferred into the solid YM medium at intervals of 1, 4, 8, 12, and 14 days to conveniently observe the growth of the strain with the naked eye. Two replicates were operated the same as above. The domesticated strains were preserved at −80 °C in glycerol medium.

2.4. Bioremediation Procedure

Laboratory experiments explored the effect of metal stress on the bioremediation of CCA-treated wood. The OD_{600nm} value of the growth medium in flasks was adjusted to 0.05, 0.1, 0.3, 0.6, and 0.9 by adding the activated *Y. lipolytica*. The pH of OD_{600nm} = 0.05, 0.1, 0.3, 0.6, and 0.9 was 4.7, 4.64, 4.62, 4.59, and 4.55, respectively. These strains were cultivated in flasks in a shaker at a rate of 125 rpm for 24 h at 27 °C. CCA-treated wood wastes of 5g were then placed into fermented liquid media of 200 mL. CCA wood samples treated with distilled water and growth medium served as controls. The flasks inoculated with different strain concentrations were sampled at 1, 3, 6, 9, 15, and 21 days compared to the controls. Two replicates of each sample with different OD_{600nm} values and control samples were performed. Flask liquids were dumped into 50 mL tubes to measure their pH using a pH meter (HI-2002, HANNA, Padova, Italy). The wood particles were separated from the solution through a strainer. Furthermore, the treated wood particle was washed with distilled water three times and then placed in an oven overnight at 50 °C for characterization and reuse.

The following experiment was performed with different amounts of wood sawdust instead of strain concentrations. Four different quantities of wood (3, 5, 7, and 9 g) were used, where the OD_{600nm} was 0.6, and the other conditions were kept the same. Two replicates of each sample with different amounts of CCA wood were performed.

The domesticated strains of the sample (A), (B), (C), (D), (E), and (F) were selected to perform the bioremediation tests to investigate the effects of Cu^{2+} stress on heavy metal removal from CCA-treated wood wastes (Table 1). Samples (A) and (B) were stressed by copper for 4 and 8 days, respectively, with Cu^{2+} concentrations of 600 $mg \cdot LTab^{-1}$. Samples (C) and (D) were stressed for 12 and 14 days, respectively, with Cu^{2+} concentrations of

700 mg·L^{-1}. Samples (E) and (F) were stressed for 12 days with Cu^{2+} concentrations of 800 and 1000 mg·L^{-1}, respectively. The OD_{600nm} was 0.6, and the other conditions were the same as the abovementioned parameters. The strains with OD_{600nm} = 0.6 without adding Cu^{2+} were used as a control. The growth medium without any strains was set as a blank control. All samples had two replicates.

Table 1. Domesticated *Y. lipolytica* strains for bioremediation.

Sample	Concentrations of Cu^{2+} (mg·L^{-1})	Days of Metal Stress
A	600	4
B	600	8
C	700	12
D	700	14
E	800	12
F	1000	12
Control	No addition	No

2.5. Characterization of CCA-Treated Wood

After bioremediation, the surface chemical properties of the CCA-treated wood samples were analyzed using a Fourier-transform infrared spectrophotometer (IR_Tracer-100, Shimadzu, Kyoto, Japan) with an attenuated total reflection. Wood samples were scanned at wavelengths in the range of 500–4000 cm^{-1} with a resolution of 4 cm^{-1}.

The surficial morphology of the wood samples was characterized using scanning electron microscopy (SEM) (SU1510, HITACHI, Kyoto, Japan) with an accelerating voltage of 10 kV. The wood residue was sputter-coated with gold powder before testing.

Organic acids produced by fungi in the solution were determined by HPLC with a Biorad Aminex HPX-87 H column (300 × 7.8 mm) (Alltech Assoc. Inc., Deerfield, IL, USA) at 40 °C using ultraviolet detection (210 nm). The results are the average of three measurements. The eluent was 5 mM H_2SO_4 at a flow rate of 0.6 mL·min^{-1}. Liquid supernatants containing Cu and Cr were pretreated with the resin Chelex-100 (Sigma Aldrich, St. Louis, MO, USA) to prevent analytical interference by heavy metals. The tested samples were diluted with an equal volume of phosphate buffer (0.08 M, pH 6.5) and supplied with Chelex 100 (0.1 g). The mixture solution was shaken before the HPLC test and passed through a membrane filter (0.45 μm) based on previous literature [24].

The heavy metal content of wood wastes before and after bioremediation was detected to analyze the metal removal efficiency of *Y. lipolytica*. An Agilent 4100 Microwave Plasma Atomic Emission Spectrometer (Agilent Technologies, Melbourne, Australia) equipped with a standard glass concentric nebulizer and cyclonic spray chamber (Agilent Technologies, Melbourne, Australia) was used to analyze the heavy metals in the CCA-treated wood wastes before and after bioremediation. Before each test, 20 s torch stabilization and 10 s uptake were applied. Meanwhile, a 5 s read time with three replicates was used for the emission measurement of each sample. The metal content is the average of three measurements.

3. Results and Discussion

3.1. Morphology of Bioremediation Wood

The surfaces of waste CCA-treated wood displayed a relatively smooth and intact structure (Figure 1). The microbial strains and medium treatments still need to change the structure of CCA wood, as depicted in Figure 1A,B. The original structure of the wood was preserved in CCA-treated wood after bioremediation with *Y. lipolytica*. This result contradicts previous findings where the lignin and cellulose of wood were decomposed [25]. The intact structure of wood facilitates its recycling. However, microbial strains covered the wood samples' surface after bioremediation. Strains adhered to the CCA wood surface increased with the increase in cultivation time (Figure 2A–C). The SEM results also showed that the cells of *Y. lipolytica* adhered to CCA-treated wood were ellipsoid with bipolar budding patterns. The size of the ellipsoidal cells was around 5 μm. Meanwhile, some characteristic features of filamentous hyphae were also observed, especially for longer cultivation times.

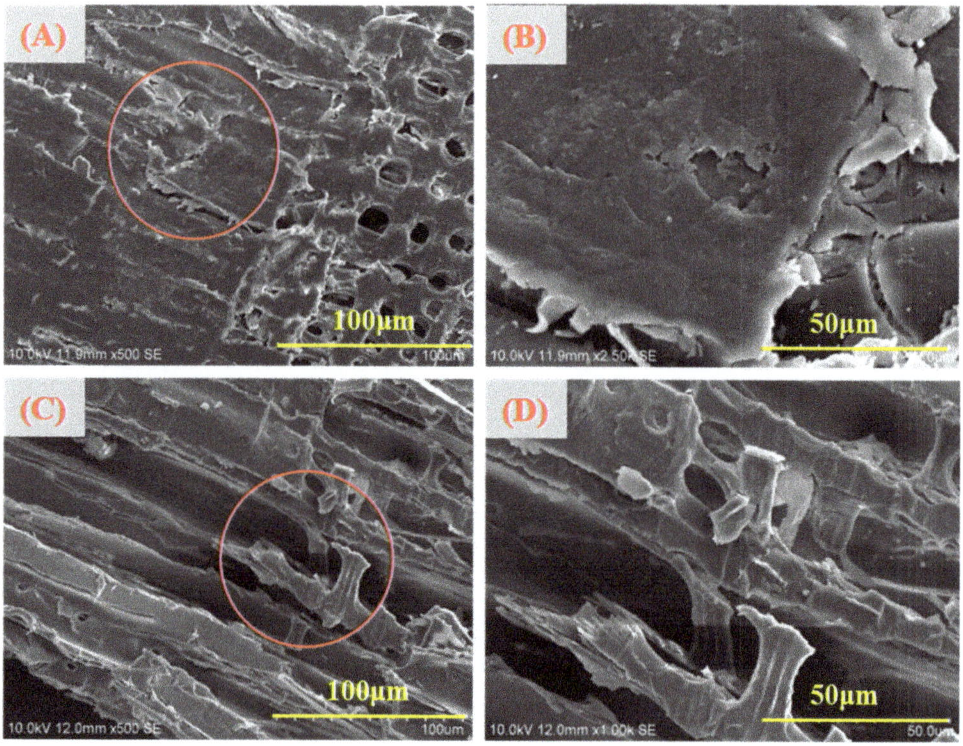

Figure 1. Micro-morphology of CCA-treated wood before treatment at 100 μm (**A**) and 50 μm scale (**B**); after treatment with uninoculated medium for 21 d at 100 μm (**C**) and 50 μm scale (**D**).

Compared to the original strains, the Cu^{2+}-domesticated strains had an ellipsoidal shape and partially changed into much smaller particles (Figure 2D–F). These observations agreed with the results of digital images of the fungi, as shown in Figure 3 and Table 2. The morphological characteristics of strain (B) after Cu^{2+} stress changed from large, oval, and round cells to small ones (Figure 3). These characteristics mainly occurred when the concentration of Cu^{2+} ranged from 500 to 700 $mg \cdot L^{-1}$, especially with an increase in cultivation days. This result indicated that the domesticated strains with round cells and small sizes are more adaptable to higher copper ion concentrations than the original strain. A previous report drew a similar conclusion that morphological changes of *Y. lipolytica* were in response to different heavy metal stress [20]. Like the original strain, microbial remediation with the domesticated strains did not destroy the surface structure of CCA-treated wood. Furthermore, many strains grew on the surface of CCA-treated wood. This observation indicated that the changes in morphological characteristics of strains hardly change the reaction mechanism of bioremediate wood waste.

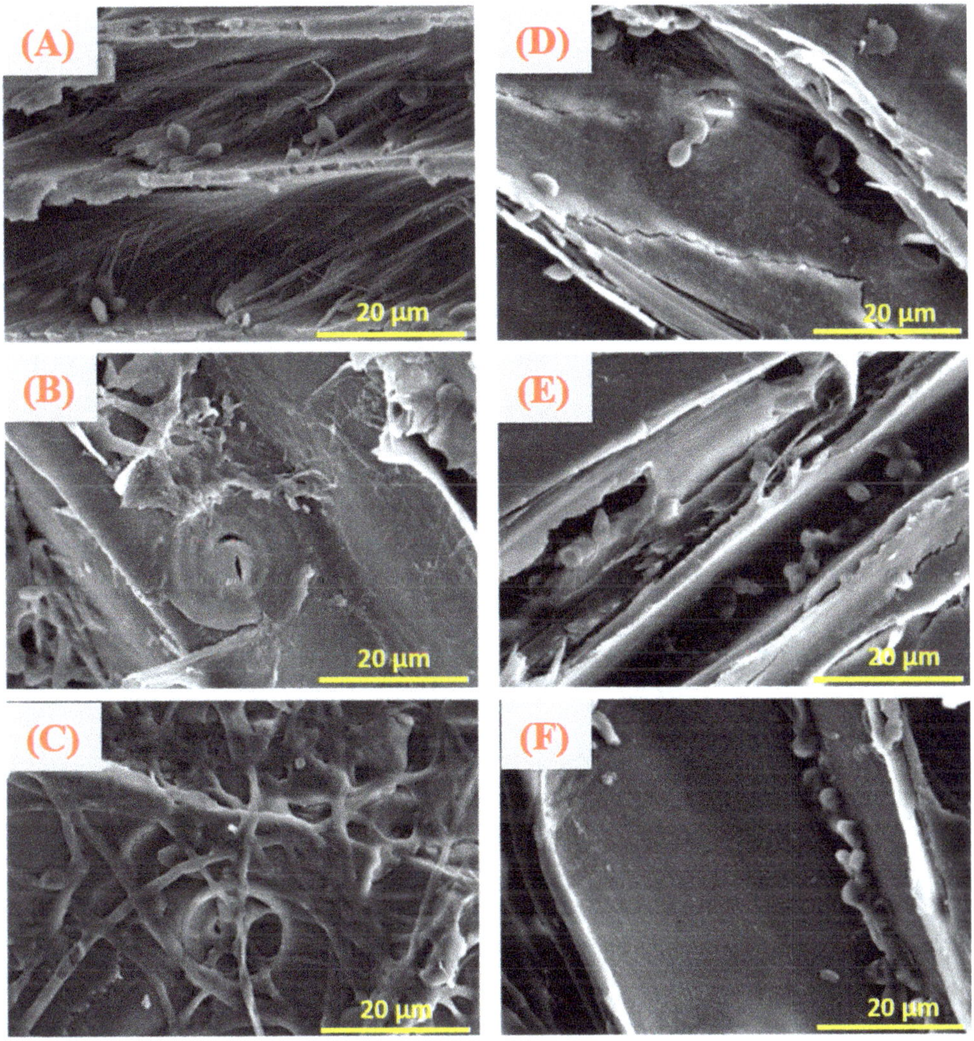

Figure 2. Micro-morphology of CCA-treated wood treated with the *Y. lipolytica*: CCA wood treated with OD_{600nm} = 0.05 for 1 d (**A**), 9 d (**B**), and 21 d (**C**), and CCA wood treated with domesticated strains of sample B for 1 d (**D**), 9 d (**E**), and 21 d (**F**).

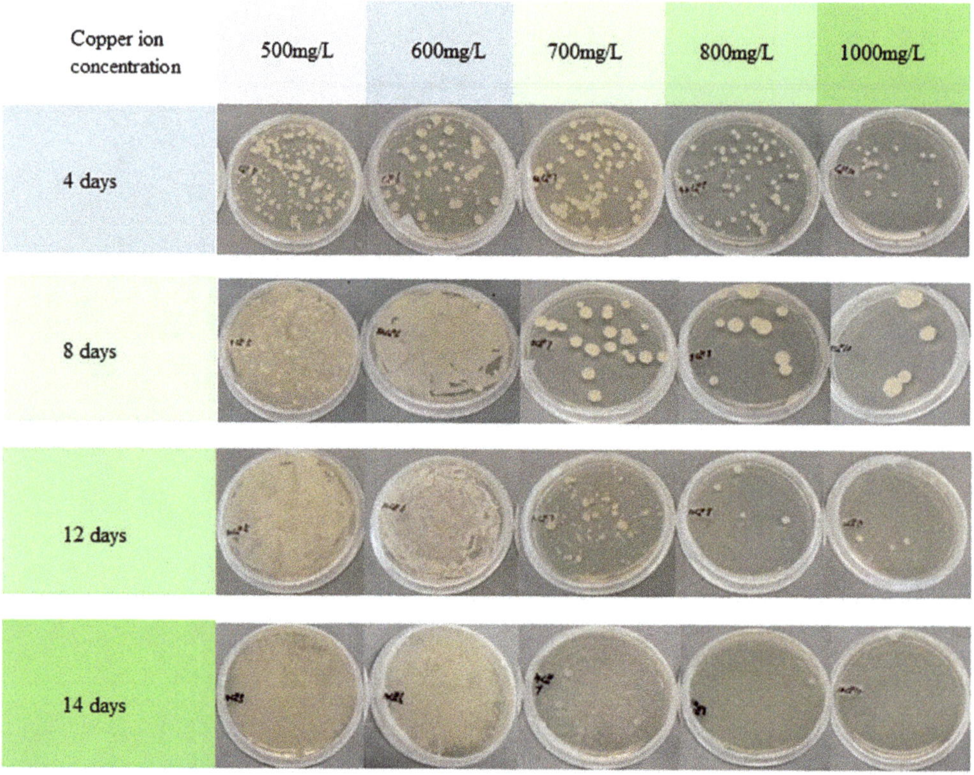

Figure 3. Digital photos of the growth status of the domesticated *Y. lipolytica*.

Table 2. The number of *Yarrowia lipolytica* colonies domesticated by different Cu^{2+} concentrations grew on the YM solid medium (obtained by counting).

Concentrations of Cu^{2+}	Number of Days of Metal Stress				
	1 Day	4 Days	8 Days	12 Days	14 Days
500 $mg \cdot L^{-1}$	116 (2.65 *)	167 (8.50)	–	–	–
600 $mg \cdot L^{-1}$	229 (42.77)	179 (6.51)	–	–	–
700 $mg \cdot L^{-1}$	170 (13.65)	92 (7.94)	21 (2.65)	99 (2.65)	–
800 $mg \cdot L^{-1}$	74 (14.36)	80 (6.81)	8 (3.51)	6 (1.15)	1 (0.58)
1000 $mg \cdot L^{-1}$	84 (15.72)	20 (2.31)	7 (2.89)	5 (3.06)	0 (0.58)

– Uncountable, (*) standard deviation.

3.2. Surface Chemical Properties of CCA-Treated Wood Samples

The surface chemical groups of the CCA-treated wood before and after bioremediation were characterized by FTIR, as shown in Figure 4. A broad and intensely sharp band at 3444 to 3330 cm^{-1} was observed. This band can be attributed to the stretching vibration of the hydroxyl group of bonded water in the CCA-treated wood. The asymmetric and symmetric vibrations of the C–H bonds were located at the 2922 and 2853 cm^{-1} wavenumbers, respectively. Those results demonstrated the presence of aliphatic structures in the CCA-treated wood. The peaks at 1240 and 1147 to 1031 cm^{-1} and the continuation spectra between 900 and 700 cm^{-1} were attributed to C–O, C–C, or C–OH vibrations of aromatic structures and C–H bonds of aromatic structures, respectively. The peaks at about 1650 to 1750 cm^{-1} indicated the presence of C = O vibrations. Nadaroglu et al. (2015) also observed these characteristic peaks in the FTIR spectrum of CCA-treated wood [26]. Furthermore, the wide absorption band at 1650–1730 cm^{-1} was attributed to the stretching vibrations of the C = O group in the CCA-treated wood. However, this band was split into two peaks after bioremediation. This result can be explained by copper (II) ions complexed with the acetyl or carboxylic acid bonds in the CCA-treated wood, causing a broad absorption band [27]. The copper ions in the CCA-treated wood were removed by microbial remediation with *Y. Lipolytica*, releasing carboxylic acid.

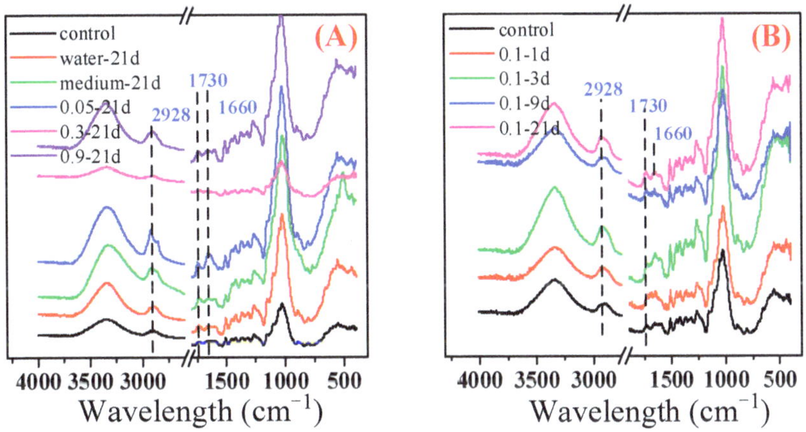

Figure 4. FTIR spectra of CCA wood treated with different conditions for 21 days (**A**) and with the concentrations of *Y. lipolytica* (OD_{600nm} = 0.1) for different cultivation days (**B**).

3.3. Bioremediation of CCA-Treated Wood

3.3.1. Influence of Initial Strain Concentrations

The CCA-wood was treated with different concentrations of *Y. lipolytica* to remove heavy metals. An uninoculated medium and distilled water were used as controls, as shown in Figure 5A–C. After 21 days, the minimum content of copper, arsenic, and chromium in the wood was 0.417, 0.522, and 2.55 mg/g, respectively, with the treatment of yeast at OD = 0.05 concentration (Figure S1). The maximum Cu removal (82.8%) was attained after 21 days of incubation with a strain concentration of OD_{600nm} = 0.05. It was observed that almost the same amount of Cu was removed after 21 days when the strain's concentration was OD_{600nm} = 0.1, 0.3, and 0.6. However, OD_{600nm} = 0.9 showed less Cu removal under the same condition. This was attributed to the fact that organic acids have a greater impact on metal removal than colony adsorption. In the early stage of treatment, the strain multiplies rapidly, consuming nutrients. During this period, the removal effect of heavy metals by high-concentration fungi was better than that of low-concentration fungi due to adsorption. The content of oxalic acid increased rapidly after 6 days of treatment (Figure 5C), playing a major role in removing heavy metals. The strains depleted nutrients quickly when the strains' concentration was high. For strains with OD_{600nm} = 0.9, the lack of nutrients led to low oxalic acid production after 6 days, resulting in a low metal removal rate. This result is consistent with Table 3, where heavy metal mainly passed into the culture medium rather than the yeast, especially for low initial strains concentration. The amount of metals adsorbed by yeast increases with the initial strain concentration of the treatment solution. This increase is because the yield of organic acid is low in the initial stage of treatment, and the removal of metals is mainly by the adsorption of strains. Over time, the metal content in the yeast cells decreased. This decrease is because the organic acids produced in the late fermentation stage, especially oxalic acid, have a stronger binding ability to metals excreted from the cells to the culture solution. However, no matter the initial strain concentration, the metal content in the treated supernatant is higher than in the yeast cells. This finding suggests that metal removal is mainly due to cell secretions (organic acids) rather than the cells' adsorption. The interactions of strains and Cu elements include extracellular complexation through organic acids, precipitation, and adsorption onto the cell wall [28]. The *Y. lipolytica* strain treatment led to extensive solubilization of Cu with increasing cultivation days under the same strains' concentrations. It is worth mentioning that more than half of the Cu ions in the wood were removed after three days in this system. The heavy metal removal was much higher than the control samples (14.3% with water and 34.4% with medium). CCA-treated wood is impregnated with CuO, CrO_3, and As_2O_5 under high pressure. These chemical compositions are water soluble, which can be leached with water or organic acids [16]. Furthermore, the medium in this study mainly consists of glucose that contains hydroxyl functional groups. These hydroxyl groups combine with heavy metals to decontaminate CCA-treated wood [16]. *Aspergillus niger,* having mature cultivation systems, has also been extensively studied in the bioremediation of treated wood. However, its removal efficiency for Cu is less than 90% after 10 days of remediation [29]. Another study showed that the Cu removal rates from CCA-treated wood treated with *Fomitopsis palustris*, *Coniophora puteana*, and *Laetiporus sulphureus* for 10 days were 71.9%, 66.5% and 50.1%, respectively [30]. Researchers have also explored other fungi to remove heavy metals from CCA-treated wood wastes, as shown in Table 4. Those results indicated the removal ability of copper ions varies with the fungal species. Those observations showed that *Y. lipolytica* is promising to obtain an excellent Cu removal rate by optimizing the medium compositions and cultivation conditions.

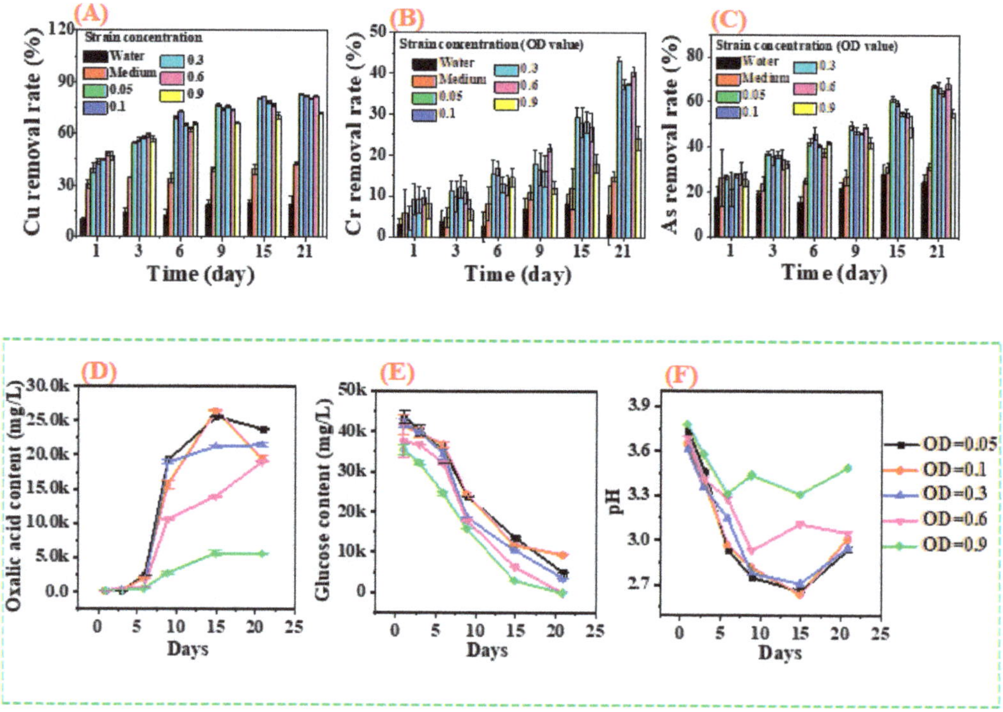

Figure 5. Removal rates of Cu (**A**), Cr (**B**), and As (**C**) from CCA-treated wood using *Y. lipolytica*, and the amounts of oxalic acid (**D**) and glucose (**E**) in the fermentation medium, the pH value of the fermentation medium (**F**).

Table 3. The cu ion content in the culture supernatant and cells of *Y. lipolytica* after treatment.

Time \ OD Values	0.05	0.6	0.6 (Fungi under Copper Ion Stress)	0.05 (Supernatant)	0.6 (Supernatant)
1 day	36.8×10^{-3} ($mg{\cdot}g^{-1}$)	1.53 ($mg{\cdot}g^{-1}$)	1.55 ($mg{\cdot}g^{-1}$)	46.0×10^{-3} ($mg{\cdot}g^{-1}$)	32.3×10^{-3} ($mg{\cdot}g^{-1}$)
15 days	36.8×10^{-3} ($mg{\cdot}g^{-1}$)	1.02 ($mg{\cdot}g^{-1}$)	1.19 ($mg{\cdot}g^{-1}$)	-	-

Table 4. Comparison of decontamination effects of different fungi on CCA wood.

Microbes	Maximum Metal Removal Rate (%)			Reference
	Cu	Cr	As	
Yarrowia lipolitica	83	43	68	This study
Aspergillus niger	49	55	97	[15]
Aureobacterium barkeri	50	68	37	[31]
Pseudomonas fluorescens	50	-	48	[31]
Fomitopsis palustris	72	87	100	[30]
Coniophora puteana	67	19	18	[30]
Laetiporus sulphureus	50	69	85	[30]

Compared with Cu, Cr, and As, elements in CCA-treated wood are difficult to remove. A previous study indicated that less than 20% of Cr and 30% of As were removed using *Alternaria alternata* or *Cladosporium herbarium* treated for 10 days [29]. In the current study, the removal rate of Cr at different strain concentration gradients was lower than that for Cu, where the maximum removal was 43.1%. Meanwhile, the maximum removal rate of As was 68.3%. Like Cu, the Cr and As removal rates increased with increasing incubation time (Figure 5B,C). However, Sierra-Alvarez (2009) found that the removal rate of Cu from the solid-state fermentation of CCA-treated wood by several brown rot fungi was very low (<10.9%), while a high removal rate for Cr was observed in the same system [24]. Similar results were also observed for *Fomitopsis palustris* and *Laetiporus sulphureus*, as shown in Table 4. Compared with copper ions, the removal of arsenic ions by these two fungi is more obvious. These observations are different from what has been found in this study. This observation suggests that microorganisms exhibit different tolerances to different metals. Thus, removing heavy metals from treated wood using a combination of different species of microorganisms is a promising exploration strategy. For the CCA-treated wood with uninoculated medium for 21 days, this medium was mainly composed of glucose, which contained hydroxyl structural groups. These hydroxyl groups combine with metal ions, thereby removing the metal ions from the CCA-treated wood [32].

Figure 5F shows that the pH of the medium decreased sharply before 15 days of domesticating, except for strains with OD_{600nm} = 0.6 and 0.9. The decrease in pH was accompanied by a significant increase in oxalic acid accumulation for all the samples with cultivation times up to 15 days. However, the pH values of the mediums increased from 15 to 21 days. This observation is attributed to the lack of a carbon source (i.e., glucose), leading to stagnant growth and strains' low organic acid production. Oxalic acid in the medium complexed with copper ions further increases Cu removal under the same concentrations. Therefore, organic acids play an important role in the bioremediation process of preservative-treated wood by microorganisms [24,29]. The removal rate of heavy metals (Cu, Cr, and As) from CCA-treated wood increased with a decreased organic acid concentration in the biological system.

The organic acid composition and pH determination were analyzed to explore the metal removal mechanism for CCA-treated wood by *Y. lipolytica*. Results showed that *Y. lipolytica* produced oxalic, citric, malic, and acetic acids, as depicted in the HPLC chromatograms (Figure 6).

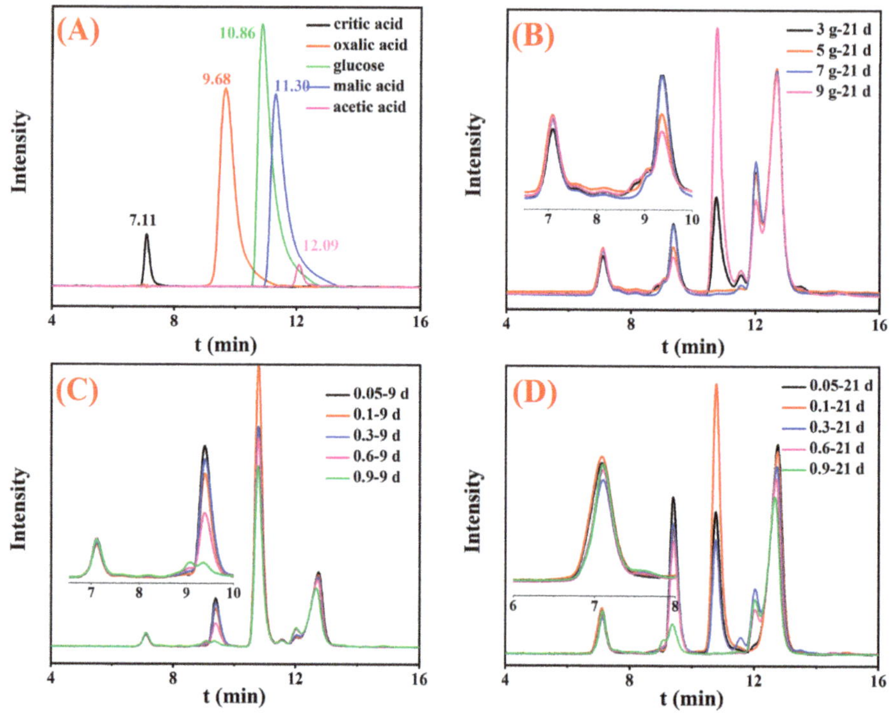

Figure 6. HPLC results of the standard samples (**A**) and the medium supernatant under different conditions: different contents of CCA-treated wood for 21 days (**B**), different concentrations of *Y. lipolytica* for 9 days (**C**), and different concentrations of *Y. lipolytica* for 21 days (**D**).

Inoculum concentration had an important influence on the removal rate of heavy metals from the CCA-treated wood. For instance, during the first 15 days of cultivation, the removal efficiencies of Cu were shifted from low to high as follows, OD_{600nm} = 0.9 < 0.6 < 0.3 < 0.05 < 0.1. Generally, the higher the strain concentration, the increased ability to remove metal. The experimental results of this study were consistent with this rule when the treatment time was less than 6 days. However, this study showed a contrary result after 21 days of treatment. The reason is that organic acids play a major role in metal removal. In the early stage of treatment, the biosorption mechanism is dominant, so the more colonies, the more metals will be adsorbed, and the removal rate will be higher. The medium's space and nutrients are insufficient to support the normal growth of high-concentration strains to produce oxalic acid after 6 days. The strains with low concentrations are more active and produce more contents of organic acids after 6 days, thereby possessing a higher metal removal efficiency. The removal efficiency of Cr and As was lower than that of Cu. This observation may be explained by oxalic acid being more sensitive to copper than As and Cr metals. Oxalic acid was predominant in the fermentation medium of the *Y. lipolytica* strains (Figures 6 and 7). Unlike Cu, the acid that can act on chromium or arsenic was not observed. In addition, the amounts and types of organic acids produced by microorganisms vary with operating conditions, such as pH, temperature, and nutrient contents [29]. Thus, the external conditions should be optimized to adjust the removal of heavy metals from preservative-treated wood.

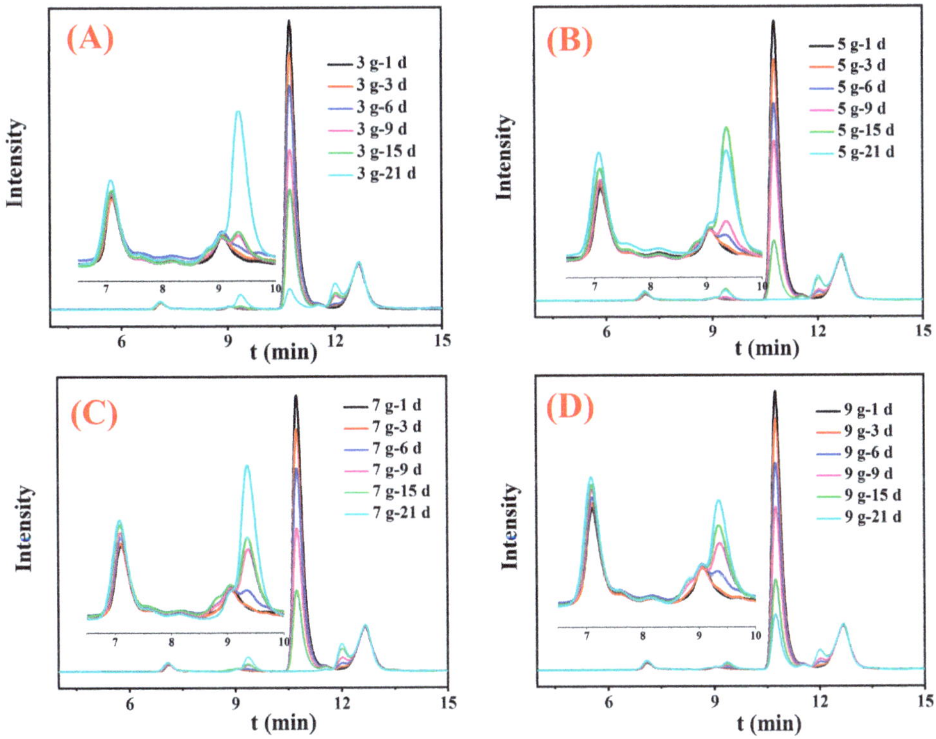

Figure 7. HPLC results for the medium supernatant with different contents of CCA-treated wood: CCA-treated wood of 3 g (**A**), CCA-treated wood of 5 g (**B**), CCA-treated wood of 7 g (**C**), and CCA-treated wood of 9 g (**D**).

3.3.2. Influence of Wood Amount

The effect of CCA-treated wood dose on the removal of heavy metals using the liquid broth medium of *Y. lipolytica* (OD_{600nm} = 0.6) was explored. The number of wood wastes affected the removal rate of heavy metals, especially copper (Figure 8A). The percentage of removed Cu gradually decreased with increased CCA-treated wood contents (Figure 8A). Moreover, increasing the cultivation period from 1 to 21 days enhanced heavy metal removal while the pH value decreased. The pH increased with increasing wood amounts (Figure 8D). It was likely attributed to the enhancement of inorganic metal salts in liquid with the increase in CCA-treated wood dose. These observations indicated that copper removal mainly depended on the amount of organic acid. Of note, the effect of wood contents on the concentration of organic acids could have been more obvious.

The effect of wood content on the removal of Cr and As showed a different tendency compared to Cu (Figure 8B,C). The amounts of CCA-treated wood did not affect the removal rates of Cr and As. Nevertheless, they were influenced to some extent by cultivation time. In contrast, a previous study revealed that the initial concentration of ions was highly affected by removing Cr (VI) using *Y. lipolytica* in aqueous solutions [33]. In addition to the organic acids produced by the yeast, mycelium and proteins effectively decontaminated toxic metals by either biosorption or bioaccumulation [34]. Biomass residues have numerous functional groups, such as hydroxyl and carboxyl, which help absorb metal ions in aqueous solutions compared to a solid matrix. These observations reveal that the ability of yeast to decontaminate CCA-treated wood varies with metal species and the state of the pollutant.

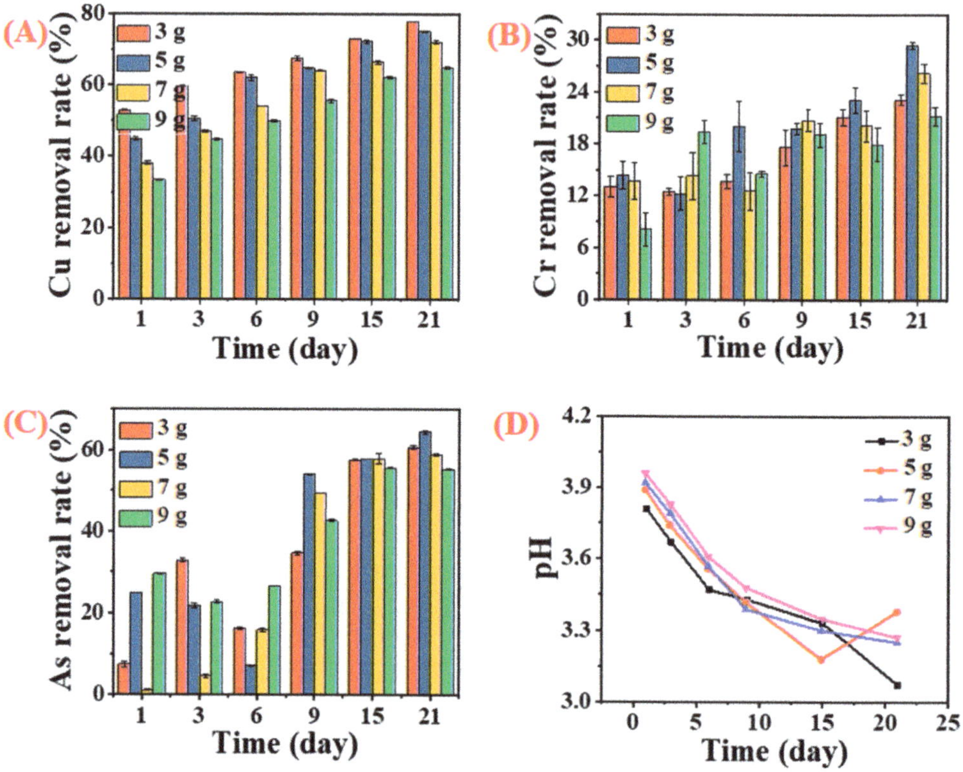

Figure 8. Removal rate of Cu (**A**), Cr (**B**), and As (**C**) from CCA-treated wood by *Y. lipolytica* (OD_{600nm} = 0.6) with different wood contents; pH of the fermentation medium with varying contents of CCA-treated wood (**D**).

3.3.3. Influence of Copper Ion Stress

A previous study reported that the metal-tolerant properties of *Y. lipolytica* increased after strains were stressed by Cu^{2+} solutions [33]. Based on the above report, *Y. lipolytica* was domesticated by Cu ions to change its form to remove heavy metals in this study. High concentrations of Cu^{2+} were used to domesticate *Y. lipolytica* at different intervals to enhance its metal tolerance. Table 2 shows the number of colonies of *Y. lipolytica* grown on a YM solid medium domesticated by different Cu^{2+} concentrations. The number of colonies was obtained by counting. The amounts of the strains increased when the concentrations of Cu^{2+} in solution were 500 and 600 $mg \cdot L^{-1}$ with the increase in metal stress days. The morphology of the domesticated strains changed from large and round colonies to smaller ones. Kolhe obtained an analogous result that the dimorphic behavior of *Y. lipolytica* from yeast to mycelial forms was altered in response to metal stress [28]. In addition, biofilm formation is another way to resist metal ion stress. Moreover, the extracellular synthesis of Cu-nanoparticles of *Y. lipolytica* was also a pathway in response to Cu stress [35].

The change in morphology was significant when the strains were cultivated in a copper concentration of 700 $mg \cdot L^{-1}$. The number of strains first reduced within eight days and afterward increased. This can be explained by the fact that heavy metals partly inhibited the strains at an early stress stage, and the survived strains were adapted to the metal-based environment after eight days. When the concentrations of Cu^{2+} were over 800 $mg \cdot L^{-1}$, it was difficult for the strains to survive more than four days. Even worse, all strains died after 14 days of stress. The results indicated that the change in morphology of *Y. lipolytica* varied with concentrations of heavy metals. Similarly, previous studies found that different types of *Y. lipolytica* have differential responses to heavy metals stress [20]. A few eugenic strains were selected to treat CCA-treated wood to determine the tolerance of the domesticated strains to heavy metals.

The metal stress treatment promoted the removal of copper from CCA-treated wood by *Y. lipolytica* (Figure 9). The metal adsorption performance of the seven domesticated yeasts varied with the processing parameters. The Cu^{2+} removal rate of the domesticated strains of samples (C) and (D) increased by around 20% compared to the original ones within 1 day. Meanwhile, sample C's removal rate was 95.9% higher within one day than for the blank control (growth medium). Even though glucose and phosphate in the medium may aid in the diffusion of metals, it is difficult for them to compete for copper ions from cellulose molecules. Both glucose and wood contain a large number of hydroxyl groups. Moreover, wood contains carboxyl groups, which are easier to complex with metal ions than hydroxyl groups. The significant increase in the removal rate of copper ions indicates that *Y. lipolytica* has played an important role in promoting the removal of copper ions. Phosphatases, metallothionein, and organic acids are reported to play essential roles in overcoming copper ion stress in *Y. lipolytica* [28]. *Y. lipolytica* exhibited higher phosphatase levels in a copper-supplemented medium, which increased with metal ion concentrations [36]. The strains' tolerance to Cu^{2+} was attributed to the increased phosphatase activity, resulting in the efflux of Cu^{2+} as phosphate complexes. The presence of phosphatase ensures the survival of fungi in the environment of high metal ions so that more organic acids can be produced to remove heavy metals from wood. Thiol groups that were in cysteine residues connected metallothionein and metal ions. Organic acids, transport proteins, and extracellular proteins secreted by the strains were imperative to the bioremediation of CCA-treated wood. They remove heavy metals by biosorption, biotransformation, bioaccumulation, and organic acid complexation [5]. Thus, the enhancement of metal removal by Cu^{2+}-domesticated strains was attributed to the synergistic effect of multiple enzymes and organic acids.

The domesticated strains did not show regular growth with increasing cultivation time. The effect of sample (C) on metal removal from treated wood showed more efficiency than the original ones for different cultivation intervals. The domesticated strains (B) showed incremental patterns of metal removal with an increase in cultivation days. The metal removal rate from CCA-treated wood using strains (B) was similar to the control sample within nine days of cultivation. It overtook the control sample removal rate after 15 days of cultivation. This may be due to the original strain being adaptable to high copper ion concentrations over time. On the other hand, metal stress treatments make strains better adapt to harsh conditions than the original strain, then shorten their metal removal time.

The response of yeasts to toxic metals depends on the concentrations of heavy metal ions. Generally, yeasts adapt to the pressure of the metal environment in different ways, such as transformation, extracellular precipitation, intracellular compartmentalization, crystallization, organic acid complexation, and adsorption onto cell walls [33,37,38]. Bankar et al. (2018) investigated *Y. lipolytica's* tolerance to heavy metals and found that the metals influenced biofilms' formation [33]. A previous study revealed that changes in its morphological behavior demonstrated the adaption of *Y. lipolytica* to the stresses of different heavy metals [20]. In this study, strains' morphology changed when the concentrations of Cu^{2+} were lower while the strains' growth was inhibited at a high Cu^{2+} concentration (Table 2). Moreover, the organic acid produced by strains plays a vital role in heavy metal removal. Therefore, other stress approaches, such as changes in cultivation conditions (pH, temperature, etc.) [39], electrochemical stress [21], UV light radiation [40], and genetic engineering [41], will be further investigated to develop selective strains with better characteristics for heavy metal removal.

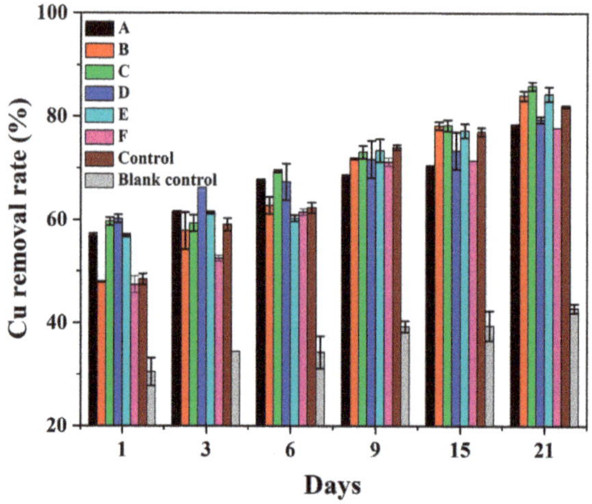

Figure 9. Cu removal rate from CCA-treated wood using domesticated strains compared with untreated ones; samples (A) and (B) were stressed for 4 and 8 days, respectively, with Cu^{2+} concentrations of 600 $mg{\cdot}L^{-1}$; samples (C) and (D) were stressed for 12 and 14 days, respectively, with Cu^{2+} concentrations of 700 $mg{\cdot}L^{-1}$; samples (E) and (F) were stressed for 12 days with Cu^{2+} concentrations of 800 and 1000 $mg{\cdot}L^{-1}$, respectively; the control was strains with OD_{600nm} = 0.6 without the addition of Cu^{2+}; the blank control was the growth medium without any strains.

4. Conclusions

Y. lipolytica was efficient in the removal of toxic metals from CCA-treated wood. The copper stress of microbial strains enhanced their metal tolerance, further increasing their heavy metal removal abilities. Bioremediation changed the surface morphology and the chemicals of CCA-treated wood because heavy metals and strain residues were removed. The organic acids produced by *Y. lipolytica* played an important role in removing heavy metals, especially copper. The Cu, Cr, and As removal rates by *Y. lipolytica* were 82.8%, 43.1%, and 68.3%, respectively. Since *Y. lipolytica* does not destroy the structure of wood particles, the decontaminated CCA wood particles can be mixed with the original wood particles in a certain proportion to prepare wood pellets. In contrast to the original strains, the removal rate of copper for domesticated strains increased by 20% under fixed conditions. This result indicated that metal stress changed the nature of the strains, increasing their heavy metal removal efficiency for CCA-treated wood. This study proved that *Y. lipolytica* strains could remove metals from CCA-treated wood, advancing research on the bioremediation of solid waste pollutants.

Supplementary Materials: The following supporting information can be downloaded at: https://www.mdpi.com/article/10.3390/jof9040469/s1, Figure S1: The content of copper, chromium and arsenic in wood, treated with different concentrations of yeast for different time.

Author Contributions: D.X.: Conceptualization, Methodology, Original draft preparation, Software; J.Z.: Data curation, Visualization, Reviewing and Editing; A.K. and S.M.: Supervision, Validation, Reviewing and Editing; H.B.: Reviewing and Editing. A.K. and H.B. Funding acquisition. All authors have read and agreed to the published version of the manuscript.

Funding: The Quebec Consortium for Industrial Bioprocess Research and Innovation (CRIBIQ), Mitacs (IT11796), the Canada Research Chair Program, the Natural Science Foundation of Shandong Province (ZR2022QC101), and the Ministry of Education (SWZ-ZD202101) funded this research.

Institutional Review Board Statement: Not applicable.

Informed Consent Statement: Not applicable.

Data Availability Statement: The data used in this study is available upon request from the corresponding author.

Acknowledgments: The authors thank the Industrial Waste Technology Centre and the Key Laboratory of Bio-based Material Science & Technology (Northeast Forestry University) for access to their facilities and technical assistance.

Conflicts of Interest: The authors declare no conflict of interest.

References

1. Morais, S.; Fonseca, H.M.; Oliveira, S.M.; Oliveira, H.; Gupta, V.K.; Sharma, B.; de Lourdes Pereira, M. Environmental and Health Hazards of Chromated Copper Arsenate-Treated Wood: A Review. *Int. J. Environ. Res. Public Health* **2021**, *18*, 5518. [CrossRef] [PubMed]
2. Civardi, C.; Schwarze, F.W.M.R.; Wick, P. Micronized copper wood preservatives: An efficiency and potential health risk assessment for copper-based nanoparticles. *Environ. Pollut.* **2015**, *200*, 126–132. [CrossRef] [PubMed]
3. Dhillon, G.S.; Lea Rosine, G.M.; Kaur, S.; Hegde, K.; Brar, S.K.; Drogui, P.; Verma, M. Novel biomaterials from citric acid fermentation as biosorbents for removal of metals from waste chromated copper arsenate wood leachates. *Int. Biodeterior. Biodegrad.* **2017**, *119*, 147–154. [CrossRef]
4. Townsend, T.; Tolaymat, T.; Solo-Gabriele, H.; Dubey, B.; Stook, K.; Wadanambi, L. Leaching of CCA-treated wood: Implications for waste disposal. *J. Hazard. Mater.* **2004**, *114*, 75–91. [CrossRef]
5. Zango Usman, U.; Mukesh, Y.; Vandana, S.; Sharma, J.; Sanjay, P.; Sidhartha, D.; Sharma Anil, K. Microbial bioremediation of heavy metals: Emerging trends and recent advances. *Res. J. Biotechnol. Vol.* **2020**, *15*, 164–178.
6. Gmar, M.; Bouafif, H.; Bouslimi, B.; Braghiroli, F.L.; Koubaa, A. Pyrolysis of Chromated Copper Arsenate-Treated Wood: Investigation of Temperature, Granulometry, Biochar Yield, and Metal Pathways. *Energies* **2022**, *15*, 5071. [CrossRef]
7. Ribeiro, A.; Rodriguez-Maroto, J.; Mateus, E.; Velizarova, E.; Ottosen, L.M. Modeling of electrodialytic and dialytic removal of Cr, Cu and As from CCA-treated wood chips. *Chemosphere* **2007**, *66*, 1716–1726. [CrossRef]
8. Alam, M.; Alshehri, T.; Wang, J.; Singerling, S.A.; Alpers, C.N.; Baalousha, M. Identification and quantification of Cr, Cu, and As incidental nanomaterials derived from CCA-treated wood in wildland-urban interface fire ashes. *J. Hazard. Mater.* **2023**, *445*, 130608. [CrossRef]
9. Kato, T.; Hatakeyama, T.; Sugawara, K. Release behavior of arsenic, chromium, and copper during heat treatments of CCA-treated wood. *J. Mater. Cycles Waste Manag.* **2021**, *23*, 1636–1645. [CrossRef]
10. Costa, L.G.d.; Brocco, V.F.; Paes, J.B.; Kirker, G.T.; Bishell, A.B. Biological and chemical remediation of CCA treated eucalypt poles after 30 years in service. *Chemosphere* **2022**, *286*, 131629. [CrossRef]
11. Blazhenko, O.V.; Zimmermann, M.; Kang, H.A.; Bartosz, G.; Penninckx, M.J.; Ubiyvovk, V.M.; Sibirny, A.A. Accumulation of cadmium ions in the methylotrophic yeast Hansenula polymorpha. *BioMetals* **2006**, *19*, 593–599. [CrossRef] [PubMed]
12. Wang, J.; Chen, C. Biosorption of heavy metals by Saccharomyces cerevisiae: A review. *Biotechnol. Adv.* **2006**, *24*, 427–451. [CrossRef]
13. Singh, P.; Raghukumar, C.; Parvatkar, R.R.; Mascarenhas-Pereira, M. Heavy metal tolerance in the psychrotolerant *Cryptococcus* sp. isolated from deep-sea sediments of the Central Indian Basin. *Yeast* **2013**, *30*, 93–101. [CrossRef] [PubMed]
14. Ilyas, S.; Rehman, A.; Ilyas, Q. Heavy metals induced oxidative stress in multi-metal tolerant yeast, *Candida* sp. PS33 and its capability to uptake heavy metals from wastewater. *Pak. J. Zool.* **2017**, *49*, 769. [CrossRef]
15. Kartal, S.N.; Kakitani, T.; Imamura, Y. Bioremediation of CCA-C treated wood by Aspergillus niger fermentation. *Holz Als Roh-Und Werkst.* **2004**, *62*, 64–68. [CrossRef]
16. Xing, D.; Magdouli, S.; Zhang, J.; Koubaa, A. Microbial remediation for the removal of inorganic contaminants from treated wood: Recent trends and challenges. *Chemosphere* **2020**, *258*, 127429. [CrossRef]
17. Bankar, A.V.; Kumar, A.R.; Zinjarde, S.S. Removal of chromium (VI) ions from aqueous solution by adsorption onto two marine isolates of Yarrowia lipolytica. *J. Hazard. Mater.* **2009**, *170*, 487–494. [CrossRef]
18. Ren, B.; Jin, Y.; Cui, C.; Song, X. Enhanced Cr (VI) adsorption using chemically modified dormant Aspergillus niger spores: Process and mechanisms. *J. Environ. Chem. Eng.* **2022**, *10*, 106955. [CrossRef]
19. Cavallo, E.; Charreau, H.; Cerrutti, P.; Foresti, M.L. Yarrowia lipolytica: A model yeast for citric acid production. *FEMS Yeast Res.* **2017**, *17*, fox084. [CrossRef]
20. Bankar, A.; Zinjarde, S.; Telmore, A.; Walke, A.; Ravikumar, A. Morphological response of Yarrowia lipolytica under stress of heavy metals. *Can. J. Microbiol.* **2018**, *64*, 559–566. [CrossRef]
21. Oliveira, A.; Sousa, T.; Amaral, P.; Coelho, M.; Araujo, O. Study of morphological and physiological parameters of cultures of Yarrowia lipolytica undergone electrochemical stress. *Chem. Eng. Trans.* **2010**, *20*, 133–138.
22. Pandharikar, G.; Claudien, K.; Rose, C.; Billet, D.; Pollier, B.; Deveau, A.; Besserer, A.; Morel-Rouhier, M. Comparative Copper Resistance Strategies of Rhodonia placenta and Phanerochaete chrysosporium in a Copper/Azole-Treated Wood Microcosm. *J. Fungi* **2022**, *8*, 706. [CrossRef] [PubMed]
23. Liu, X.; Xu, J.; Xia, J.; Lv, J.; Wu, Z.; Deng, Y. Improved production of citric acid by Yarrowia lipolytica using oleic acid as the oxygen-vector and co-substrate. *Eng. Life Sci.* **2016**, *16*, 424–431. [CrossRef]

24. Sierra-Alvarez, R. Removal of copper, chromium, and arsenic from preservative-treated wood by chemical extraction-fungal bioleaching. *Waste Manag.* **2009**, *29*, 1885–1891. [CrossRef] [PubMed]
25. Schubert, M.; Volkmer, T.; Lehringer, C.; Schwarze, F. Resistance of bioincised wood treated with wood preservatives to blue-stain and wood-decay fungi. *Int. Biodeterior. Biodegrad.* **2011**, *65*, 108–115. [CrossRef]
26. Nadaroglu, H.; Kalkan, E.; Celik, H. Equilibrium studies of copper ion adsorption onto a modified kernel of date (*Fructus dactylus*). *Int. J. Environ. Sci. Technol.* **2015**, *12*, 2079–2090. [CrossRef]
27. Gołofit, T.; Zielenkiewicz, T.; Gawron, J. FTIR examination of preservative retention in beech wood (*Fagus sylvatica* L.). *Eur. J. Wood Wood Prod.* **2012**, *70*, 907–909. [CrossRef]
28. Kolhe, N.; Damle, E.; Pradhan, A.; Zinjarde, S. A comprehensive assessment of Yarrowia lipolytica and its interactions with metals: Current updates and future prospective. *Biotechnol. Adv.* **2022**, *59*, 107967. [CrossRef]
29. Kartal, S.N.; Katsumata, N.; Imamura, Y. Removal of copper, chromium, and arsenic from CCA-treated wood by organic acids released by mold and staining fungi. *For. Prod. J.* **2006**, *56*, 33–37.
30. Kartal, S.N.; Munir, E.; Kakitani, T.; Imamura, Y. Bioremediation of CCA-treated wood by brown-rot fungi Fomitopsis palustris, Coniophora puteana, and Laetiporus sulphureus. *J. Wood Sci.* **2004**, *50*, 182–188. [CrossRef]
31. Clausen, C.A. Isolating metal-tolerant bacteria capable of removing copper, chromium, and arsenic from treated wood. *Waste Manag. Res.* **2000**, *18*, 264–268. [CrossRef]
32. Wang, X.; Li, J.; Guo, Q.; Xu, M.; Zhang, X.; Li, T. Cation-π interaction in Mg (OH) 2@ GO-coated activated carbon fiber cloth for rapid removal and recovery of divalent metal cations by flow-through adsorption. *Resour. Conserv. Recycl.* **2023**, *188*, 106648. [CrossRef]
33. Bankar, A.; Zinjarde, S.; Shinde, M.; Gopalghare, G.; Ravikumar, A. Heavy metal tolerance in marine strains of Yarrowia lipolytica. *Extremophiles* **2018**, *22*, 617–628. [CrossRef] [PubMed]
34. Hattori, T.; Hisamori, H.; Suzuki, S.; Umezawa, T.; Yoshimura, T.; Sakai, H. Rapid copper transfer and precipitation by wood-rotting fungi can affect copper removal from copper sulfate-treated wood blocks during solid-state fungal treatment. *Int. Biodeterior. Biodegrad.* **2015**, *97*, 195–201. [CrossRef]
35. El-Sayed, M.T. Bioremediation and Extracellular Synthesis of Copper Nanoparticles from Wastewater using vYarrowia lipolytica AUMC 9256. *Egypt. J. Bot.* **2018**, *58*, 563–579.
36. Ito, H.; Inouhe, M.; Tohoyama, H.; Joho, M. Effect of copper on acid phosphatase activity in yeast Yarrowia lipolytica. *Z. Für Nat. C* **2007**, *62*, 70–76. [CrossRef]
37. Adamo, G.M.; Brocca, S.; Passolunghi, S.; Salvato, B.; Lotti, M. Laboratory evolution of copper tolerant yeast strains. *Microb. Cell Factories* **2012**, *11*, 1. [CrossRef]
38. Hosiner, D.; Gerber, S.; Lichtenberg-Fraté, H.; Glaser, W.; Schüller, C.; Klipp, E. Impact of Acute Metal Stress in Saccharomyces cerevisiae. *PLoS ONE* **2014**, *9*, e83330. [CrossRef]
39. Rywińska, A.; Wojtatowicz, M.; Żarowska, B.; Rymowicz, W. Biosynthesis of citric acid by yeast Yarrowia lipolytica A-101-1.31 under repeated batch cultivation. *Electron. J. Pol. Agric. Univ.* **2008**, *11*, #07.
40. Kostina, E.; Wulff, A.; Julkunen-Tiitto, R. Growth, structure, stomatal responses and secondary metabolites of birch seedlings (Betula pendula) under elevated UV-B radiation in the field. *Trees* **2001**, *15*, 483–491. [CrossRef]
41. Rzechonek, D.A.; Dobrowolski, A.; Rymowicz, W.; Mirończuk, A.M. Aseptic production of citric and isocitric acid from crude glycerol by genetically modified Yarrowia lipolytica. *Bioresour. Technol.* **2019**, *271*, 340–344. [CrossRef] [PubMed]

Journal of *Fungi*

MDPI

Article

Risk Assessment of the Wild Edible *Leccinum* Mushrooms Consumption According to the Total Mercury Content

Marek Šnirc [1], Ivona Jančo [2], Martin Hauptvogl [3], Silvia Jakabová [1], Lenka Demková [4] and Július Árvay [1,*]

1 Institute of Food Sciences, Faculty of Biotechnology and Food Sciences, Slovak University of Agriculture in Nitra, Tr. A. Hlinku 2, 949 76 Nitra, Slovakia
2 AgroBioTech Research Centre, Slovak University of Agriculture in Nitra, Tr. A. Hlinku 2, 949 76 Nitra, Slovakia
3 Institute of Environmental Management, Faculty of European Studies and Regional Development, Slovak University of Agriculture in Nitra, Tr. A. Hlinku 2, 949 76 Nitra, Slovakia
4 Department of Ecology, Faculty of Humanities and Natural Sciences, University of Prešov, 081 16 Prešov, Slovakia
* Correspondence: julius.arvay@uniag.sk; Tel.: +421-37-641-4497

Abstract: Wild-growing edible mushrooms contain many biologically valuable substances. However, they are considered a risk commodity due to their extremely high capacity for bioaccumulation of potential risk elements and pollutants from the environment. Four bolete mushrooms from the genus *Leccinum* were collected from 16 forested areas of Slovakia from June to October 2019. The total mercury content in soil and fruiting body parts was determined by an AMA-254 Advanced Mercury Analyzer. Soil pollution by total mercury was evaluated by contamination factor (C_f^i). Bioaccumulation factor (BCF), translocation factor (Qc/s), percentage of provisional tolerable weekly intake (%PTWI), and target hazard quotient (THQ) were used to describe and compare uptake and transition abilities of mushrooms, and the health risk arising from consumption of the mushrooms. Total mercury content varied between 0.05 to 0.61 mg kg^{-1} DW in the soil/substrate samples, and between 0.16 and 5.82 (caps), and 0.20 and 3.50 mg kg^{-1} DW (stems) in fruiting body samples. None of the analyzed locations represented a health risk based on %PTWI values, however, three locations may pose a significant health risk from the perspective of THQ values.

Keywords: mercury; risk assessment; contamination; bioaccumulation; *Leccinum*

Citation: Šnirc, M.; Jančo, I.; Hauptvogl, M.; Jakabová, S.; Demková, L.; Árvay, J. Risk Assessment of the Wild Edible *Leccinum* Mushrooms Consumption According to the Total Mercury Content. *J. Fungi* **2023**, *9*, 287. https://doi.org/10.3390/jof9030287

Academic Editors: Miha Humar and Ivan Širić

Received: 1 February 2023
Revised: 16 February 2023
Accepted: 20 February 2023
Published: 22 February 2023

1. Introduction

Mushrooms are an old group of heterotrophic organisms. Their body is composed of hyphae, and a mature hypha forms fructifications—fruiting bodies which are composed of two basic parts: the cap and the stem (both can have various shapes, sizes, and colors, depending on the species) [1]. In Europe, they are collected for consumption as a good source of digestible proteins, amino acids (e.g., leucine, lysine, methionine, tryptophan), carbohydrates, fibers (mainly in the form of chitin), vitamins (B1, B2, B12, C, D, niacin, folic acid), phenolics, organic acids, sterols, alkaloids, and terpenoids [2]. Among wild fungi collected worldwide, the most commonly consumed and traded species are brittlegills (*Russula* spp.), milk caps (*Lactarius* spp.), chanterelles (*Cantharellus* spp.), agarics (*Amanita* spp.), scaber stalks (*Leccinum* spp.), and boletes (*Boletus* spp.) [3]. One of the most economically and ecologically important mushroom species is *Leccinum* [4]. Fruiting bodies of the genus *Leccinum* are especially popular in Slovakia [5]. They are easily identified by their prominent squamulose stem ornamentation, whitish, brownish, or light-yellow pores, and white context that never changes color or that stains grey, blue, or reddish tints when damaged [4,6].

Risk elements contamination in the environment is a universal threat, which was exacerbated by intensive agriculture and rapid urbanization. Risk elements tend to persist in the environment for centuries [7]. Soil is the principal repository of these toxic elements. Although parent material is the major contributor of risk metals to the soil environment, these elements become bioavailable more slowly than those originating from anthropogenic activities [8]. The released risk elements accumulate in the environment and may alter the microbial processes which can cause an increase in their availability and their toxicity to higher plants, mushrooms, and other organisms [9–12], as well as physicochemical properties of soils leading to loss of fertility, disturbing plant metabolism, and reducing biomass production and crop yields [13]. Saprophytic mushrooms have high decomposition ability and increased activity of catalase that multiplies the concentration of these elements [14].

Mercury is a global pollutant that has raised great concerns worldwide [15]. Unlike other risk metals, mercury can remain in the atmosphere for a long period and migrate long distances. Eventually, about 93.7% of mercury enters the land and water ecosystem through dry and wet deposition [16]. Sources of mercury pollution in soil include atmospheric deposition, sewage irrigation, livestock manure, discarded mercury-containing appliances, etc., while the use of pesticides, lime fertilizer and seed coating with organic mercury also increases the risk of mercury pollution during crop plantation [17]. Mercury can also be organically or inorganically bound in coal, but it could also be present in its elementary form as Hg^0. The high combustion temperatures cause the primary occurrence of elementary mercury. During the cooling of the flue gas, mercury might react with other components [18]. Inorganic mercury can be converted to highly neurotoxic methyl mercury (MeHg), which is bioaccumulated and biomagnified in the food chain and thus endangers human health. Even at a low level, MeHg may pose chronic toxicity to a certain population due to long-term exposure. The presence and accumulation of MeHg in the environment are mainly a result of mercury methylation driven by anaerobic microorganisms [19]. The usual contents of mercury in wild-growing mushrooms range between <0.5 and 5.0 mg kg^{-1} DW and some species, e.g., *Macrolepiota procera* have been reported to be mercury bioaccumulators [20]. From an environmental point of view, mushrooms have a positive impact on increasing soil fertility through their ability to break down and dissolve complex compounds into simple ones, and their ability to reduce or eliminate environmental pollutants [21,22]. In this regard, mushrooms and some higher plants have an important role in the ecosystem due to their bioremediation ability [23]. Nevertheless, several studies from different parts of the world have proved that mushrooms can accumulate high amounts of potentially harmful elements, especially when collected from heavily contaminated regions (mining sites, industrial areas) or soils with high metal content [8,9,24]. In such conditions, the mushrooms are toxic and non-edible [25].

The aim of the study was to determine the level of mercury contamination of the genus *Leccinum* and soil/substrate samples. Ecological risks of mercury were evaluated by calculating the contamination and bioaccumulation factors. By using a provisional tolerable weekly intake and target hazard quotient, the health risk associated with the consumption of the investigated mushroom genus was evaluated.

2. Materials and Methods

2.1. Study Areas, Sampling and Sample Preparation

The samples (n = 249) of 4 mushroom species of the genus *Leccinum* (*Leccinum scabrum* (Bull.) Gray, *Leccinum pseudoscabrum* (Kallenb.) Šutara, *Leccinum albostipitatum* (den Bakker & Noordel) and *Leccinum piceinum* (Pilát & Dermek) Singer) were collected from 16 forested areas of Slovakia from June to October 2019 (Figure 1). The number of individual species collected in the individual study areas is shown in Table 1. Directly after the sampling, all mushroom samples were cleaned up from any organic and inorganic debris, and the bottom part of the stem was cut off. After that, they were divided into two parts: cap and stem. The individual cap and stem samples were sliced into pieces using a ceramic knife and dried to a constant weight at 40 °C in a laboratory dry heat oven with forced air circulation (Memmert GmbH & Co. KG, Schwabach, Germany) for 22 h. The dried samples were pulverized in the rotary homogenizer (IKA Mills A 10 basic—Werke GmbH & Co.

KG, Staufen, Germany) and stored in polyethene bags until further analysis. Underlying soil/substrate samples (n = 249) were collected together with the mushroom samples at the same location from a depth of approximately 0.10 m. Under laboratory conditions, the samples were air-dried at room temperature for 3 weeks. Afterwards, they were sieved through a mesh sieve (2 mm) and stored in paper bags until the analysis.

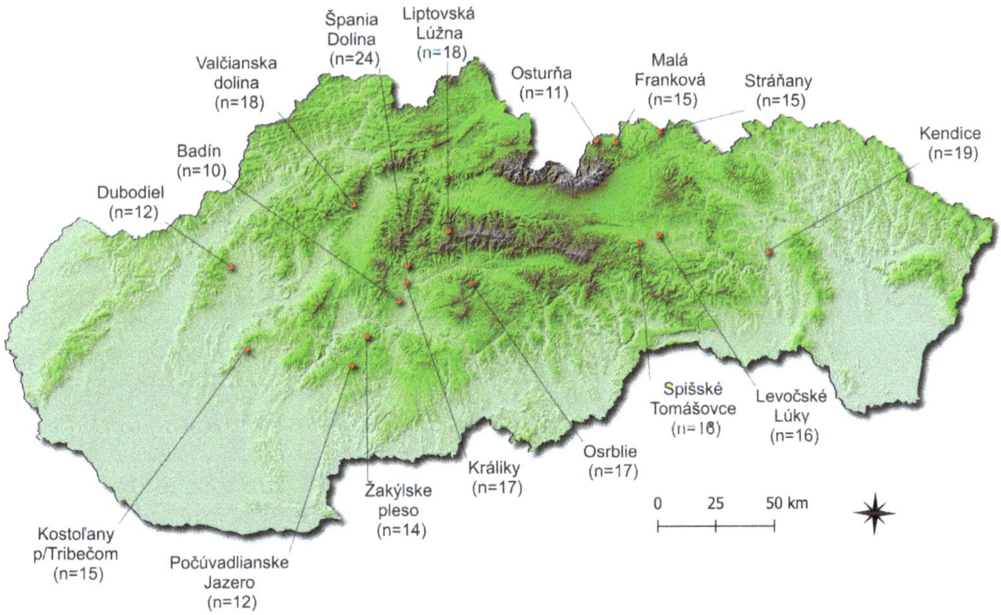

Figure 1. Mushroom and its underlying substrate sampling areas.

Table 1. The number of collected samples of individual mushroom species in the sampling localities.

Locality	*L. albostipitatum*	*L. piceinum*	*L. pseudoscabrum*	*L. scabrum*
Badín			10	
Dubodiel				12
Kendice			19	
Kostoľany p/Tribečom			15	
Králiky			17	
Levočské Lúky			6	10
Malá Franková	2		1	12
Liptovská Lúžna	18			
Osrblie	17			
Osturňa		9		2
Počúvadlianske Jazero			12	
Spišské Tomášovce	10			6
Stráňany	7			8
Špania Dolina	17	7		
Valčianska dolina				18
Žakýlske pleso			14	

2.2. Sample Analysis

Total mercury content was determined by cold-vapor AAS analyzer AMA 254 (Al-tec, Prague, Czech Republic), in all types of dried and homogenized samples. The limit of detection for Hg was set at 1.5×10^{-6} mg kg^{-1} DW and the limit of quantification at 4.45×10^{-6} mg kg^{-1} DW. Two Certified Reference Materials (CRM) from the Institute for Reference Materials and Measurements were used to check the accuracy and precision of the analytical method. The recovery value varied between 0.06 and 0.07 mg kg^{-1} DW for the loam soil (ERM-CC141) and between 0.05 and 0.06 mg kg^{-1} DW for the Mussel tissue (ERM-CE278k).

2.3. *Risk Assessment*

2.3.1. Contamination Factor (C_f^i)

To assess the level of ecological load of the monitored localities, the Hg content in the soil/substrate was evaluated.

The contamination factor described by Hakanson [26] was used to express the level of soil/substrate pollution by mercury. It is calculated as follows:

$$C_f^i = \frac{C_{0-1}^i}{C_n^i}$$

where: C_{0-1}^i is the total Hg content in soil and C_n^i is the background Hg level, which is (0.06 mg kg^{-1}) [21]. The contamination factor values were divided into four categories: low contamination factor ($C_f^i < 1$); moderate contamination factor ($1 \leq C_f^i < 3$); considerable contamination factor ($3 \leq C_f^i < 6$); very high contamination factor ($C_f^i \geq 6$).

2.3.2. Bioaccumulation Factor (BCF) and Translocation Quotient (Qc/s)

The bioaccumulation factor (BCF) was calculated to assess the level of transition and accumulation of Hg from soil/substrate to the above-ground parts (fruiting body). It was calculated as follows:

$$BCF = \frac{C_{Hg}}{C_s}$$

where: C_{Hg} is the measured mercury content in mushroom samples and C_s is the measured mercury content in soil/substrate. BCF < 1 indicates excluders, BCF > 1 indicates accumulators [27].

The translocation quotient (Qc/s) was evaluated to compare the level of Hg translocation within the fruiting body.

$$Q_{c/s} = \frac{Hg_{cap}}{Hg_{stem}}$$

where, Hg_{cap} is the concentration of mercury in mushroom caps, and Hg_{stem} is the mercury concentration in mushroom stems.

2.3.3. Provisional Tolerable Weekly Intake (PTWI)

The percentage of the provisional tolerable weekly intake (%PTWI) was used to consider the potential risk arising from the long-term consumption of the studied mushrooms. The tolerable weekly intake per adult person weighing 70 kg for Hg was established at 0.28 mg per person per week [28]. Taking into account the average consumption of "Other vegetables including mushrooms" which was 0.18 kg FW per person per week in Slovakia in 2020 [29] the %PTWI was calculated as follows:

$$\%PTWI = \frac{BS_{Hg} \times 0.18}{0.28} \times 100$$

where: BS_{Hg} is the measured content of Hg in the biological sample (mg kg^{-1} of fresh weight (FW) in mushrooms). The fresh weight of the mushrooms was calculated providing that the dry matter represented 10% of the mushroom fruiting body [20,30]. If the detected value was greater than 100%, the consumption of mushroom samples from the area would be potentially hazardous.

2.3.4. Target Hazard Quotient (THQ)

With the purpose of a comprehensive assessment of the dangers arising from the long-term consumption of mushrooms, the target hazard quotient (THQ) was used. THQ considered numerous parameters, which can influence the health of consumers. THQ can be expressed as the ratio of toxic element exposure and the highest reference dose at which no adverse effects on human health are expected [30]. THQ was calculated as follows:

$$THQ = \frac{Efr \times ED \times ADC \times C_E}{RfDo \times BW \times ATn} \times 10^{-3}$$

where: Efr is the frequency of exposure (365 days), ED is the exposure duration (70 years), ADC is the average daily consumption of mushrooms, which was according to the Statistical Office of the Slovak Republic [29] estimated to be 25.7 g/day, C_E is the average Hg concentration in mushroom samples (mg kg^{-1} FW), and $RfDo$ is the oral reference dose for mercury (0.0003 mg kg^{-1} day^{-1}) [20]. BW is the average body weight (70 kg) and ATn is the average exposure time (365 days × 70 years = 25,550 days), 10^{-3} is a factor considering the unit's conversion. If the THQ is lower than 1, carcinogenic health effects are not expected; if the THQ is bigger than 1, there is a serious possibility that adverse health effects can be experienced.

2.4. Statistical Analysis

Descriptive statistical analysis, normality tests and Spearman correlation test were performed using Jamovi software version 2.3.9. The distribution of the analyzed quantitative variables (based on the Kolmogorov–Smirnov test and the Shapiro–Wilk test) was non-normal, therefore, the non-parametric ANOVA test (Kruskal–Wallis) and Wilcoxon test were used for the comparison of mercury content among the tested localities. For a better understanding and interpretation of the results, each locality was compared with the median value (horizontal line) using the Wilcoxon test. Spearman correlation was used to determine the relationships between soil/substrate and fruiting body parts of the tested mushrooms. Non-parametric ANOVA was performed using the RStudio software, version 2022.07.2 [31–37].

3. Results and Discussion

3.1. Soil/Substrate Samples Analysis

The average Hg content is 0.06 mg kg^{-1} in the soil in Slovakia [38]. The total Hg content in the soil/substrate in the study areas varied from 0.05 to 0.61 mg kg^{-1} (Table 2). The highest average Hg content was detected in Levočské Lúky (0.61 ± 0.12 mg kg^{-1}) and the lowest in Osrblie (0.05 ± 0.02 mg kg^{-1}) and Králiky (0.05 ± 0.06 mg kg^{-1}). The limit value of Hg for soils in Slovakia, which is set to 0.50 mg kg^{-1} DW [39], was exceeded at 2 sampling sites (Levočské Lúky; Spišské Tomášovce). The concentrations of Hg that exceeded the permissible limit values were detected in forest soils in Slovakia, and also at localities that were not influenced by anthropogenic pollution sources. The level of ecological load of the monitored localities varied from 0.77 (Osrblie) to 10.13 (Levočské Lúky). According to Hakanson [26], Osrblie (C^i_f = 0.77), Liptovská Lúžna (C^i_f = 0.8) and Králiky (C^i_f = 0.85) belong to the low contamination degree (C^i_f < 1). A total of 9 localities (Kostoľany pod Tríbečom (C^i_f 1.11), Valčianska dolina (C^i_f = 1.18), Žakýlske pleso (C^i_f = 1.22), Kendice (C^i_f = 1.34), Počúvadlianske Jazero (C^i_f = 1.58), Badín (C^i_f = 2.1), Osrurňa (C^i_f = 2.1), Stráňany (C^i_f = 2.43) and Malá Franková (C^i_f = 2.63) belong to the moderate contamination degree (moderate contamination factor 1 ≤ C^i_f < 3). Dubodiel (C^i_f = 3.49) and Špania Dolina (C^i_f = 5.8) belong to the considerable contamination degree (considerable contamination factor 3 ≤ C^i_f < 6). A very high contamination factor (C^i_f ≥ 6) was observed in the Spišské Tomášovce (C^i_f = 8.76) and Levočké lúky (C^i_f = 10.13). Mercury soil contamination in Slovakia has received attention in earlier studies. Serious Hg pollution has been confirmed around the former mine sites [30]. Additionally, some studies have confirmed long-distance air pollution and its serious consequences for environmental quality, even in localities where there is no direct source of pollution [30,40]. High values of Hg content in mushroom samples from the east of Slovakia in soil/substrate were also confirmed by Jančo et. al. [41], specifically in the locality Snina (0.68 mg kg^{-1}). Lower values of Hg content in soil/substrate from Central Slovakia were recorded by Árvay et al. [42] where the content of total Hg in the soil/substrate varied between 0.05 and 0.27 mg kg^{-1} DW. Another study by Árvay et al. [43] investigated soil/substrate of a historical mining area near Banská Bystrica (Slovakia). The content of total Hg in the underlying substrate ranged between 0.05 and 0.27 mg kg^{-1} (n = 33). Falandysz et al. [44] determined the contents of Hg in the fruiting bodies of 15 higher mushroom species and soil/substrate collected from Wieluńska Upland in the northern part of the Sandomierska Valley in south-central Poland. A total of 227 soil samples were analyzed. Mean mercury contents in the underlying soil/substrate of the 15 mushroom species (17 samples of *M. procera*) were between 0.03 and 0.09 mg kg^{-1}. Another study by Mleczek et al. [45] showed the mean Hg content of 0.06 mg kg^{-1} DW in the soil/substrate from Polish forests. According to Falandysz and Gucia [46], the range of Hg content was between 0.01 and 0.54 mg kg^{-1} DW in topsoil samples (Poland).

Table 2. Mercury content in soil/substrate (mg kg^{-1} DW) and contamination factor of soil in the studied localities.

Locality	AVG ± SD	Min–Max	C_f^i
Badín	0.13 ± 0.02	0.09–0.15	2.10
Dubodiel	0.21 ± 0.05	0.18–0.33	3.49
Kendice	0.08 ± 0.04	0.04–0.18	1.34
Kostol'any p/Tribečom	0.07 ± 0.04	0.05–0.18	1.11
Králiky	0.05 ± 0.06	0.02–0.25	0.85
Levočské Lúky	0.61 ± 0.12	0.38–0.81	10.1
Malá Franková	0.16 ± 0.05	0.09–0.25	2.63
Liptovská Lúžna	0.18 ± 0.02	0.15–0.23	3.80
Osrblie	0.05 ± 0.02	0.02–0.09	0.77
Osturňa	0.13 ± 0.21	0.06–0.81	2.10
Počúvadlianske Jazero	0.10 ± 0.01	0.07–0.11	1.58
Spišské Tomášovce	0.53 ± 0.27	0.02–0.85	8.76
Stráňany	0.15 ± 0.02	0.09–0.16	2.43
Valčianska dolina	0.07 ± 0.01	0.07–0.08	1.18
Špania Dolina	0.30 ± 0.14	0.01–0.55	5.80
Žakýlske pleso	0.07 ± 0.01	0.06–0.10	1.22

The Kruskal–Wallis test confirmed significant differences among individual locations (p < 2.2e-16). Subsequently, the individual locations were compared with the median value using the Wilcoxon test (Figure 2). Significantly higher Hg contents were detected in the locations Dubodiel (p < 0.01), Levočské Lúky (p < 0.0001), Liptovská Lúžna (p < 0.01), Spišské Tomášovce (p < 0.0001) and Špania Dolina (p < 0.001). On the other hand, significantly lower mercury content in the soil was recorded in Kendice (p < 0.05), Kostol'any pod Tribečom (p < 0.01), Králiky (p < 0.001), Osrblie (p < 0.0001), Valčianska dolina (p < 0.001) and Žakýlske pleso (p < 0.01). In the case of Badín, Malá Franková, Osturňa, Počúvadlianske Jazero and Stráňany, there were no significant differences compared to the median value.

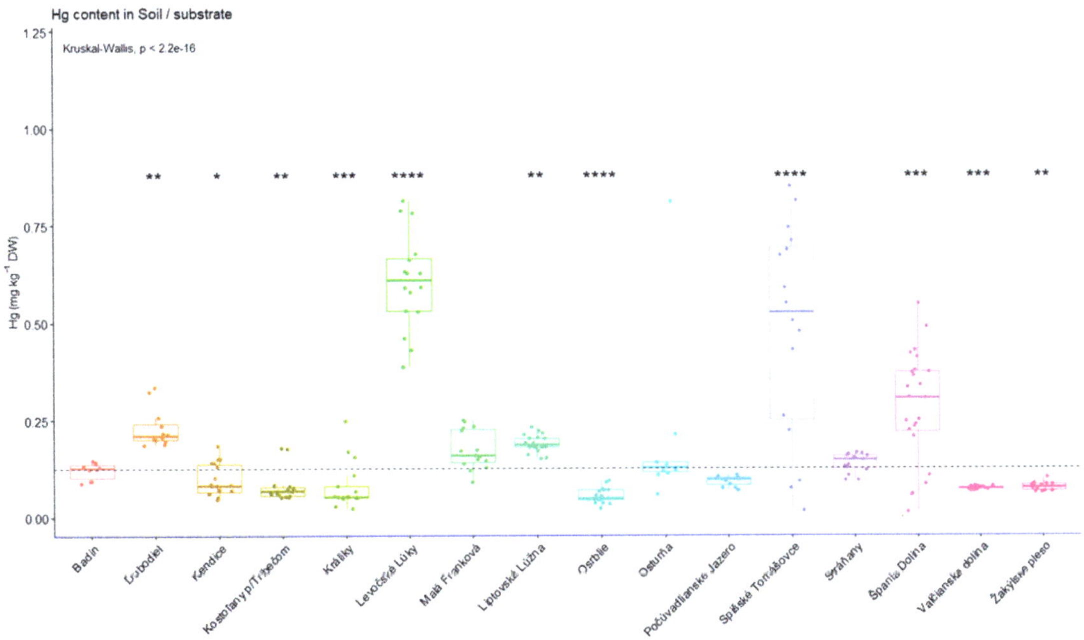

Figure 2. Significant differences in soils/substrate mercury concentrations in mg kg^{-1}, concerning localities (* $p < 0.05$; ** $p < 0.01$; *** $p < 0.001$; **** $p < 0.0001$).

3.2. Mercury Concentration in Fruiting Bodies

The element accumulation by wild-growing mushrooms has been the subject of numerous scientific papers around the world. Depending on the collection site, a higher, lower, or significantly differentiated abilities of mushrooms to accumulate some toxic elements were reported. However, the efficiency of the element accumulation does not always depend on their content in soil/substrate, but the element contents in such cases depend on mushroom species, genus, or the families to which they belong. Increasing age of mycelium, up to decades in wild-growing species, and a protracted interval between fructifications significantly elevate the contents of many elements in fruiting bodies, and usually higher levels occur in caps than in stems [20,41,45].

In the present study, the fruiting bodies were divided into two edible mushroom parts: caps and stems. Average Hg contents are shown in Table 3. There were no significant differences in the Hg content among the tested *Leccinum* species based on the Kruskal–Wallis test ($p = 0.304$ for caps; $p = 0.057$ for stems). Therefore, the mushroom samples were examined with the exception of the *Leccinum* genus. In the case of caps, the Hg concentration varied between 0.16 ± 0.05 mg kg^{-1} DW (Králiky) and 5.82 ± 2.28 mg kg^{-1} DW (Spišské Tomášovce). The EU limit value in the fruiting bodies of edible mushrooms for Hg is 0.75 mg kg^{-1} FW [41]. The dry matter content in all samples was approximately 10%. The results showed that none of the tested samples exceeded the maximum allowed limit. In general, there was a strong significant difference among the mercury concentrations in caps (p – 2.2e-16) (Figure 3). Significantly higher mercury concentrations were in Levočské Lúky ($p < 0.0001$), Liptovská Lúžna ($p < 0.0001$) and Spišské Tomášovce ($p < 0.0001$). On the other hand, there were also significantly lower Hg concentrations in caps in Králiky ($p < 0.0001$), Malá Franková ($p < 0.05$), Počúvadlianske Jazero ($p < 0.001$) and Žakýlske pleso ($p < 0.05$).

Table 3. Mercury content in fruiting bodies (mg kg^{-1} DW) in the studied localities.

Locality	Cap		Stem	
	AVG ± SD	Min–Max	AVG ± SD	Min–Max
Badín	0.97 ± 0.25	0.39–1.90	0.48 ± 0.18	0.13–0.62
Dubodiel	0.55 ± 0.06	0.48–0.69	0.43 ± 0.08	0.26–0.56
Kendice	0.62 ± 0.25	0.09–0.96	0.70 ± 0.37	0.16–1.55
Kostoľany p/Tribečom	0.78 ± 0.43	0.27–1.66	0.60 ± 0.23	0.21–0.95
Králiky	0.16 ± 0.05	0.08–0.27	0.20 ± 0.82	0.06–2.86
Levočské Lúky	1.89 ± 2.60	0.89–8.99	1.18 ± 1.35	0.31–5.94
Malá Franková	0.34 ± 0.51	0.1–1.97	0.22 ± 0.18	0.06–0.68
Liptovská Lúžna	1.27 ± 0.41	0.78–2.13	0.97 ± 0.21	0.50–1.18
Osrblie	0.71 ± 0.37	0.07–1.10	0.53 ± 0.25	0.05–0.87
Osturňa	0.98 ± 0.95	0.02–3.15	0.53 ± 0.35	0.22–1.26
Počúvadlianske Jazero	0.36 ± 0.04	0.27–0.38	0.32 ± 0.03	0.28–0.36
Spišské Tomášovce	5.82 ± 2.28	2.50–9.61	3.50 ± 1.23	0.48–4.43
Stráňany	0.64 ± 0.90	0.32–3.26	0.44 ± 0.45	0.19–1.53
Valčianska dolina	0.52 ± 0.03	0.47–0.56	0.40 ± 0.08	0.29–0.58
Špania Dolina	0.58 ± 0.23	0.28–1.80	0.34 ± 0.09	0.20–0.50
Žakýlske pleso	0.41 ± 0.17	0.18–0.71	0.35 ± 0.14	0.12–0.64

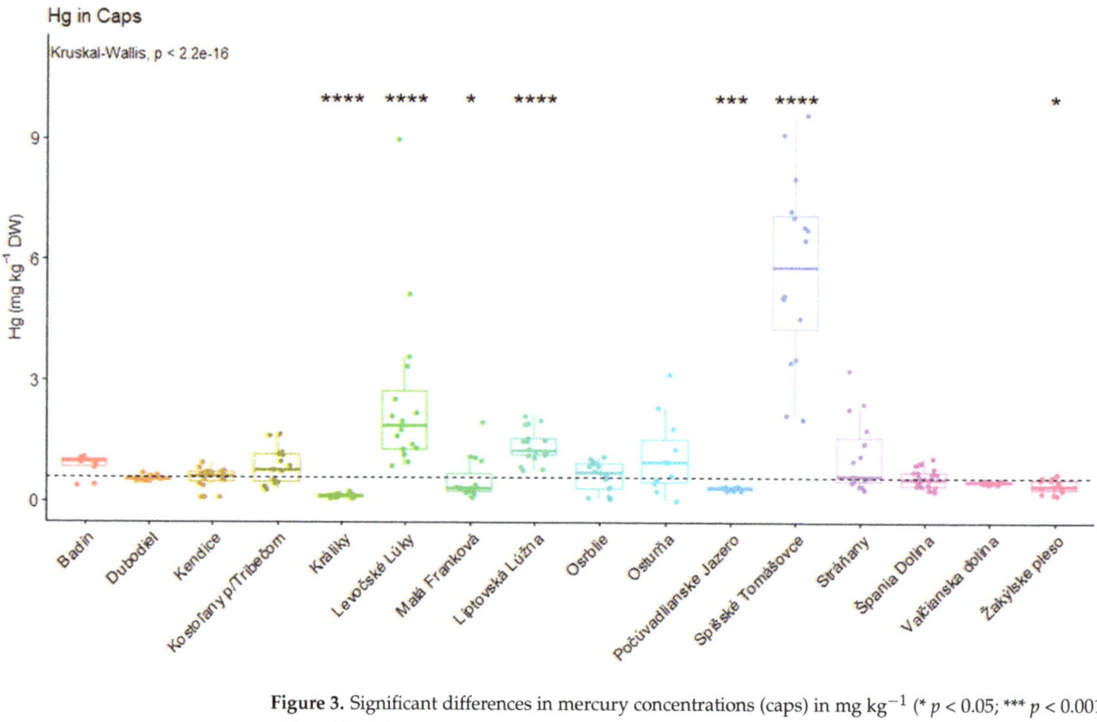

Figure 3. Significant differences in mercury concentrations (caps) in mg kg^{-1} (* $p < 0.05$; *** $p < 0.001$; **** $p < 0.0001$).

In the case of stems, high total Hg concentrations were observed in Spišské Tomášovce (3.50 ± 1.23 mg kg^{-1} DW). On the other hand, the lowest total mercury concentrations were observed in Králiky (0.20 ± 0.82 mg kg^{-1} DW). Figure 4 displays significant differences in Hg concentrations in stems among the localities. There were significantly higher concentrations in Kendice ($p < 0.05$), Levočské lúky ($p < 0.0001$), Liptovská Lúžna ($p < 0.001$) and Spišské Tomášovce ($p < 0.0001$). Significantly lower Hg concentrations in stems were observed in Králiky, Levočské lúky ($p < 0.05$), Malá Franková ($p < 0.001$), Počúvadlianske

Jazero ($p < 0.01$), Špania Dolina ($p < 0.01$) and Žakýlske pleso ($p < 0.05$). Árvay et al. [43] recorded comparable results when analyzing total Hg content in *M. procera* from a historical mining area of Banská Bystrica. The detected average content of total Hg in stems was 1.40 (0.12–1.75) mg kg^{-1} DW. The highest Hg content was measured in *M. procera* cap, and the average value was 1.98 (between 0.41 and 3.20 mg kg^{-1} DW). Parasol Mushroom contained the greatest (compared to other species) mean Hg contents in the study of Falandysz et al. [47] both in caps (between 4.50 ± 1.70 and 4.40 ± 2.40 mg kg^{-1} DW) and stems (between 2.80 ± 1.30 and 3.00 ± 2.00 mg kg^{-1} DW). The Parasol Mushroom also showed a great potential to bioaccumulate Hg from the soil. Mleczek et al. [48] studied 34 elements in four edible mushroom species: *Boletus edulis, Imleria badia, Leccinum scabrum,* and *Macrolepiota procera*, and underlying soil substrate collected from Polish forests between 1974 and 2019. The average Hg content detected in *Leccinum scabrum* was 1.41 (0.674–2.94) mg kg^{-1} DW. Jančo et al. [41,49] also confirmed the significant effect of the locality on the total Hg content in edible parts of mushrooms. Some previous studies showed that *Macrolepiota procera* collected from Central Slovakia (Banská Bystrica) had an average Hg concentration of 1.98 mg kg^{-1} (from 0.41 to 3.20 mg kg^{-1}) in the caps and 1.40 mg kg^{-1} (from 0.12 to 1.75 mg kg^{-1}) in stems. Falandysz et al. [50,51] determined Hg content in *Leccinum* caps and stems collected from China ranging from 0.54 to 4.80 g kg^{-1} DW and 0.32 to 2.80 mg kg^{-1} DW, respectively, and in *Leccinum* sp. (caps and stems) collected from Poland ranging from 0.180 to 1.50 mg kg^{-1} DW and from 0.04 to 0.65 mg kg^{-1} DW, respectively. The concentration of Hg in edible mushrooms was lower than the legislation limits in all samples. The observed Hg concentrations in *Leccinum* sp. were comparable with other reports from Europe and lower than those from China [52–54].

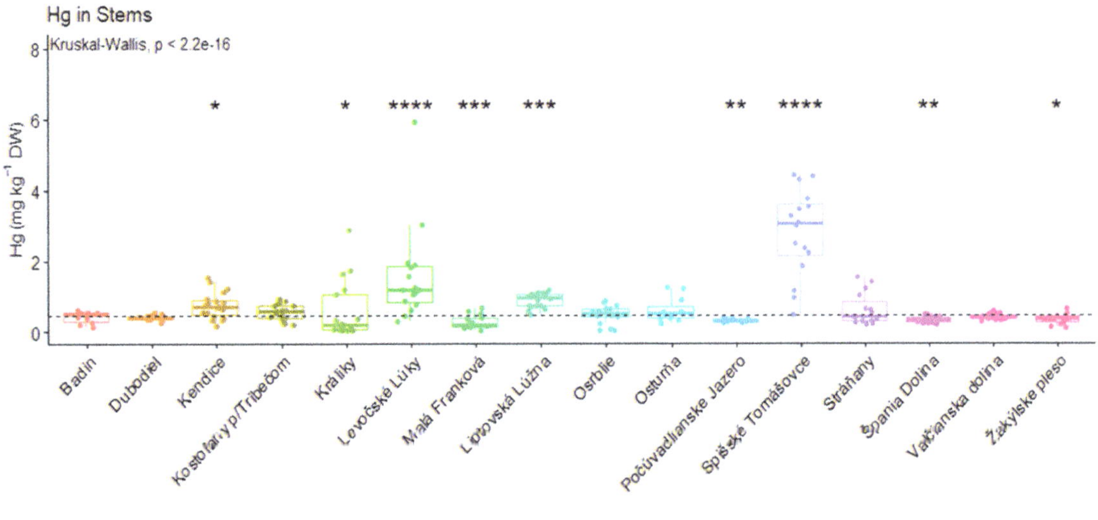

Figure 4. Significant differences in mercury concentrations (stems) in mg kg^{-1}, (* $p < 0.05$; ** $p < 0.01$; *** $p < 0.001$; **** $p < 0.0001$).

The results of Spearman's correlation coefficient are displayed in Figure 5. Correlation analysis was made for each mushroom fruiting body parts and soil/substrate (with the exception of the genus *Leccinum*). The strongest relationship was observed between fruiting body parts (cap-stem). This finding suggests that if the mushroom absorbs risk elements from its environment, all the fruiting body parts are influenced. A significant relationship between soil/substrate pollution and Hg content in the fruiting body parts was also observed. This was confirmed by other authors [5,30,53], however, the ability of different mushroom species to accumulate elements from the soil substrate differ significantly [24].

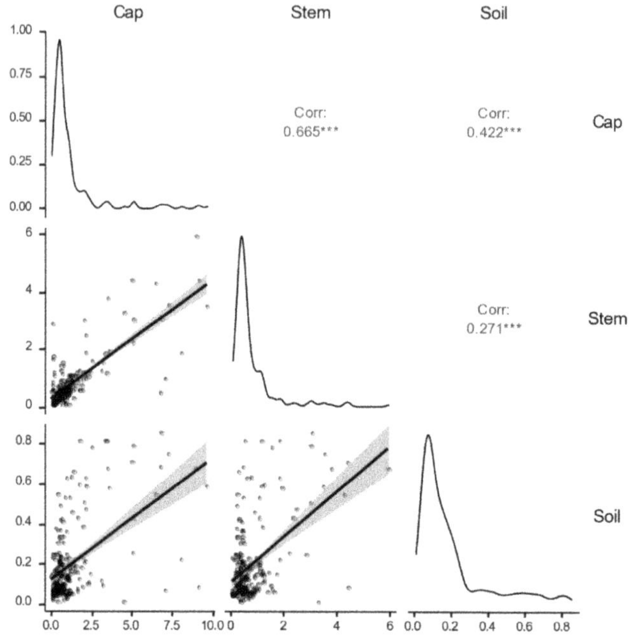

Figure 5. Spearman's correlation matrix between the soil/substrate and the fruiting body parts (cap, stem). (*** $p < 0.001$).

3.3. Bioaccumulation Factor (BCF) and the Translocation Quotient (Qc/s)

The content of Hg in mushroom species regarding the pollution of the environment (soil/substrate) was expressed using the bioaccumulation factor (Table 4). The bioaccumulation factor ranged from 4.27 (*L. piceinum*) to 7.15 (*L. albostipitatum*) in the case of caps, and from 2.61 (*L. scabrum*) to 4.78 (*L. albostipitatum*) in the case of stems. Fruiting body parts of mushrooms are significantly affected by soil quality. A mushroom (and/or plant) is considered an accumulator or hyperaccumulator if BCF > 1 [55]. All studied mushroom species could be considered accumulators. These results are closely correlated with the findings of other authors, who found high BCF values in caps and stems [30,47,56]. The results of the Wilcoxon signed-rank test (Table 4) showed significantly higher BCF values in caps compared to stems.

Table 4. Bioaccumulation factor (BCF) values in caps and stems, p-values of the Wilcoxon test between caps and stems within the species and translocation quotient (Qc/s) values determined for four *Leccinum* species.

Mushroom Species	BCF (Cap)	BCF (Stem)	*p*-Value	Qc/s
L. albostipitatum	7.15	4.78	<0.0001	1.59
L. piceinum	4.27	3.35	0.0186	1.66
L. pseudoscabrum	4.55	4.48	0.0230	1.18
L. scabrum	3.79	2.61	<0.0001	1.50

The translocation quotient (*Qc/s*) in the mushroom fruiting body is expressed by the cap/stem ratio, and it represents the mobility of metals including mercury in the mushroom fruiting body [30,57]. *Qc/s* values ranged from 1.18 (*L. pseudoscabrum*) to 1.66 (*L. piceinum*). Comparable Qc/s values in the genus *Leccinum* were also published in other studies [58–61]. Translocation quotient values greater than 1 indicate that the concentration of Hg in the caps of analyzed mushrooms is greater than the concentration of Hg in the stems. Due to the high mobility of heavy metals in the fruiting body, there is a risk of bioaccumulation of large amounts of these elements in the superior part of mushrooms, which is more commonly consumed by humans [62]. In the present study, all the analyzed samples had Qc/s values higher than 1.

3.4. Health Risk Assessment

Percentage of the Provisional Tolerable Weekly Intake (%PTWI) and Target Hazard Quotient (THQ)

Toxic metals, including Hg, present in food, directly affect the health of consumers. The Food and Agriculture Organization and World Health Organization establish safe levels of metal intake in terms of provisional tolerable daily intake (PTWI). The comparison of %PTWI values between individual *Leccinum* species in caps showed significant differences between *L. albostipitatum* and *L. pseudoscabrum* ($p < 0.001$), *L. albostipitatum* and *L. scabrum* ($p = 0.003$). In the case of stems, there were also significant differences between *L. albostipitatum* and *L. pseudoscabrum* ($p = 0.038$), *L. albostipitatum* and *L. scabrum* ($p = 0.025$). According to the geographical origin, none of the analyzed locations represented a health risk from the perspective of PTWI values. The %PTWI ranged from 1.04% (Králiky) to 37.4% (Spišské Tomášovce) in caps and from 1.27% (Králiky) to 19.6% (Spišské Tomášovce) in stems. Krasińska and Falandysz [63] also did not record increased values of %PTWI (20–60% of PTWI) in *Leccinum* species collected in Poland. In a previous study [5], %PTWI values for caps ranged from 2.64% to 48.3%, and for stems from 2.28% to 18.7 % in mushrooms of the genus *Leccinum* collected from Slovakia.

Another frequently used indicator of human health risk associated with food consumption is the target hazard quotient (THQ). It connects the risk element concentrations in food with their toxicity, quantity and quality of food consumption, and consumers' body mass. Oral exposure to mushroom samples with THQ values less than 1 poses no significant health risk. THQ values higher than 1 were observed in Levočské Lúky and Spišské Tomášovce (for both caps and stems), as well as Badín and Osturňa (only for caps). Čéryová et. al. [5] reported THQ values greater than 1 (genus *Leccinum*) in Mníšek nad Popradom (Eastern Slovakia). PTWI and THQ values are presented in Table 5. Mushroom consumption is considered safe concerning health risks if consumers avoid a long-term consumption of mushrooms from explicitly polluted areas.

Table 5. THQ values and %PTWI of the studied mushrooms based on Hg content according to the geographical origin.

Locality	%PTWI		THQ	
	Cap	Stem	Cap	Stem
Badín	6.27	3.90	1.19	0.59
Dubodiel	3.56	2.75	0.68	0.52
Kendice	3.99	4.50	0.76	0.86
Kostoľany p/Tribečom	5.40	3.87	0.96	0.74
Králiky	1.40	1.27	0.20	0.24
Levočské Lúky	12.1	7.62	2.31	1.45
Malá Franková	2.16	1.41	0.41	0.27
Liptovská Lúžna	8.20	6.21	1.56	1.18
Osrblie	4.59	3.37	0.87	0.64
Osturňa	6.33	3.38	1.20	0.64
Počúvadlianske Jazero	2.31	2.70	0.44	0.39
Spišské Tomášovce	37.4	19.6	7.13	3.74
Stráňany	4.14	2.83	0.79	0.54
Valčianska dolina	3.32	2.57	0.63	0.49
Špania Dolina	3.72	2.18	0.71	0.41
Žakýlske pleso	2.61	2.26	0.50	0.43

4. Conclusions

The present study was carried out to investigate the accumulation of total mercury in the edible wild-growing mushrooms of the genus *Leccinum* and underlying substrate from 16 forested locations in Slovakia. The limit value of Hg for soils in Slovakia was exceeded at two locations. Three studied locations had a low contamination factor, nine locations had a moderate contamination factor, two locations had a considerable contamination factor and there were two localities with a very high contamination factor. There were statistically significant differences among the investigated locations in the total Hg content in the substrate. The significant relationship of the location to the content of total Hg in the substrate and mushroom fruiting bodies was confirmed. The significant positive correlation was found among total Hg content in the substrate and mushroom fruiting body parts. Significantly higher values of total Hg were found in caps compared to stems. There were no significant differences in the total Hg content among the tested *Leccinum* species. Fruiting body parts of mushrooms are significantly affected by the soil quality and the high mercury bioaccumulative potential of the studied *Leccinum* species. According to the geographical origin, none of the analyzed locations poses a health risk from the perspective of %PTWI. However, some locations with a THQ higher than 1 were observed, and long-term consumption of mushrooms from these areas can pose a health risk. Consumption of mushrooms in Slovakia can be safe in terms of health risks if consumers avoid long-term consumption of mushrooms from polluted areas.

Author Contributions: Conceptualization, J.Á.; Data curation, M.Š.; Formal analysis, I.J. and J.Á.; Funding acquisition, M.Š.; Investigation, M.Š., I.J., M.H. and J.Á.; Methodology, I.J. and M.H.; Resources, L.D.; Visualization, M.H.; Writing—original draft, M.Š. and J.Á.; Writing—review & editing, S.J. and L.D. All authors have read and agreed to the published version of the manuscript.

Funding: This work was supported by the Grant Agency of the Slovak University of Agriculture in Nitra, the project number 17-GASPU-2021 and project of Ministry of Education, Science, Research and Sport of the Slovak Republic VEGA No. 1/0602/22.

Institutional Review Board Statement: Not applicable.

Informed Consent Statement: Not applicable.

Data Availability Statement: The datasets analyzed during the current study are available from the corresponding author on reasonable request.

Acknowledgments: This project was supported by the Grant Agency of the Slovak University of Agriculture in Nitra, the project number 17-GASPU-2021: Possibilities of reducing the level of contamination of edible wild growing mushrooms from environmentally polluted areas of Slovakia while preserving their valuable biologically active substances and by project of Ministry of Education, Science, Research and Sport of the Slovak Republic project VEGA No. 1/0602/22. Special thanks go to Vladimír Kunca for his help in identifying of mushroom samples and Rastislav Sabo for his assistance in the mushroom sampling.

Conflicts of Interest: The authors declare no conflict of interest.

References

1. Bernaś, E.; Jaworska, G. Vitamins profile as an indicator of the quality of frozen Agaricus bisporus mushrooms. *J. Food Compost. Anal.* **2016**, *49*, 1–8. [CrossRef]
2. Kumar, P.; Kumar, V.; Eid, E.M.; Al-Huqail, A.A.; Adelodun, B.; Fayssal, S.A.; Goala, M.; Kumar-Arya, A.; Bachheti, A.; Andabaka, Ž.; et al. Spatial Assessment of Potentially Toxic Elements (PTE) Concentration in Agaricus bisporus Mushroom Collected from Local Vegetable Markets of Uttarakhand State, India. *J. Fungi* **2022**, *8*, 452. [CrossRef]
3. Wang, X.-M.; Zhang, J.; Wu, L.-H.; Zhao, Y.-L.; Li, T.; Li, J.-Q.; Wang, Y.-Z.; Liu, H.-G. A mini-review of chemical composition and nutritional value of edible wild-grown mushroom from China. *Food Chem.* **2014**, *151*, 279–285. [CrossRef]
4. Meng, X.; Wang, G.-S.; Wu, G.; Wang, P.-M.; Yang, Z.L.; Li, Y.-C. The genus Leccinum (Boletaceae, Boletales) from China based on morphological and Molecular Data. *J. Fungi* **2021**, *7*, 732. [CrossRef]
5. Čeryová, N.; Jančo, I.; Šnirc, M.; Lidiková, J.; Sabo, R.; Wassenaar, T.; Meroro, A.; Amwele, H.; Árvay, J. Mercury content in the wild edible Leccinum mushrooms growing in Slovakia: Environmental and Health Risk Assessment. *J. Microbiol. Biotechnol. Food Sci.* **2022**, *12*, e9455. [CrossRef]

6. Den Bakker, H.C.; Zuccarello, G.C.; Kuyper, T.H.W.; Noordeloos, M.E. Evolution and host specificity in the ectomycorrhizal genus Leccinum. *New Phytol.* **2004**, *163*, 201–215.
7. Demková, L.; Árvay, J.; Bobuľská, L.; Hauptvogl, M.; Michalko, M. Activity of the soil enzymes and moss and lichen biomonitoring method used for the evaluation of soil and air pollution from tailing pond in Nižná Slaná (Slovakia). *J. Environ. Sci. Health A* **2019**, *54*, 495–507. [CrossRef]
8. Širić, I.; Kumar, P.; Adelodun, B.; Fayssal, S.A.; Bachheti, K.R.; Bachheti, A.; Ajibade, O.F.; Kumar, V.; Taher, M.A.; Eid, E.M. Risk Assessment of Heavy Metals Occurrence in Two Wild Edible Oyster Mushrooms (*Pleurotus* spp.) Collected from Rajaji National Park. *J. Fungi* **2022**, *8*, 1007. [CrossRef]
9. Árvay, J.; Demková, L.; Hauptvogl, M.; Michalko, M.; Bajčan, D.; Stanovič, R.; Tomáš, J.; Hrstková, M.; Trebichalský, P. Assessment of environmental and health risks in former polymetallic ore mining and smelting area, Slovakia: Spatial Distribution and accumulation of mercury in four different ecosystems. *Ecotoxicol. Environ. Saf.* **2017**, *144*, 236–244. [CrossRef]
10. Kovacik, A.; Arvay, J.; Tusimova, E.; Harangozo, L.; Tvrda, E.; Zbynovska, K.; Cupka, P.; Andrascikova, S.; Tomas, J.; Massanyi, P. Seasonal variations in the blood concentration of selected heavy metals in sheep and their effects on the biochemical and hematological parameters. *Chemosphere* **2017**, *168*, 365–371.
11. Ruan, X.; Ge, S.; Jiao, Z.; Zhan, W.; Wang, Y. Bioaccumulation and risk assessment of potential toxic elements in the soil-vegetable system as influenced by historical wastewater irrigation. *Agric. Water Manag.* **2023**, *279*, 108197. [CrossRef]
12. Yang, L.; Ren, Q.; Zheng, K.; Jiao, Z.; Ruan, X.; Wang, Y. Migration of heavy metals in the soil-grape system and potential health risk assessment. *Sci. Tot. Environ.* **2022**, *806*, 150646.
13. Asad, S.A.; Farooq, M.; Afzal, A.; West, H. Integrated Phytobial Heavy Metal Remediation Strategies for a sustainable clean environment—A Review. *Chemosphere* **2019**, *217*, 925–941. [CrossRef]
14. Árvay, J.; Hauptvogl, M.; Šnirc, M.; Gažová, M.; Demková, L.; Bobuľská, L.; Hrstková, M.; Bajčan, D.; Harangozo, Ľ.; Bilčíková, J.; et al. Determination of elements in wild edible mushrooms: Levels and risk assessment. *J. Microbiol. Biotechnol. Food Sci.* **2019**, *8*, 999–1004. [CrossRef]
15. Selin, N.E. A proposed global metric to aid Mercury pollution policy. *Science* **2018**, *360*, 607–609. [CrossRef]
16. Hsu-Kim, H.; Eckley, C.S.; Achá, D.; Feng, X.; Gilmour, C.C.; Jonsson, S.; Mitchell, C.P. Challenges and opportunities for managing aquatic mercury pollution in altered landscapes. *AMBIO* **2018**, *47*, 141–169. [CrossRef]
17. Xu, J.; Zhang, J.; Lv, Y.; Xu, K.; Lu, S.; Liu, X.; Yang, Y. Effect of soil mercury pollution on Ginger (zingiber officinale Roscoe): Growth, product quality, health risks and Silicon Mitigation. *Ecotoxicol. Environ. Saf.* **2020**, *195*, 110472. [CrossRef]
18. Schwieger, A.-C.; Gebauer, K.; Ohle, A.; Beckmann, M. Determination of mercury binding forms in humic substances of lignite. *Fuel* **2020**, *274*, 117800. [CrossRef]
19. Tang, W.-L.; Liu, Y.-R.; Guan, W.-Y.; Zhong, H.; Qu, X.-M.; Zhang, T. Understanding Mercury methylation in the changing environment: Recent advances in assessing microbial methylators and Mercury bioavailability. *Sci. Tot. Environ.* **2020**, *714*, 136827. [CrossRef]
20. Kalač, P. *Mineral Composition and Radioactivity of Edible Mushrooms*; Academic Press: Cambridge, MA, USA, 2019; pp. 299–326. ISBN 9780128175651.
21. Kunca, V.; Čiliak, M. Habitat preferences of hericium erinaceus in Slovakia. *Fungal Ecol.* **2017**, *27*, 189–192. [CrossRef]
22. Kunca, V.; Pavlik, M. Fruiting body production of, and suitable environmental ranges for, growing the umbrella polypore medicinal mushroom, polyporus umbellatus (agaricomycetes), in natural conditions in Central Europe. *Int. J. Med. Mushrooms* **2019**, *21*, 121–129. [CrossRef]
23. Demo, M.; Bako, A.; Húska, D. Biomass production potential of different willow varieties (*Salix* spp.) grown in soil-climatic conditions of south-western Slovakia. *Wood Res.* **2013**, *58*, 651–661.
24. Liu, S.; Liu, H.; Li, J.; Wang, Y. Research Progress on Elements of Wild Edible Mushrooms. *J. Fungi* **2022**, *8*, 964. [CrossRef]
25. Bahadori, M.B.; Sarikurkcu, C.; Yalcin, O.U.; Cengiz, M.; Gungor, H. Metal concentration, phenolics profiling, and antioxidant activity of two wild edible melanoleuca mushrooms (M. Cognata and M. Stridula). *Microchem. J.* **2019**, *150*, 104172. [CrossRef]
26. Hakanson, L. An ecological risk index for aquatic pollution control.a sedimentological approach. *Water Res.* **1980**, *14*, 975–1001. [CrossRef]
27. Dryżałowska, A.; Falandysz, J. Bioconcentration of Mercury by mushroom Xerocomus Chrysenteron from the spatially distinct locations: Levels, possible intake and safety. *Ecotox. Environ. Saf.* **2014**, *107*, 97–102. [CrossRef]
28. Joint, F.A.O.; World Health Organization; WHO Expert Committee on Food Additives. Safety Evaluation of Certain Food Additives/Prepared by the by the Seventy Fourth Meeting of the Joint FAO/WHO Expert Committee on Food Additives (JECFA). 2012. Available online: https://books.google.ro/books?hl=ro&lr=&id=YFAMU9qYD_YC&oi=fnd&pg=PP7&ots=c5okGh31Ln&sig=W99eNu8GUWHlWsG-ooJzM1dlkws&redir_esc=y#v=onepage&q&f=false (accessed on 9 May 2022).
29. Statistical Organization of Slovak Republic 2021. Food Consumption in the Slovak Republic 2020. Available online: www.statistics.sk (accessed on 9 May 2022).
30. Demková, L.; Árvay, J.; Hauptvogl, M.; Michalková, J.; Šnirc, M.; Harangozo, Ľ.; Bobuľská, L.; Bajčan, D.; Kunca, V. Mercury content in three edible wild-growing mushroom species from different environmentally loaded areas in Slovakia: An Ecological and human health risk assessment. *J. Fungi* **2021**, *7*, 434. [CrossRef]
31. The Jamovi Project Jamovi. (Version 2.2) [Computer Software]. 2021. Available online: https://www.jamovi.org (accessed on 21 December 2022).

32. R Core Team. R: A Language and Environment for Statistical Computing. (Version 4.0) [Computer software]. 2021. Available online: https://cran.r-project.org (accessed on 1 April 2021).
33. Patil I Ggstatsplot: 'ggplot2' Based Plots with Statistical Details. [R package]. 2018. Available online: https://CRAN.R-project.org/package=ggstatsplot (accessed on 21 December 2022).
34. Brunson, J.C. ggalluvial: Alluvial Plots in 'ggplot2'. [R package]. 2018. Available online: https://CRAN.Rproject.org/package=ggalluvial (accessed on 21 December 2022).
35. Koneswarakantha, B. Easyalluvial: Generate Alluvial Plots with a Single Line of Code. [R package]. 2019. Available online: https://CRAN.R-project.org/package=easyalluvial (accessed on 20 December 2022).
36. Wickham, H.; Chang, W.; Henry, L.; Pedersen, T.L.; Takahashi, K.; Wilke, C.; Woo, K. RStudio ggplot2: Create Elegant Data Visualisations Using the Grammar of Graphics. [R package]. 2018. Available online: https://CRAN.Rproject.org/package=ggplot2 (accessed on 20 December 2022).
37. RStudio Team. *RStudio: Integrated Development for R*; RStudio Inc.: Boston, MA, USA, 2015; Available online: http://www.rstudio.com/ (accessed on 12 December 2022).
38. Šefčík, P.; Pramuka, S.; Gluch, A. Assessment of soil contamination in Slovakia according index of geoaccumulation. *Agriculture* **2008**, *54*, 119–130.
39. Act of the National Council of the Slovak Republic, No. 220/2004 Coll. Available online: http://www.podnemapy.sk/portal/verejnost/konsolidacia/z_220_2004.pdf (accessed on 29 October 2022).
40. Li, X.; Yang, H.; Zhang, C.; Zeng, G.; Liu, Y.; Xu, W.; Wu, Y.; Lan, S. Spatial distribution and transport characteristics of heavy metals around an antimony mine area in central China. *Chemosphere* **2017**, *170*, 17–24. [CrossRef]
41. Jančo, I.; Šnirc, M.; Hauptvogl, M.; Demková, L.; Franková, H.; Kunca, V.; Lošák, T.; Árvay, J. Mercury in Macrolepiota procera (scop.) singer and its underlying substrate—Environmental and Health Risks Assessment. *J. Fungi* **2021**, *7*, 772. [CrossRef]
42. Árvay, J.; Tomáš, J.; Hauptvogl, M.; Massányi, P.; Harangozo, Ľ.; Tóth, T.; Stanovič, R.; Bryndzová, Š.; Bumbalová, M. Human exposure to heavy metals and possible public health risks via consumption of wild edible mushrooms from Slovak Paradise National Park, Slovakia. *J. Environ. Sci. Health B* **2015**, *50*, 833–843. [CrossRef]
43. Árvay, J.; Záhorcová, Z.; Tomáš, J.; Hauptvogl, M.; Stanovič, R.; Harangozo. Mercury in edible wild-grown mushrooms from historical mining area—Slovakia: Bioaccumulation and risk assessment. *J. Microbiol. Biotechnol. Food Sci.* **2015**, *4*, 1–4. [CrossRef]
44. Falandysz, M.; Gucia, B.; Skwarzec, A.J. Total Mercury in mushrooms and underlying soil substrate from the Borecka Forest, northeastern Poland. *Arch. Environ. Contam. Toxicol.* **2002**, *42*, 145–154. [CrossRef]
45. Mleczek, M.; Budka, A.; Kalač, P.; Siwulski, M.; Niedzielski, P. Family and species as determinants modulating mineral composition of selected wild-growing mushroom species. *Environ. Sci. Pollut. Res.* **2020**, *28*, 389–404. [CrossRef]
46. Falandysz, J.; Gucia, M. Bioconcentration factors of Mercury by Parasol Mushroom (Macrolepiota procera). *Environ. Geochem. Health* **2008**, *30*, 121–125. [CrossRef]
47. Falandysz, J.; Bielawski, L.; Kawano, M.; Brzostowski, A.; Chudzyński, K. Mercury in mushrooms and soil from the Wieluńska Upland in south-central Poland. *J. Environ. Sci. Health A* **2002**, *37*, 1409–1420. [CrossRef]
48. Mleczek, M.; Siwulski, M.; Budka, A.; Mleczek, P.; Budzyńska, S.; Szostek, M.; Kuczyńska-Kippen, N.; Kalač, P.; Niedzielski, P.; Gąsecka, M.; et al. Toxicological risks and nutritional value of wild edible mushroom species -a half-century monitoring study. *Chemosphere* **2021**, *263*, 128095. [CrossRef]
49. Jančo, I.; Šnirc, M.; Hauptvogl, M.; Franková, H.; Čeryová, N.; Štefániková, J.; Árvay, J. Arsenic, cadmium and mercury in the Macrolepiota procera (scop.) singer fruiting bodies. *J. Microbiol. Biotechnol. Food Sci.* **2021**, *11*, e4764. [CrossRef]
50. Jančo, I.; Trebichalský, P.; Bystrická, J.; Tirdiľová, I.; Štefániková, J. Determination of cadmium, lead and mercury content in Macrolepiota procera in selected areas of Slovakia.19th international multidisciplinary scientific geoconference SGEM2019, water resources. forest, Marine and Ocean Ecosystems. *SGEM Int. Multidiscip. Sci. GeoConference EXPO Proc.* **2019**, *19*, 681–688. [CrossRef]
51. Árvay, J.; Tomáš, J.; Hauptvogl, M.; Kopernická, M.; Kováčik, A.; Bajčan, D.; Massányi, P. Contamination of wild-grown edible mushrooms by heavy metals in a former mercury-mining area. *J. Environ. Sci. Health B* **2014**, *49*, 815–827. [CrossRef]
52. Falandysz, J.; Zhang, J.; Wang, Y.; Krasińska, G.; Kojta, A.; Saba, M.; Shen, T.; Li, T.; Liu, H. Evaluation of the mercury contamination in mushrooms of genus Leccinum from two different regions of the world: Accumulation, distribution and probable dietary intake. *Sci. Tot. Environ.* **2015**, *537*, 470–478. [CrossRef]
53. Melgar, M.J.; Alonso, J.; García, M.A. Mercury in edible mushrooms and underlying soil: Bioconcentration factors and toxicological risk. *Sci. Tot. Environ.* **2009**, *407*, 5328–5334. [CrossRef]
54. Svoboda, L.; Havlíčková, B.; Kalač, P. Contents of cadmium, mercury and lead in edible mushrooms growing in a historical silver-mining area. *Food Chem.* **2006**, *96*, 580–585. [CrossRef]
55. Scragg, A. *Environmental Biotechnology*; Oxford University Press: New York, NY, USA, 2005; pp. 218–258.
56. Falandysz, J.; Borovička, J. Macro and trace mineral constituents and radionuclides in mushrooms: Health benefits and risks. *Appl. Microbiol. Biotechnol.* **2012**, *97*, 477–501. [CrossRef]
57. Bidar, G.; Garçon, G.; Pruvot, C.; Dewaele, D.; Cazier, F.; Douay, F.; Shirali, P. Behavior of trifolium repens and lolium perenne growing in a heavy metal contaminated field: Plant metal concentration and phytotoxicity. *Environ. Pollut.* **2007**, *147*, 546–553. [CrossRef]

58. Krasińska, G.; Falandysz, J. Mercury in Hazel Bolete Leccinum griseum and soil substratum: Distribution, bioconcentration and dietary exposure. *J. Environ. Sci. Health A* **2015**, *50*, 1259–1264. [CrossRef]
59. Falandysz, J.; Kowalewska, I.; Nnorom, I.C.; Drewnowska, M.; Jarzyńska, G. Mercury in Red Aspen boletes (*Leccinum aurantiacum*) mushrooms and the soils. *J. Environ. Sci. Health A* **2012**, *47*, 1695–1700. [CrossRef]
60. Falandysz, J.; Bielawski, L. Mercury and its bioconcentration factors in Brown Birch scaber stalk (*Leccinum scabrum*) from various sites in Poland. *Food Chem.* **2007**, *105*, 635–640. [CrossRef]
61. Falandysz, J.; Kunito, T.; Kubota, R.; Bielawski, L.; Mazur, A.; Falandysz, J.J.; Tanabe, S. Selected elements in Brown Birch scaber stalk Leccinum scabrum. *J. Environ. Sci. Health A* **2007**, *42*, 2081–2088. [CrossRef]
62. Busuioc, G.; Elekes, C.C.; Stihi, C.; Iordache, S.; Ciulei, S.C. The bioaccumulation and translocation of Fe, Zn, and Cu in species of mushrooms from Russula genus. *Environ. Sci. Pollut. Res.* **2011**, *18*, 890–896.
63. Krasińska, G.; Falandysz, J. Mercury in Orange Birch Bolete leccinum versipelle and soil substratum: Bioconcentration by mushroom and probable dietary intake by consumers. *Environ. Sci. Pollut. Res.* **2015**, *23*, 860–869. [CrossRef]

Journal of Fungi

MDPI

Article

Risk Assessment of Heavy Metals Occurrence in Two Wild Edible Oyster Mushrooms (*Pleurotus* spp.) Collected from Rajaji National Park

Ivan Širić [1], Pankaj Kumar [2,*], Bashir Adelodun [3,4], Sami Abou Fayssal [5,6], Rakesh Kumar Bachheti [7,8], Archana Bachheti [9], Fidelis O. Ajibade [10], Vinod Kumar [2], Mostafa A. Taher [11,12] and Ebrahem M. Eid [13,14,*]

1 University of Zagreb, Faculty of Agriculture, Svetosimunska 25, 10000 Zagreb, Croatia
2 Agro-Ecology and Pollution Research Laboratory, Department of Zoology and Environmental Science, Gurukula Kangri (Deemed to Be University), Haridwar 249404, Uttarakhand, India
3 Department of Agricultural and Biosystems Engineering, University of Ilorin, PMB 1515, Ilorin 240003, Nigeria
4 Department of Agricultural Civil Engineering, Kyungpook National University, Daegu 37224, Korea
5 Department of Agronomy, Faculty of Agronomy, University of Forestry, 10 Kliment Ohridski Blvd, 1797 Sofia, Bulgaria
6 Department of Plant Production, Faculty of Agriculture, Lebanese University, Beirut 1302, Lebanon
7 Department of Industrial Chemistry, College of Applied Science, Addis Ababa Science and Technology University, Addis Ababa P.O. Box 16417, Ethiopia
8 Centre of Excellence in Nanotechnology, Addis Ababa Science and Technology University, Addis Ababa P.O. Box 16417, Ethiopia
9 Department of Environmental Science, Graphic Era (Deemed to be University), Dehradun 248002, Uttarakhand, India
10 Department of Civil and Environmental Engineering, Federal University of Technology, PMB 704, Akure 340001, Nigeria
11 Biology Department, Faculty of Science and Arts, King Khalid University, Mohail Assir 61321, Saudi Arabia
12 Botany Department, Faculty of Science, Aswan University, Aswan 81528, Egypt
13 Biology Department, College of Science, King Khalid University, Abha 61321, Saudi Arabia
14 Botany Department, Faculty of Science, Kafrelsheikh University, Kafr El-Sheikh 33516, Egypt
* Correspondence: rs.pankajkumar@gkv.ac.in (P.K.); ebrahem.eid@sci.kfs.edu.eg (E.M.E.)

Citation: Širić, I.; Kumar, P.; Adelodun, B.; Abou Fayssal, S.; Bachheti, R.K.; Bachheti, A.; Ajibade, F.O.; Kumar, V.; Taher, M.A.; Eid, E.M. Risk Assessment of Heavy Metals Occurrence in Two Wild Edible Oyster Mushrooms (*Pleurotus* spp.) Collected from Rajaji National Park. *J. Fungi* **2022**, *8*, 1007. https://doi.org/10.3390/jof8101007

Academic Editor: Laurent Dufossé

Received: 3 September 2022
Accepted: 21 September 2022
Published: 25 September 2022

Abstract: This study aimed at assessing the concentration of six heavy metals (Cd, Cr, Cu, Fe, Mn, and Zn) in two wild edible oyster mushrooms (*Pleurotus ostreatus* and *Pleurotus djamor*) collected from Rajaji National Park in Haridwar, India. For this purpose, mushroom samples were collected from selected locations (forest, residential, tourist, industrial areas, and transportation activities) from June 2021 to July 2022 and subsequently analyzed for selected heavy metals using atomic absorption spectroscopy (AAS). Results showed that both *Pleurotus* spp. had significantly varying ($p < 0.05$) concentrations of heavy metals. However, *P. ostreatus* showed relatively higher concentration levels of these metals compared to *P. djamor*. The mean concentrations (mg/kg dry weight) of the Cd, Cr, Cu, Fe, Mn, and Zn in *P. ostreatus* and *P. djamor* were 0.10 and 0.08, 0.87 and 0.64, 16.19 and 14.77, 28.49 and 27.15, 9.93 and 8.73, and 18.15 and 15.76, respectively. As indicated by the multivariate analysis tools i.e., principal component analysis (PCA) and hierarchical cluster analysis (HCA), the locations near the residential, industrial, and transportation activities had higher concentration levels of heavy metals. Moreover, the health risk studies using the target hazard quotient (THQ < 1) showed no significant health risk as the consumption of both *Pleurotus* spp., except for at one location, had high-traffic activities. The findings of this study provide vital information about the occurrence of potentially toxic heavy metals in wild edible *Pleurotus* spp. in Rajaji National Park in Haridwar, India representing a safeguard for mushroom consumers.

Keywords: food quality; forest biodiversity; health risk; metal elements; traditional foraging; wild mushrooms

1. Introduction

Mushrooms are protein-rich and low-fat sources of food, making them nutritionally equivalent to meat for the rising vegan community around the world [1]. According to a recent report, the global mushroom market accounted for 15.25 million tons in 2021 and is projected to reach 24.05 million tons by 2028 with a compound annual growth rate of 6.74% during the 2021–2028 period [2]. India accounts for around 2% of the world's commercial mushroom production (around 0.31 million tons), however, several communities still depend on foraging wild mushrooms [3]. Being rich in vitamin D, riboflavin, and essential amino and fatty acids [4], mushroom consumption increased during the COVID-19 pandemic, showing positive impacts on human immunity. Whereas, only a little increase was observed in India as the consumption of mushrooms recently faced with a lack of appetite in some local areas due to its strangeness to traditional cuisine [5]. Despite this local behavior, this consumption is becoming trendier among youths due to the increasing awareness vis-à-vis its health benefits and low-cost production, as tremendous amounts of raw material based on agro-industrial wastes could be easily found [6,7].

Wild mushroom foraging is a cultural tradition of European countries, mainly in Central and Eastern Europe. Locals tend to forage collectively in the autumn–spring season, in which a diversity of delicious edible mushrooms flourish [3]. Yet, these knowledge and differentiation skills are not popular globally. For instance, some Indian locals tend to avoid wild mushroom foraging and consumption as they found it hard to differentiate between edible, non-edible, and poisonous species. Moreover, a lack of research on wild mushroom species in the Indian sub-continent constitutes an additional limiting factor for this exciting adventure [8]. Additional constraints of wild mushroom consumption include their growth in polluted zones near mines and factories. These sites are sources of potentially toxic heavy metals such as cadmium (Cd), cobalt (Co), copper (Cu), chromium (Cr), iron (Fe), lead (Pb), manganese (Mn), mercury (Hg), nickel (Ni), selenium (Se) and zinc (Zn). In this context, a recent report revealed a high potential of some wild edible mushroom species such as *Tricholoma* spp. to accumulate Cd concentrations with a bioconcentration factor (BCF) higher than 1 [9]. However, the authors noted no potential health risk, as the latter's index (HRI) besides the daily intake of heavy metals was lower than 1 and 0.426, respectively. Although these results were very promising, the high bioaccumulation capacity is a matter of concern for researchers as many sources of pollution may be concealed in other locations. Earlier, wild *Agaricus campestris*, *Boletus edulis*, and *Boletus reticulatus* were found to be high accumulators of Cd and Hg, respectively [10]. Most of these potentially toxic heavy metals were detected in mushrooms' caps rather than stipes. Being the most-desired part of mushrooms, cap accumulation of highly undesired elements is a worrying issue. Therefore, the need to study the environmental ecosystem (mainly soils) where wild edible mushrooms grow is always a needed fore step. It was reported that high levels of Cd were thought to be the main cause of breast, pancreas, and kidney cancers. Other potentially toxic heavy metals such as Hg can lead to the disruption of growth factor synthesis and affect reproduction, respectively [11]. Recently, the health risk and heavy-metal composition of *Agaricus bisporus* mushrooms sold in Uttarakhand State local markets were assessed [12]. Authors found a decreasing order of heavy metals as: Fe > Zn > Mn > Cr > Cu > Ni > Cd, with no potential hazard in consumption. However, further research should be performed on other local wild and edible species found in markets in order to assure safe mushroom consumption with no short or long-term health risks.

Oyster mushroom species (*Pleurotus* spp.) are mainly grown on lignocellulosic-rich materials or dead wood. These species are a great source of human protein, natural amino acids, and essential poly- and monounsaturated fatty acids [13,14]. Recently, it was acknowledged that *P. ostreatus* grown on agro-industrial residues such as olive pruning residues and spent coffee grounds had the potential to accumulate some desirable and undesirable metals. However, the reported concentrations were within the same limits set by the European Commission (EC), World Health Organization (WHO), and United States Food and Drug Administration (FDA) [15]. In the same vein, *P. ostreatus*, grown on spent mushroom substrate, supplemented with nano-amino additives outlined high Pb levels, (>0.3 mg/kg) exceeding limits set by the guidelines, which stipulate the need to reconsider consuming the produce regularly [14]. However, wild *P. ostreatus* mushrooms were acknowledged to enclose higher concentrations of heavy metals compared to cultivated ones. In this context, this species collected from mine-polluted soils delineated Zn, Cd, and Cr concentrations higher than the safe limits, with concentrations in stalks higher than those in caps [16]. Although the authors did not detect human carcinogenic risks, they simulated possible and serious health hazards if continuously consumed by locals. *P. djamor* is well appreciated for its high antioxidants and protein content. This species has been reported to accumulate the highest amounts of heavy metals among *Pleurotus* spp., and mainly has the highest Pb concentrations [17]. These authors also acknowledged the following decreasing order in terms of total metal bioaccumulation: *P. cirinopileatus* > *P. djamor* > *P. eryngii* > *P. ostreatus*.

Rajaji National Park is a famous natural reserve located on both sides of the Ganges River, Uttarakhand (820 km^2). It is well known for its high biodiversity and is home to several rare plant and animal species [18]. This park is also a reserve for elephants and tigers, being the first park in Uttarakhand to offer the most prestigious wildlife conservation. Rajaji National Park hosts wild Safaris for tourists and locals all around the year [19]. In this context, wild mushroom populations settling in Rajaji National Park were very scantly investigated and clumsily identified in the literature [20]. Therefore, the current study investigates the heavy metal accumulation of two identified wild edible *Pleurotus* spp. and assesses their potential risk for human consumption. The findings of this study suggest a heavy-metal accumulation potential of wild edible *Pleurotus* spp. that could potentially be harmful to consumers.

2. Materials and Methods

2.1. Description of Study Area and Mushroom Sampling

For the present study, samples of two oyster mushroom species i.e., *P. ostreatus* and *P. djamor* were obtained from ten (10) locations around the Chilla Forest Range of Rajaji National Park in Haridwar city, Uttarakhand, India. Table 1 and Figure 1 show the list and map of sampling locations for the current study, respectively. The elevation of sampling locations ranged from 286–515 m. Moreover, the perimeter and total area of the sampling area were 42 km and 100 km^2, respectively. The area receives an average annual rainfall of 2136.7 mm with August as the wettest month. The average temperature and humidity of the area are recorded as 23 °C and 68%, respectively [21]. The monsoon season in the region begins in July and lasts up to September. During this period, a wide variety of wild mushrooms appear in the forest areas, including *Pleurotus* spp. The sampling locations are easily accessible to the local communities and do not need any special permission from the forest authorities. Therefore, wild mushrooms are commonly collected and consumed by the local people. The mushroom samples were collected from June 2021 to July 2022 following standard field sampling methods [22]. The mushroom bodies were carefully picked and immediately transported to the laboratory for further analysis. However, the number of samples collected (*P. ostreatus*, $n = 49$; *P. djamor*, $n = 39$) for each species varied due to their availability. Herein, the *Pleurotus* spp. were morphologically identified using the field guide of Kumari et al. [23] and several online databases such as https://www.mushroom.world/ and http://www.mycokey.com/ (accessed on 1 September 2022).

2.2. Heavy-Metal Analysis

The collected mushroom samples were oven dried at 60 °C until a constant weight was achieved. The dried samples were again converted into fine powder form using a mechanical grinder (JMG3345, 450W, Usha, Gurgaon, India). Analysis of six heavy metals (Cd, Cr, Cu, Fe, Mn, Zn) was performed using atomic absorption spectroscopy (AAS, ECIL, Hyderabad, India). For this, 1 g of powdered mushroom sample was placed in a 150 mL-capacity conical flask and dissolved in 10 mL of a di-acid mixture having a 3:1 ratio of concentrated HNO_3 and $HClO_4$. The mixture was placed on a rotary shaker and self-digested for 12 h. After the completion of self-digestion, a 3% HNO_3 solution was added to make the final volume of the contents 50 mL. Again, the flask was placed on a hot plate and digested for 1 h at 150 °C until 10 mL of residual volume was left. Finally, more HNO_3 solution was added to make the liquid volume 50 mL and usable for heavy metal analysis using AAS. The analytical conditions of the AAS instrument such as detection limits, wavelength (nm), cathode lamp current (mA), slit width (0.2–0.7 mm), and flame (air-acetylene mixture) were adjusted based on the heavy metals as given in the previous study [24–26]. The standard recovery percentage of heavy metals ranged from 98.50–99.10%. All reagents used for heavy metal analysis were of laboratory grade.

Table 1. Description of the *Pleurotus* spp. sampling locations in Rajaji National Park, Haridwar, India.

Site Name	Code	Samples Collected in Two Years	Latitude (N) ^	Longitude (E) ^	Elevation (m)
Chilla Forest Colony	CF	*P. ostreatus* (*n* = 5); *P. djamor* (*n* = 4)	29°57′46.57″	78°11′41.59″	296
Chandi Devi Forest	CD	*P. ostreatus* (*n* = 5); *P. djamor* (*n* = 5)	29°56′10.38″	78°10′56.72″	517
Bheemgoda Barrage Forest	BF	*P. ostreatus* (*n* = 4); *P. djamor* (*n* = 5)	29°57′20.73″	78°11′10.16″	333
Sureshwari Devi Forest	SD	*P. ostreatus* (*n* = 5); *P. djamor* (*n* = 5)	29°58′70.49″	78° 6′26.45″	315
Rishikesh Canal Road	RC	*P. ostreatus* (*n* = 5); *P. djamor* (*n* = 5)	29°58′58.16″	78°14′20.12″	328
Mansa Devi Forest	MD	*P. ostreatus* (*n* = 5); *P. djamor* (*n* = 5)	29°57′24.01″	78° 9′56.09″	440
Sapt Rishi Ghat	SR	*P. ostreatus* (*n* = 5); *P. djamor* (*n* = 2)	29°58′40.43″	78°11′18.52″	299
BHEL Forest Colony	BF	*P. ostreatus* (*n* = 5); *P. djamor* (*n* = 4)	29°57′9.92″	78° 6′22.65″	304
Devpura Forest Colony	DF	*P. ostreatus* (*n* = 5); *P. djamor* (*n* = 2)	29°56′25.96″	78° 8′14.40″	322
Bijnor Canal Road	BC	*P. ostreatus* (*n* = 5); *P. djamor* (*n* = 2)	29°55′90.43″	78°10′27.73″	286

^: Coordinates and referencing system (CRS) is rendered as EPSG:4326–WGS 84 projection; Source: Google Earth Pro.

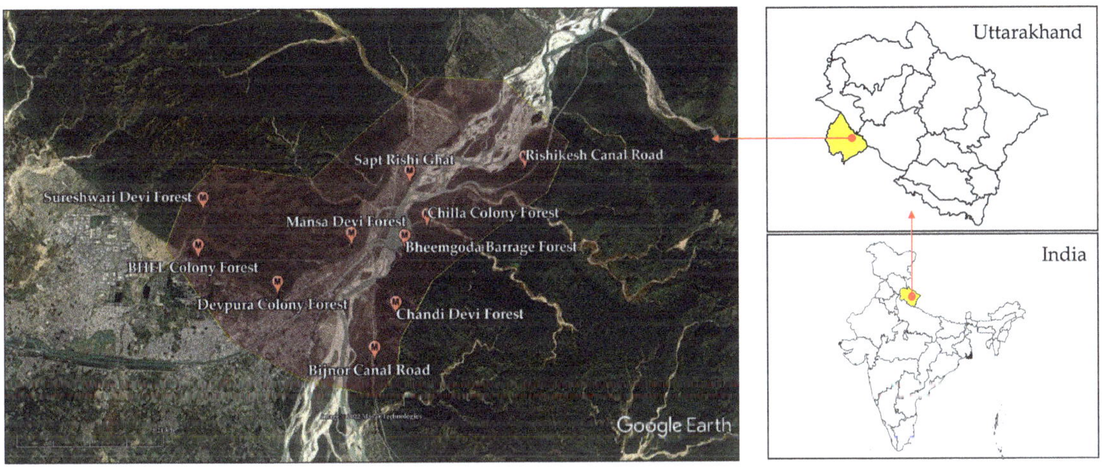

Figure 1. Map showing the sampling locations of *Pleurotus* spp. collection in and around Rajaji National Park at Haridwar, Uttarakhand, India (Source: Google Earth Pro).

2.3. Data Analysis

In the present study, the health risk of heavy metals contents in two *Pleurotus* spp. was analyzed using the target hazard quotient (THQ) method [26,27]. The cumulative health risk of heavy metals in child and adult human groups was assessed using the THQ index, as given in Equation (1):

$$\text{THQ} = 10^{-3} \times \frac{Ef \times Ed \times Ir \times HMc}{Bw \times Cp \times Rd} \quad (1)$$

Moreover, the THQ sum of six heavy metals was used to present the cumulative health risk index (HRI) value. Following Equation (2) were used to calculate HRI:

$$HRI = \sum THQ_{Cd+Cr+Cu+Fe+Mn+Zn} \quad (2)$$

where, *Ef* is exposure frequency of heavy metals, *Ed* is exposure duration (365 days), *Ir* is the mushroom ingestion rate, *HMc* is the heavy-metal concentration in the mushroom sample (mg/kg), *Bw* is the average body weight (70 kg for adult and 16 kg for child), *Cp* is the consumption period (25,550 days for adult and 5840 days for child), *Rd* is the reference doses (Cd: 5.0×10^{-4}; Cr: 3.0×10^{-3}; Cu: 4.2×10^{-2}; Fe: 7.0×10^{-1}; Mn: 1.4×10^{-2}, and Zn: 3.0×10^{-1}) [28], and 10^{-3} is the conversion factor, respectively. Moreover, the data were analyzed using principal component analysis (PCA) and hierarchical clustering (CL) to derive the relationship between heavy-metal concentration in *Pleurotus* spp. and their sampling locations (correlation matrix). CL is used to create clusters that have a predetermined ordering from top to bottom. Based on their similarity, nodes are compared with each other where larger groups are built by joining groups of nodes. The data were tested using a one-way analysis of variance (ANOVA) and a Tukey's post hoc test. The level of statistical significance was adjusted as the probability (p) < 0.05.

2.4. Software and Tools

Microsoft Excel 2019 (Microsoft, Redmond, WA, USA) and OriginPro 2022b (OriginLab, Northampton, MA, USA) software packages were used for the data analysis and visualization.

3. Results and Discussion

3.1. Heavy Metal Levels in Collected Pleurotus spp.

The results of heavy-metals concentration in *Pleurotus* spp. collected from different locations of Rajaji National Park are shown in Table 2. One-way ANOVA revealed significantly varying ($p < 0.05$) heavy-metal contents among each species between selected locations. Both *Pleurotus* spp. showed the following mean heavy-metals compositional order: Fe > Zn > Cu > Mn > Cr > Cd. The contents of Cd ranged between 0.03 and 0.16, and between 0.05 and 0.11 for *P. ostreatus* and *P. djamor*, respectively, in all studied locations. Although, the mean Cd contents found in *P. ostreatus* and *P. djamor* were 0.10 and 0.08 mg/kg, respectively, noting no risk to human health; numerous locations (except Chilla Forest Colony, Chandi Devi Forest, Sureshwari Devi, and Mansa Devi Forest) outlined Cd contents in *P. ostreatus*, approaching the safe limit set by Indian Standards (0.10 mg/kg) [28]. Conversely, *P. djamor* showed a value exceeding the safe limit only in Bijnor Canal Road (0.11 mg/kg). Moreover, the skewness test of heavy-metal contents in *Pleurotus* spp. showed both positive (0.05 to 1.76) and negative (–0.02 to –1.88) values, depicting random symmetric distribution throughout the sampling locations. Also, the Kurtosis tests showed both positive (0.23 to 1.07) and negative (–0.16 to –0.85) values, indicating that the data tend to have fewer outliers.

In a previous study, *P. ostreatus* was acknowledged to have a high affinity to bioaccumulate Cd, especially when grown widely [30]; this is concordant with the findings of the current study. Other wild mushroom species (*Trichloma spp.*) denoted high Cd contents in mushroom caps and stipes, with the former higher than the latter (Cd in caps > C in stipes) [9]. In Cd-contaminated locations, mass residential colonies were associated with mass tourism for which numerous rest houses are thriving for the last decades releasing their wastes to the nearby areas. Moreover, several famous temples attracting locals for praying and religious movements could be a source of contamination due to significant amounts of liberated wastes [31]. Furthermore, Bijnor Canal Road is famous for its mass transport between trees without any preliminary landscape planning. Tourist stations have been considered as main sources of organic and inorganic pollution, including heavy metals [32]. Similarly, toxic heavy-metal contents tend to accumulate in forest soils in the vicinity of roadways [33].

Table 2. Heavy-metal concentration (mean ± SD; n = 2 to 5) in two *Pleurotus* spp. collected from different locations of Rajaji National Park, Haridwar, India.

Sampling Site	*Pleurotus* spp.	Heavy-Metal Concentration (mg/kg Dry Weight)					
		Cd	Cr	Cu	Fe	Mn	Zn
Chilla Forest Colony	*P. ostreatus*	0.06 ± 0.02 ab	0.81 ± 0.14 b	14.10 ± 1.20 b	29.76 ± 1.53 c	9.33 ± 0.20 c	13.56 ± 0.62 a
	P. djamor	0.05 ± 0.01 b	0.73 ± 0.05 b	12.08 ± 0.08 a	30.20 ± 0.86 c	8.14 ± 0.15 b	14.09 ± 0.40 a
Chandi Devi Forest	*P. ostreatus*	0.10 ± 0.03 bc	0.57 ± 0.07 a	17.90 ± 1.04 cd	25.01 ± 3.17 ab	10.36 ± 1.10 c	18.64 ± 1.56 c
	P. djamor	0.07 ± 0.02 ab	0.42 ± 0.08 a	15.05 ± 0.49 bc	23.19 ± 2.02 a	8.99 ± 0.72 b	16.03 ± 1.10 b
Bheemgoda Barrage Forest	*P. ostreatus*	0.11 ± 0.03 c	0.75 ± 0.09 b	14.43 ± 0.32 b	30.86 ± 1.38 c	7.80 ± 0.43	19.70 ± 0.98 c
	P. djamor	0.09 ± 0.01 c	0.38 ± 0.02 a	15.09 ± 0.75 b	27.74 ± 0.84 bc	8.64 ± 0.27 b	13.50 ± 2.15 a
Sureshwari Devi Forest	*P. ostreatus*	0.03 ± 0.01 a	0.84 ± 0.10 bc	13.52 ± 0.61 ab	28.04 ± 2.03 c	10.93 ± 1.09 c	16.27 ± 1.53 b
	P. djamor	*Bdl*	0.45 ± 0.15 a	14.19 ± 0.87 b	24.82 ± 4.20 ab	7.55 ± 0.14 a	15.01 ± 1.07 b
Rishikesh Canal Road	*P. ostreatus*	0.13 ± 0.02 c	0.92 ± 0.03 c	15.06 ± 0.16 bc	31.56 ± 1.64 c	10.70 ± 0.81 c	18.65 ± 1.94 c
	P. djamor	0.08 ± 0.02 ab	0.64 ± 0.07 b	14.15 ± 0.90 b	32.07 ± 1.98 c	9.07 ± 0.29 c	15.71 ± 0.62 b
Mansa Devi Forest	*P. ostreatus*	0.09 ± 0.01 c	0.89 ± 0.08 c	16.21 ± 0.19 c	21.88 ± 0.43 a	8.31 ± 0.37 b	17.10 ± 0.78 b
	P. djamor	0.07 ± 0.01 b	0.77 ± 0.04 c	13.62 ± 1.35 a	25.29 ± 1.05 ab	7.18 ± 0.11 a	19.58 ± 0.25 c
Sapt Rishi Ghat	*P. ostreatus*	0.11 ± 0.03 bc	0.96 ± 0.05 c	18.40 ± 0.46 d	29.67 ± 0.57 c	10.11 ± 0.22 c	16.39 ± 2.38 b
	P. djamor	0.05 ± 0.01 b	0.62 ± 0.12 b	15.10 ± 1.10 b	30.16 ± 1.32 c	9.75 ± 0.45 c	17.22 ± 0.70 b
BHEL Forest Colony	*P. ostreatus*	0.13 ± 0.02 cd	1.04 ± 0.02 d	17.21 ± 0.84 cd	25.64 ± 1.88 b	9.42 ± 0.60 c	20.84 ± 0.58 c
	P. djamor	0.10 ± 0.03 c	0.85 ± 0.03 c	16.12 ± 1.32 c	22.92 ± 3.05 a	8.09 ± 0.44 a	15.58 ± 1.39 b
Devpura Forest Colony	*P. ostreatus*	0.11 ± 0.02 c	0.79 ± 0.14 b	15.88 ± 0.78 b	26.78 ± 1.19 b	10.25 ± 0.97 c	18.24 ± 0.66 c
	P. djamor	0.08 ± 0.01 c	0.57 ± 0.09 b	16.26 ± 0.35 c	23.08 ± 3.60 a	8.58 ± 0.48 b	12.02 ± 2.04 a
Bijnor Canal Road	*P. ostreatus*	0.16 ± 0.02 d	1.14 ± 0.17 c	19.20 ± 0.40 d	35.66 ± 0.94 cd	12.05 ± 0.51 d	22.10 ± 0.71 d
	P. djamor	0.11 ± 0.04 c	0.95 ± 0.09 c	16.07 ± 0.69 c	32.02 ± 1.55 c	11.32 ± 0.35 bc	18.90 ± 1.20 c
Minimum	*P. ostreatus*	0.03	0.57	13.52	21.88	7.80	13.56
	P. djamor	0.05	0.38	12.08	22.92	7.18	12.02
Maximum	*P. ostreatus*	0.16	1.14	19.20	35.66	12.05	22.10
	P. djamor	0.11	0.95	16.26	32.07	11.32	19.58
Median	*P. ostreatus*	0.11	0.87	16.05	28.86	10.18	18.44
	P. djamor	0.08	0.63	15.07	26.52	8.61	15.65
Mean	*P. ostreatus*	0.10	0.87	16.19	28.49	9.93	18.15
	P. djamor	0.08	0.64	14.77	27.15	8.73	15.76
SD	*P. ostreatus*	0.04	0.16	1.94	3.89	1.26	2.46
	P. djamor	0.02	0.19	1.30	3.74	1.18	2.34
Skewness	*P. ostreatus*	−0.68	−0.16	0.17	0.13	−0.21	−0.23
	P. djamor	0.05	0.17	−0.85	0.18	1.07	0.22
Kurtosis	*P. ostreatus*	0.82	0.62	1.31	0.31	−0.02	0.23
	P. djamor	−0.70	−0.97	0.62	−1.88	1.76	−0.35
Safe Limit [28,29]	-	0.10	2.30	40.00	425.00	30.00	50.00

The same letters (a–d) indicate no significant difference between the sampling location values at $p < 0.05$. *Bdl*: below detectable limits.

Annually, the Rajaji National Park witnesses spring water overflowing on the surface; this type of water often transports debris enclosing contaminants and other pollutants according to an earlier report from Penn State University [34], which may explain the unsafe Cd content found in several samples. Cd is considered one of the most dangerous metals found in mushrooms [35] and can seriously affect kidneys, and the human respiratory and reproductive systems [36,37]. Cr, Cu, Fe, and Zn contents found in *P. ostreatus* and *P. djamor* samples collected from all locations were far below the safe limits; [28,29] being in the following respective ranges: 0.57–1.14 mg/kg and 0.38–0.95 mg/kg, 13.52–19.20 mg/kg and 12.08–16.26 mg/kg, 21.88–35.66 mg/kg and 22.92–32.07 mg/kg, 7.80–12.05 mg/kg and 7.18–11.32 mg/kg, 13.56–22.10 mg/kg and 12.02–19.58 mg/kg. These metals are essential for human body nutrition and development when found in low contents. Skewness and Kurtosis values were all between −2 and +2 which shows an acceptable asymmetry and a normal univariate distribution [38].

3.2. Multivariate Analysis Results

The multivariate analysis provides better insights into the understanding of randomly distributed data sets [23]. In this study, principal component and hierarchal cluster analyses were used to derive the relationship and similarities between the heavy metal contents in two *Pleurotus* spp. and their sampling locations in and around Rajaji National Park, Haridwar, India. Results revealed that by using the PCA tool, the randomly distributed data was orthogonally transformed onto two principal components namely PC1 and PC2 for both *Pleurotus* spp. The standardized score plot given in Figure 2a showed the relative dominance of different PCs within the sampling locations. In particular, the PCs of *P. ostreatus* had eigenvalues of 16.50 and 6.90 covering the variances of 62.06 and 25.96%, respectively Figure 2b. For *P. ostreatus*, the axial vector length of Fe was highly correlated for Rishikesh Canal Road, Bheemgoda Barrage Forest, and Sapt Rishi Ghat, sampling sites while Zn and Cu dominated the Bijnor Canal Road location. Herein, the Mansa Devi Forest location showed the lowest heavy-metal contents For *P. ostreatus*, which suggested the safest location for *P. ostreatus* cultivation. Figure 2c shows the hierarchal cluster dendrogram with heatmap for heavy-metal contents in *P. ostreatus* collected from different locations of Rajaji National Park, Haridwar, India. Hierarchal cluster analysis showed that Mn–Fe and Cd–Zn had the highest similarities in terms of analyzed heavy-metals concentration in *P. ostreatus* mushroom. Also, similarities were identified between different sampling locations such as Sureshwari Devi Forest–Chilla Forest Colony and Devpura Forest Colony–Rishikesh Canal Road, respectively.

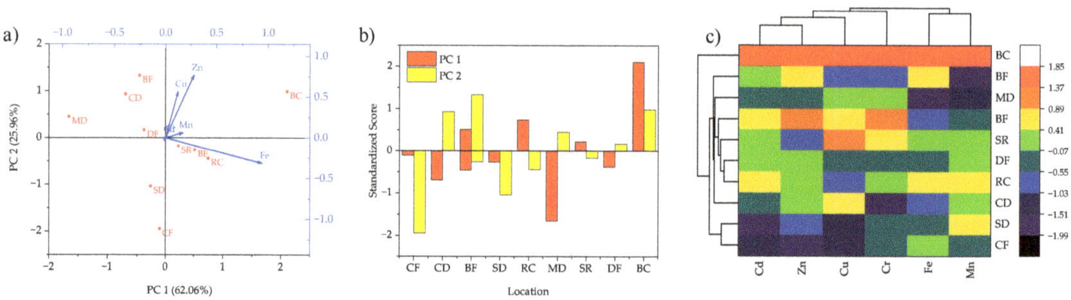

Figure 2. (**a**) Axial biplot; (**b**) PCA-standardized scores; and (**c**) HCA dendrogram with heatmap for heavy metal contents in *P. ostreatus* collected from different locations of Rajaji National Park, Haridwar, India.

On the other hand, the extracted eigenvalues of PC1 and PC2 for heavy metal contents in *P. djamor* samples were observed as 15.30 and 4.80 depicting 67.86 and 21.29% variance, respectively. As given in Figure 3a, the vector length showed the highest concentration of Fe at Bijnor Canal Road, Rishikesh Canal Road, Sapt Rishi Ghat, and Chilla Forest Colony. Moreover, Zn was observed highest at Bijnor Canal Road, while other heavy metals showed relatively similar vector lengths at different sampling locations. Figure 3b shows the distribution of PCs over different sampling locations and confirmed that values of some scores positively dominated PC1 and PC2. Similarly, Figure 3c shows that Mn–Fe, Zn–Cr, and Cd–Cu lie within the same clusters due to their similarities in terms of analyzed heavy metal concentrations in *P. djamor* samples. However, the similarities between sampling locations were not effectively established due to random distribution except for Sapt Rishi Ghat–Rishikesh Canal Road and Bheemgoda Barrage Forest–Chandi Devi Forest.

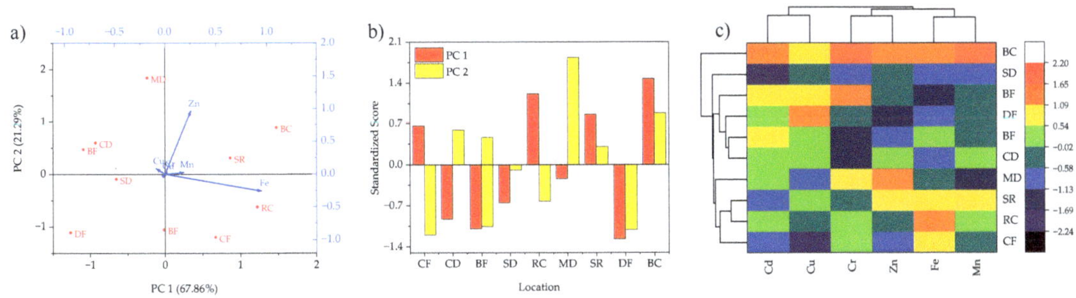

Figure 3. (**a**) Axial biplot; (**b**) PCA-standardized scores and (**c**) HCA dendrogram with heatmap for heavy metal contents in *P. djamor* collected from different locations of Rajaji National Park, Haridwar, India.

Multivariate statistical analyses are one of the most effective methods for generalizing randomly distributed large data sets. In a recent study by Kumar et al. [12], the contents of heavy metals in *Agaricus bisporus* mushroom collected from different locations in Uttarakhand State, India were analyzed using geospatial and multivariate statistical tools i.e., principal component analysis and cluster analysis. They found that both PCA and HCA helped identify the districts which had the highest heavy-metal contamination across the state. Similarly, Sarikurkcu et al. [39] also analyzed the heavy-metal concentrations in eight selected *Russula* spp. from Turkey. They found that the variability of essential and non-essential heavy metals was successfully depicted using PCA and HCA. Moreover, Barea-Sepúlveda et al. [40] also utilized PCA and HCA tools to understand the variabilities of toxic heavy metals occurrence in *Macrolepiota procera* mushrooms collected from different locations in Spain and Morocco. Thus, these reports suggested that PCA and HCA can be useful for understanding the heavy metal variabilities in mushroom samples.

3.3. Results of Health Risk Studies

As wild mushroom consumption has increased in numerous regions of the world, especially in developing countries, the need to investigate any possible health risks has become indispensable. Therefore, this can be predicted by studying the target hazard quotient (THQ) and health risk index (HRI) in adult and child groups for a better insight into any possible toxicity on human health [6]. Here, Table 3 summarizes the values of adult and child THQs for each heavy metal found in *P. ostreatus* and *P. djamor*, as well as their resultant HRIs. It was revealed that child THQs were always higher than the adult and that THQs of *P. ostreatus* were higher compared to those of *P. djamor*. A similar observation was recently acknowledged by Kumar et al. [12] regarding *Agaricus bisporus* mushrooms collected from Uttarakhand state markets. The study of HRIs on *Tricholoma spp.* revealed that they were higher in mushroom caps than in mushroom stipes [9]. Conversely, the current study evaluated THQs and HRIs of whole *Pleurotus spp.* mushrooms; therefore, further studies should be put into action to detect any possible potential risk associated with a specific part of the edible mushrooms. The average child HRIs of *P. ostreatus* and *P. djamor* was 2.33 fold higher than the average adult HRIs in the same species. Moreover, child and adult HRIs were higher by 1.1–1.4-fold in *P. ostreatus* than in *P. djamor*, as shown in Figure 4. All HRIs were < 1 outlining a safe consumption without potential health risk [41], except child HRI of *P. ostreatus*, collected from Bijnor Canal Road (1.151), which was mostly attributed to the high Cd content.

Table 3. Health risk index of heavy metal contents collected from different locations of Rajaji National Park, Haridwar, India.

Sampling Site	*Pleurotus* spp.	Target Hazard Quotient (THQ)											
		Cd		Cr		Cu		Fe		Mn		Zn	
		Child	Adult	Child	Adult	Child	Adult	Child	Adult	Child	Adult	Child	Adult
Chilla Forest Colony	*P. ostreatus*	0.065	0.028	0.141	0.060	0.182	0.078	0.023	0.010	0.360	0.154	0.024	0.010
	P. djamor	0.054	0.023	0.132	0.056	0.156	0.067	0.023	0.010	0.314	0.135	0.025	0.011
Chandi Devi Forest	*P. ostreatus*	0.108	0.046	0.099	0.042	0.230	0.099	0.019	0.008	0.400	0.171	0.034	0.014
	P. djamor	0.076	0.032	0.076	0.032	0.194	0.083	0.018	0.008	0.347	0.149	0.029	0.012
Bheemgoda Barrage Forest	*P. ostreatus*	0.119	0.051	0.130	0.056	0.186	0.080	0.024	0.010	0.301	0.129	0.036	0.015
	P. djamor	0.097	0.042	0.068	0.029	0.194	0.083	0.021	0.009	0.334	0.143	0.024	0.010
Sureshwari Devi Forest	*P. ostreatus*	0.032	0.014	0.146	0.063	0.174	0.075	0.022	0.009	0.422	0.181	0.029	0.013
	P. djamor	0.000	0.000	0.081	0.035	0.183	0.078	0.019	0.008	0.292	0.125	0.027	0.012
Rishikesh Canal Road	*P. ostreatus*	0.141	0.060	0.160	0.069	0.194	0.083	0.024	0.010	0.413	0.177	0.034	0.014
	P. djamor	0.087	0.037	0.115	0.049	0.182	0.078	0.025	0.011	0.350	0.150	0.028	0.012
Mansa Devi Forest	*P. ostreatus*	0.097	0.042	0.155	0.066	0.209	0.089	0.017	0.007	0.321	0.138	0.031	0.013
	P. djamor	0.076	0.032	0.139	0.059	0.175	0.075	0.020	0.008	0.277	0.119	0.035	0.015
Sapt Rishi Ghat	*P. ostreatus*	0.119	0.051	0.167	0.072	0.237	0.102	0.023	0.010	0.390	0.167	0.030	0.013
	P. djamor	0.054	0.023	0.112	0.048	0.194	0.083	0.023	0.010	0.377	0.161	0.031	0.013
BHEL Forest Colony	*P. ostreatus*	0.141	0.060	0.181	0.077	0.222	0.095	0.020	0.008	0.364	0.156	0.038	0.016
	P. djamor	0.108	0.046	0.153	0.066	0.208	0.089	0.018	0.008	0.312	0.134	0.028	0.012
Devpura Forest Colony	*P. ostreatus*	0.119	0.051	0.137	0.059	0.204	0.088	0.021	0.009	0.396	0.170	0.033	0.014
	P. djamor	0.087	0.037	0.103	0.044	0.209	0.090	0.018	0.008	0.331	0.142	0.022	0.009
Bijnor Canal Road	*P. ostreatus*	0.173	0.074	0.198	0.085	0.247	0.106	0.028	0.012	0.465	0.199	0.040	0.017
	P. djamor	0.119	0.051	0.171	0.073	0.207	0.089	0.025	0.011	0.437	0.187	0.034	0.015

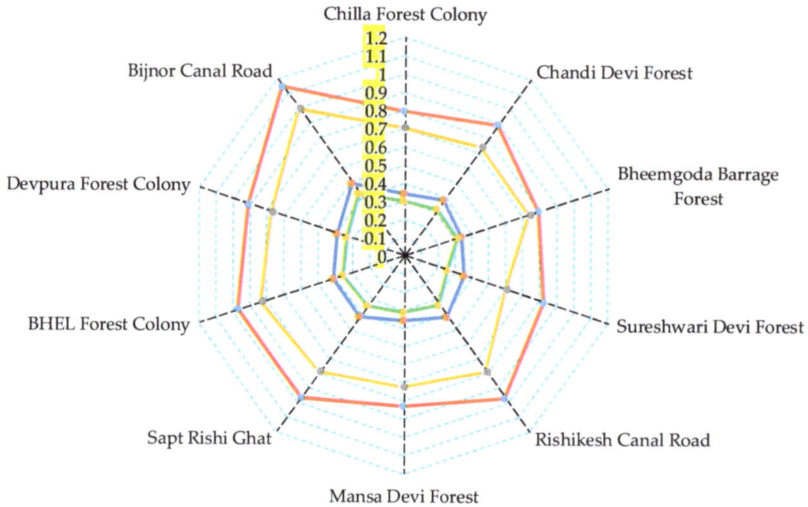

Figure 4. Spider-web diagram showing the health risk index (HRI) values of heavy metal contents in two oyster mushroom spp. (PO: *P. ostreatus* and PD: *P. djamor*) collected from different locations of Rajaji National Park, Haridwar, India.

This region is currently heavily affected by the increasing activities of neighboring mines besides the uprising water canal pollution, although a restoration process took place in 2014. Accordingly, children's intake of *P. ostreatus* from Bijnor Canal Road should be avoided as they may pose a serious effect on the development of their central nervous system leading to learning disabilities besides possible cancerogenic impacts [42]. For *P. ostreatus*, child HRIs' in studied locations followed this decreasing order: Bijnor Canal Road > Rishikesh Canal Roa –Sapt Rishi Gat > BHEL Forest Colony > Devpura Forest Colony > Chandi Devi Forest > Mansa Devi Forest > Sureshwari Devi Forest > Bheemgoda Barrage Forest > Chilla Forest Colony; whereas, adult HRIs' decreasing order was as follows: Bijnor Canal Road > Rishikesh Canal Road–Sapt Rishi Gat > BHEL Forest Colony > Devpura Forest Colony > Chandi Devi Forest > Mansa Devi Forest > Sureshwari Devi Forest > Bheemgoda Barrage Forest–Chilla Forest Colony.

For *P. djamor*, the following child HRIs trend was noted: Bijnor Canal Road > BHEL Forest Colony > Sapt Rishi Gat > Rishikesh Canal Road > Devpura Forest Colony > Bheemgoda Barrage Forest > Chandi Devi Forest > Mansa Devi Forest > Chilla Forest Colony > Sureshwari Devi Forest. Whereas, adult HRIs had the following decreasing order: Bijnor Canal Road > BHEL Forest Colony > Sapt Rishi Gat > Rishikesh Canal Road > Devpura Forest Colony > Chandi Devi Forest–Bheemgoda Barrage Forest > Mansa Devi Forest > Chilla Forest Colony > Sureshwari Devi Forest. The obtained findings simulate that HRI values were not only species-related, but also soil-related (mainly Cd contamination). Other recent studies outlined the strong inter-relationship between soil mercury (Hg) content and HRI values in wild edible mushrooms [40,43,44]. Thus, the THQ approach in the current study was helpful to estimate potential health risks associated with consumption of heavy metal contaminated *Pleurotus* spp.

4. Conclusions

This study concluded that two oyster mushrooms (*P. ostreatus* and *P. djamor*) collected from the Rajaji National Park, Haridwar, India showed the occurrence of selected heavy metals (Cd, Cu, Cr, Fe, Mn, and Zn). Comparatively, the contents of six heavy metals were found at maximum in *P. ostreatus* followed by *P. djamor*. However, the concentration varied significantly ($p < 0.05$) based on the sampling locations. Overall, the decreasing order of mean heavy-metal concentration in both *Pleurotus* spp. was Fe > Zn > Cu > Mn > Cr > Cd. The findings revealed that the samples collected from locations close to the residential, industrial areas, and transportation activities showed relatively higher heavy-metal contents as compared to those collected from the deep forest areas as indicated by the multivariate analyses. Similarly, the health risk index studies also indicated safe consumption of *Pleurotus* spp. collected from all locations except for those near heavy traffic and industrial areas. Considering the importance of wild edible mushrooms in human nutrition, proper measures should be taken in terms of heavy-metal contamination and associated health risks. Further studies on the biomonitoring of heavy metals in other edible mushroom species are highly recommended.

Author Contributions: Conceptualization, I.Š., P.K. and V.K.; Data curation, P.K.; Formal analysis, P.K.; Funding acquisition, M.A.T. and E.M.E.; Investigation, P.K.; Methodology, P.K.; Project administration, E.M.E.; Resources, V.K.; Software, P.K.; Validation, I.Š., P.K., B.A., S.A.F., R.K.B., A.B., F.O.A., V.K., M.A.T. and E.M.E.; Writing—original draft, P.K. and S.A.F.; Writing—review & editing, I.Š., P.K., B.A., R.K.B., A.B., F.O.A., V.K., M.A.T. and E.M.E. All authors have read and agreed to the published version of the manuscript.

Funding. This research was funded by King Khalid University (grant number RGP. 1/102/43), Abha, Saudi Arabia.

Institutional Review Board Statement: Not applicable.

Informed Consent Statement: Not applicable.

Data Availability Statement: Not applicable.

Acknowledgments: The authors are grateful to their host institutes for providing the necessary facilities to conduct this study. All individuals included in this section have consented to the acknowledgment. M.A.T. extends his appreciation to King Khalid University for funding this work through the Research Group Project under grant number RGP. 1/182/43.

Conflicts of Interest: The authors declare no conflict of interest.

References

1. Alam, A.; Agrawal, K.; Verma, P. Fungi and Its By-Products in Food Industry: An Unexplored Area. In *Microbial Products for Health, Environment and Agriculture*; Springer: Singapore, 2021; pp. 103–120.
2. Fortune Business Insights. *Mushroom Market Size, Share & Industry Analysis, by Type, Form and Regional Forecast 2019-2026*; Fortune Business Insights Pvt. Ltd.: Pune, India, 2020; Volume 21.
3. Patra, A.; Mukherjee, A.K. Mushroom Mycetism—A Neglected and Challenging Medical Emergency in the Indian Subcontinent: A Road Map for Its Prevention and Treatment. *Toxicon* **2022**, *217*, 56–77. [CrossRef] [PubMed]
4. Nagulwar, M.M.; More, D.R.; Mandhare, L.L. Nutritional Properties and Value Addition of Mushroom: A Review. *Pharma Innov. J.* **2020**, *9*, 395–398. [CrossRef]
5. Pieroni, A.; Houlihan, L.; Ansari, N.; Hussain, B.; Aslam, S. Medicinal Perceptions of Vegetables Traditionally Consumed by South-Asian Migrants Living in Bradford, Northern England. *J. Ethnopharmacol.* **2007**, *113*, 100–110. [CrossRef] [PubMed]
6. Kumar, P.; Kumar, V.; Adelodun, B.; Bedeković, D.; Kos, I.; Širić, I.; Alamri, S.A.M.; Alrumman, S.A.; Eid, E.M.; Abou Fayssal, S.; et al. Sustainable Use of Sewage Sludge as a Casing Material for Button Mushroom (*Agaricus bisporus*) Cultivation: Experimental and Prediction Modeling Studies for Uptake of Metal Elements. *J. Fungi* **2022**, *8*, 112. [CrossRef]
7. Kumar, P.; Eid, E.M.; Al-Huqail, A.A.; Širić, I.; Adelodun, B.; Abou Fayssal, S.; Valadez-Blanco, R.; Goala, M.; Ajibade, F.O.; Choi, K.S.; et al. Kinetic Studies on Delignification and Heavy Metals Uptake by Shiitake (*Lentinula edodes*) Mushroom Cultivated on Agro-Industrial Wastes. *Horticulturae* **2022**, *8*, 316. [CrossRef]
8. Verma, R.K.; Pandro, V.; Mishra, S.N.; Raj, D.; Asaiya, A.J.K. Sal Forest: A Source of Wild Edible Mushrooms for Livelihood Support to Tribal People of Dindori District, Madhya Pradesh, India. *Int. J. Curr. Microbiol. Appl. Sci.* **2019**, *8*, 563–575. [CrossRef]
9. Širić, I.; Kumar, P.; Eid, E.M.; Bachheti, A.; Kos, I.; Bedeković, D.; Mioč, B.; Humar, M. Occurrence and Health Risk Assessment of Cadmium Accumulation in Three Tricholoma Mushroom Species Collected from Wild Habitats of Central and Coastal Croatia. *J. Fungi* **2022**, *8*, 685. [CrossRef]
10. Širić, I.; Kasap, A.; Bedeković, D.; Falandysz, J. Lead, Cadmium and Mercury Contents and Bioaccumulation Potential of Wild Edible Saprophytic and Ectomycorrhizal Mushrooms, Croatia. *J. Environ. Sci. Health B* **2017**, *52*, 156–165. [CrossRef]
11. Orywal, K.; Socha, K.; Nowakowski, P.; Zon, W.; Kaczynski, P.; Mroczko, B.; Lozowicka, B.; Perkowski, M. Health Risk Assessment of Exposure to Toxic Elements Resulting from Consumption of Dried Wild-Grown Mushrooms Available for Sale. *PLoS ONE* **2021**, *16*, e0252834. [CrossRef]
12. Kumar, P.; Kumar, V.; Eid, E.M.; Al-Huqail, A.A.; Adelodun, B.; Abou Fayssal, S.; Goala, M.; Arya, A.K.; Bachheti, A.; Andabaka, Ž.; et al. Spatial Assessment of Potentially Toxic Elements (PTE) Concentration in Agaricus Bisporus Mushroom Collected from Local Vegetable Markets of Uttarakhand State, India. *J. Fungi* **2022**, *8*, 452. [CrossRef]
13. Abou Fayssal, S.; Alsanad, M.A.; Yordanova, M.H.; el Sebaaly, Z.; Najjar, R.; Sassine, Y.N. Effect of Olive Pruning Residues on Substrate Temperature and Production of Oyster Mushroom (*Pleurotus ostreatus*). *Acta Hortic.* **2021**, *1327*, 245–252. [CrossRef]
14. Sassine, Y.N.; Naim, L.; el Sebaaly, Z.; Abou Fayssal, S.; Alsanad, M.A.; Yordanova, M.H. Nano Urea Effects on Pleurotus Ostreatus Nutritional Value Depending on the Dose and Timing of Application. *Sci. Rep.* **2021**, *11*, 5588. [CrossRef] [PubMed]
15. Abou Fayssal, S.; el Sebaaly, Z.; Alsanad, M.A.; Najjar, R.; Yordanova, M.H.; Sassine, Y.N. Combined Effect of Olive Pruning Residues and Spent Coffee Grounds on Pleurotus Ostreatus Production, Composition, and Nutritional Value. *PLoS ONE* **2021**, *16*, e0255794. [CrossRef] [PubMed]
16. Sithole, S.C.; Agboola, O.O.; Mugivhisa, L.L.; Amoo, S.O.; Olowoyo, J.O. Elemental Concentration of Heavy Metals in Oyster Mushrooms Grown on Mine Polluted Soils in Pretoria, South Africa. *J. King Saud Univ. Sci.* **2022**, *34*, 101763. [CrossRef]
17. Mleczek, M.; Budka, A.; Siwulski, M.; Mleczek, P.; Budzyńska, S.; Proch, J.; Gąsecka, M.; Niedzielski, P.; Rzymski, P. A Comparison of Toxic and Essential Elements in Edible Wild and Cultivated Mushroom Species. *Eur. Food Res. Technol.* **2021**, *247*, 1249–1262. [CrossRef]
18. Harihar, A.; Pandav, B.; Goyal, S.P. Density of Leopards (Panthera Pardus) in the Chilla Range of Rajaji National Park, Uttarakhand, India. *Mammalia* **2009**, *73*, 68–71. [CrossRef]
19. Joshi, R.; Singh, R. Feeding Behaviour of Wild Asian Elephants (*Elephas maximus*) in the Rajaji National Park. *J. Am. Sci.* **2008**, *4*, 34–48.
20. Kothiyal, G.; Singh, K.; Kumar, A.; Juyal, P.; Guleri, S. Wild Macrofungi (Mushrooms) Diversity Occurrence in the Forest of Uttarakhand, India. *Biodiv. Res.* **2019**, *53*, 7–32.
21. Singh, G.; Mishra, N.; Thakural, L.N.; Kumar, S. Statistical Downscaling of Climate Change Scenarios of Rainfall in Haridwar District of Uttarakhand, India. In *Smart Technologies for Energy, Environment and Sustainable Development*; Springer: Singapore, 2022; pp. 131–142.
22. Rajarathnam, S.; Bano, Z. Pleurotus Mushrooms. Part I A. Morphology, Life Cycle, Taxonomy, Breeding, and Cultivation. *C R C Crit. Rev. Food Sci. Nutr.* **1987**, *26*, 157–223. [CrossRef]

23. Kumari, B.; Sharma, V.P.; Barh, A.; Upadhyay, R.C. *Mushroom Wealth of India*; M/S Bishen Singh Mahendra Pal Singh: Dehradun, India, 2022.
24. Elbagermi, M.A.; Edwards, H.G.M.; Alajtal, A.I. Monitoring of Heavy Metal Content in Fruits and Vegetables Collected from Production and Market Sites in the Misurata Area of Libya. *ISRN Anal. Chem.* **2012**, *2012*, 827645. [CrossRef]
25. Abrham, F.; Gholap, A.V. Analysis of Heavy Metal Concentration in Some Vegetables Using Atomic Absorption Spectroscopy. *Pollution* **2021**, *7*, 205–216.
26. Lion, G.N.; Olowoyo, J.O. Population Health Risk Due to Dietary Intake of Toxic Heavy Metals from Spinacia Oleracea Harvested from Soils Collected in and around Tshwane, South Africa. *South Afr. J. Bot.* **2013**, *88*, 178–182. [CrossRef]
27. Mahabadi, M. Assessment of Heavy Metals Contamination and the Risk of Target Hazard Quotient in Some Vegetables in Isfahan. *Pollution* **2020**, *6*, 69–78.
28. Sinha, S.K.; Upadhyay, T.K.; Sharma, S.K. Heavy Metals Detection in White Button Mushroom (*Agaricus bisporus*) Cultivated in State of Maharashtra, India. *Biochem. Cell Arch.* **2019**, *19*, 3501–3506. [CrossRef]
29. USEPA. Integrated Risk Information System. Available online: https://www.epa.gov/iris (accessed on 31 August 2022).
30. Zhu, F.; Qu, L.; Fan, W.; Qiao, M.; Hao, H.; Wang, X. Assessment of Heavy Metals in Some Wild Edible Mushrooms Collected from Yunnan Province, China. *Environ. Monit. Assess.* **2011**, *179*, 191–199. [CrossRef]
31. Kumari, S.; Kothari, R.; Kumar, V.; Kumar, P.; Tyagi, V.V. Kinetic Assessment of Aerobic Composting of Flower Waste Generated from Temple in Jammu, India: A Lab-Scale Experimental Study. *Environ. Sustain.* **2021**, *4*, 393–400. [CrossRef]
32. Wiłkomirski, B.; Sudnik-Wójcikowska, B.; Galera, H.; Wierzbicka, M.; Malawska, M. Railway Transportation as a Serious Source of Organic and Inorganic Pollution. *Water Air Soil Pollut.* **2011**, *218*, 333–345. [CrossRef]
33. Kupka, D.; Kania, M.; Pietrzykowski, M.; Łukasik, A.; Gruba, P. Multiple Factors Influence the Accumulation of Heavy Metals (Cu, Pb, Ni, Zn) in Forest Soils in the Vicinity of Roadways. *Water Air Soil Pollut.* **2021**, *232*, 194. [CrossRef]
34. Fergus, C. The Science of Spring Water. *Res. Penn State* **2002**, *23*. Available online: https://www.psu.edu/news/research/story/science-spring-water/ (accessed on 1 September 2022).
35. Gebrelibanos, M.; Megersa, N.; Taddesse, A.M. Levels of Essential and Non-Essential Metals in Edible Mushrooms Cultivated in Haramaya, Ethiopia. *Int. J. Food Contam.* **2016**, *3*, 2. [CrossRef]
36. WHO. *Exposure to Cadmium: A Major Public Health Concern*; WHO: Geneva, Switzerland, 2019.
37. Genchi, G.; Sinicropi, M.S.; Lauria, G.; Carocci, A.; Catalano, A. The Effects of Cadmium Toxicity. *Int. J. Environ. Res. Public Health* **2020**, *17*, 3782. [CrossRef] [PubMed]
38. Darren, G.; Paul, M. *SPSS for Windows Step by Step: A Simple Guide and Reference*; Pearson Education, Inc.: Boston, MA, USA, 1999; p. 386.
39. Sarikurkcu, C.; Akata, I.; Tepe, B. Metal Concentration and Health Risk Assessment of Eight Russula Mushrooms Collected from Kizilcahamam-Ankara, Turkey. *Environ. Sci. Pollut. Res.* **2021**, *28*, 15743–15754. [CrossRef] [PubMed]
40. Barea-Sepúlveda, M.; Espada-Bellido, E.; Ferreiro-González, M.; Bouziane, H.; López-Castillo, J.G.; Palma, M.; Barbero, F.G. Toxic Elements and Trace Elements in Macrolepiota Procera Mushrooms from Southern Spain and Northern Morocco. *J. Food Compos. Anal.* **2022**, *108*, 104419. [CrossRef]
41. Leung, A.O.W.; Duzgoren-Aydin, N.S.; Cheung, K.C.; Wong, M.H. Heavy Metals Concentrations of Surface Dust from E-Waste Recycling and Its Human Health Implications in Southeast China. *Environ. Sci. Technol.* **2008**, *42*, 2674–2680. [CrossRef] [PubMed]
42. Tchounwou, P.B.; Yedjou, C.G.; Patlolla, A.K.; Sutton, D.J. Heavy Metal Toxicity and the Environment. In *Molecular, Clinical and Environmental Toxicology*; Luch, A., Ed.; Springer: Basel, Switzerland, 2012; Volume 101, pp. 133–164.
43. Pecina, V.; Valtera, M.; Trávníčková, G.; Komendová, R.; Novotný, R.; Brtnický, M.; Juřička, D. Vertical Distribution of Mercury in Forest Soils and Its Transfer to Edible Mushrooms in Relation to Tree Species. *Forests* **2021**, *12*, 539. [CrossRef]
44. Barea-Sepúlveda, M.; Espada-Bellido, E.; Ferreiro-González, M.; Bouziane, H.; López-Castillo, J.G.; Palma, M.; Barbero, F.G. Exposure to Essential and Toxic Elements via Consumption of Agaricaceae, Amanitaceae, Boletaceae, and Russulaceae Mushrooms from Southern Spain and Northern Morocco. *J. Fungi* **2022**, *8*, 545. [CrossRef]

Journal of *Fungi*

MDPI

Article

Mycoremediation of Flotation Tailings with *Agaricus bisporus*

Sylwia Budzyńska [1,*], Marek Siwulski [2], Anna Budka [3], Pavel Kalač [4], Przemysław Niedzielski [5], Monika Gąsecka [1] and Mirosław Mleczek [1]

1 Department of Chemistry, Poznan University of Life Sciences, Wojska Polskiego 75, 60-625 Poznań, Poland
2 Department of Vegetable Crops, Poznan University of Life Sciences, Dąbrowskiego 159, 60-594 Poznań, Poland
3 Department of Mathematical and Statistical Methods, Poznan University of Life Sciences, Wojska Polskiego 28, 60-637 Poznań, Poland
4 Department of Applied Chemistry, Faculty of Agriculture, University of South Bohemia, 370 04 České Budějovice, Czech Republic
5 Faculty of Chemistry, Adam Mickiewicz University in Poznań, Uniwersytetu Poznańskiego 8, 61-614 Poznań, Poland
* Correspondence: sylwia.budzynska@up.poznan.pl; Tel.: +48-61-848-7849; Fax: +48-61-848-7824

Abstract: Due to their enzymatic and bioaccumulation faculties the use of macromycetes for the decontamination of polluted matrices seems reasonable for bioremediation. For this reason, the aim of our study was to evaluate the mycoremediation ability of *Agaricus bisporus* cultivated on compost mixed with flotation tailings in different quantities (1, 5, 10, 15, and 20% addition). The biomass of the fruit bodies and the content of 51 major and trace elements were determined. Cultivation of *A. bisporus* in compost moderately polluted with flotation tailings yielded significantly lower (the first flush) and higher (the second flush) biomass of fruit bodies, compared with the control treatment. The presence of toxic trace elements did not cause any visible adverse symptoms for *A. bisporus*. Increasing the addition of flotation tailings to the compost induced an elevated level of most determined elements. A significant increase in rare earth elements (both flushes) and platinum group elements (first flush only) was observed. The opposite situation was recorded for major essential elements, except for Na and Mg in *A. bisporus* from the second flush under the most enriched compost (20%). Nevertheless, calculated bioaccumulation factor values showed a selective accumulation capacity—limited for toxic elements (except for Ag, As, and Cd) and the effective accumulation of B, Cu, K, and Se. The obtained results confirmed that *A. bisporus* can be used for practical application in mycoremediation in the industry although this must be preceded by larger-scale tests. This application seems to be the most favorable for media contaminated with selected elements, whose absorption by fruiting bodies is the most efficient.

Keywords: accumulation; basidiomycete; bioremediation; champignon; common mushroom; edible mushroom; sludge; toxic elements; wastes

Citation: Budzyńska, S.; Siwulski, M.; Budka, A.; Kalač, P.; Niedzielski, P.; Gąsecka, M.; Mleczek, M. Mycoremediation of Flotation Tailings with *Agaricus bisporus*. *J. Fungi* **2022**, *8*, 883. https://doi.org/10.3390/jof8080883

Academic Editors: Miha Humar and Ivan Širić

Received: 24 July 2022
Accepted: 18 August 2022
Published: 22 August 2022

1. Introduction

Mycoremediation as a form of bioremediation may be an effective, eco-friendly technique for decontamination of polluted environmental matrices because of its simplicity and highly efficient implementation process [1–7]. It is also one of the least costly forms of remediation, and both micromycetes and macromycetes may be used [8,9]. Fungi-mediated remediation as a cost-effective method may use mycelium to effectively secrete extracellular enzymes, finally transforming organic pollutants into non-toxic compounds (bioaugmentation) or accumulating toxic elements [10–12].

There are numerous literature data about biodegradation, bioconversion, or biosorption for the degradation of common pollutants using different mushroom species [13,14]. Biodegradation of polycyclic aromatic hydrocarbons by *Phanerochaete chrysosporium*, *Trametes versicolor*, *Pleurotus ostreatus*, and *P. eryngii*; pesticides and herbicides by *Botryosphaeria laricina*, *Aspergillus glaucus*, *T. pavonia*, *Penicillium spiculisporus*, and *P. verruculosum*; antibiotics and pharmaceuticals by *Leptospaherulina* sp, *Irpex lacteus*, *Lentinula edodes*, *Mucor hiemalis*, and *Phanerochaete chrysosporium* were reviewed by Akhtar and Mannan [9]. Effective biosorption of heavy metals (aluminium (Al), chromium (Cr), cobalt (Co), copper (Cu), lead (Pb), iron (Fe), nickel (Ni), manganese (Mn), and zinc (Zn)) with the use of spent *Agaricus bisporus* from production has also been described in the literature [4,15,16]. In terms of heavy metals, mycoremediation by easily cultivable, fast-growing, and highly accumulating white rot fungi *P. sajor-caju*, *A. bitorquis*, and *Ganoderma lucidum* with a potential for Cr, Cu, and Pb remediation was described by Hanif and Bhatti [17]. This method may also use macromycetes to accumulate toxic elements in their biomass [18]. An example of such a species is *Pleurotus* spp., which is known to effectively accumulate selected trace elements in whole fruit bodies [19]. This method limits the initial toxicity of elements after their accumulation without the risk of the production of toxic metabolites, which are usually present in bioremediation with microbes [20].

It should be remembered that despite many advantages, biological methods of environmental remediation also have limiting factors. In the review of Boopathy [20], various factors that limit the use of bioremediation technologies were summarized. Some information can be found in the literature on the critical aspects and limitations of the use of mycoremediation. They should not be forgotten and strategies to overcome them are necessary [9]. One of the examples of limiting factors is the reduced bioavailability of pollutants. Puglisi et al. [21] observed that some fungi overcome this limitation by the production of unique proteins (hydrophobins), due to their ability to dissolve hydrophobic molecules into aqueous media. For an effective process, optimal conditions should be present (temperature: 10–28 °C, pH: about 6.5, humidity: 86–90%, and CO_2 level: 15–20%) [22,23]. In the case of the in situ process, maintaining the above-mentioned conditions can be difficult, which is another limitation. Rubichaud et al. [24] confirmed that cold environmental conditions can impede the activity of various fungal enzymes necessary to degrade toxic pollutants. Therefore, the selection of suitable macrofungi for the particular substrate is essential. A high rate of element accumulation combined with a more frequent harvest cycle is a clear argument for using this method in practice [25].

A separate issue is the efficiency of metal accumulation by fruit bodies related to their concentration in naturally and artificially polluted soil [26,27]. Sithole et al. [28] studied accumulation in mushrooms growing around three mining areas and reported that heavy metal contents can be significantly different. This confirms that both element concentration and soil chemistry influence the bioavailability of metals and their contents in fruit bodies. It seems that the enrichment of samples may be diverse, which is finally related to the potential toxicity of mushrooms and differing contents of essential inorganic elements [29].

Among the many industrial activities with negative environmental effects, the production of hazardous wastes poses serious environmental and social problems around the world. One group of such wastes is metal processing tailings. Flotation is a common mineral processing method used to upgrade copper sulfide ores where copper mineral particles are concentrated in froth, and associated gangue minerals are separated as tailings [30–33]. According to Zhai et al. [34], 60 million tons of Cu slag are generated annually worldwide during flotation and cause irreversible water and soil pollution. Finding an environmentally friendly remediation technology is crucial. There is potential for transforming tailing wastes into valuable products due to their considerable concentrations of many critical metals/metalloids. The recovery of elements and the use of the mineral residues in high- and low-value products can be very profitable from an industrial point of view (for producers of these pollutants and companies experienced in these methods). The amounts of generated wastes are so significant that a combination of several different approaches (reduce, reprocess, upcycle, downcycle) is needed [35]. One of the stages may be effective mycoremediation.

Agaricus bisporus is the most important commercially cultivated mushroom, contributing approximately 40–45% to world mushroom production [36,37]. Since it is so commonly cultivated, it would seem to be ideal for mycoremediation purposes. There are more and more reports of such use in the literature. Kryczyk et al. [38] presented in vitro cultures of *A. bisporus* demonstrating their remediation capacity for Cd and Pb from a supplemented medium. Kumar et al. [39] described an integrated approach for sustainable management of industrial wastewater and agricultural residues in *A. bisporus* production while minimizing the risks associated with their disposal. Ugya and Imam [40] described the effectiveness of *A. bisporus* in the remediation of refinery wastewater. The species showed high reduction efficiency for sulphate, phosphate, nitrate, alkalinity, electrical conductivity (EC), biological and chemical oxygen demands (BOD and COD), and heavy metals (Ag, Hg, Mn, Pb, and Zn).

In view of the above, the aim of this study was to evaluate the mycoremediation ability of *A. bisporus*. The content of 51 elements in the mushroom fruit bodies growing on compost mixed with highly polluted flotation tailings in different quantities (1, 5, 10, 15, and 20% addition) enriched with flotation tailing was determined. The biomass of the collected fruit bodies was also assessed to estimate how polluted materials affect mushroom development. An experiment was performed to show the mineral composition of fruit bodies after the mycoremediation process.

2. Materials and Methods

2.1. Experimental Design

The substrate used for the experimental cultivation came from the commercial production of compost for mushroom growing (WRONA Company, Pszczyna, Poland). The compost was based on wheat straw, chicken manure, and gypsum and was prepared using a conventional method characteristic for phase II compost (fermentation and pasteurization). The compost was mixed with flotation tailings in quantities of 0 (control), 1 (FT_1), 5 (FT_5), 10 (FT_{10}), 15 ($FT_{15,}$), and 20% (FT_{20}) by weight of the compost. Granulation [%] of flotation tailings was 11, 88, and 1 for clay, silt and sand, respectively. The pH of this component was 7.19, with an EC of 6.98 dS m^{-1}. The characteristics of element concentration in flotation tailings are described in Table 1.

The compost was inoculated with commercial *A. bisporus* mycelium in 5% of the compost weight. The strain EuroMycel 58 was used. The mixtures were placed in plastic containers (15 × 18 × 14 cm) at 1 kg per container for each particular experimental system (Figure 1). The compost layer thickness was 8 cm. Eight containers for each treatment were prepared. Incubation was carried out in a growing chamber at a temperature of 24–25 °C and 85–90% of humidity. Once the compost was completely overgrown with mycelium, a 3.5 cm layer of casing soil was applied to the substrate surface. The moisture content of the casing soil was 75%. The casing soil came from the Wokas Company (Łosice, Poland) and was prepared based on sphagnum peat moss with chalk (pH 6.7). Incubation was continued until the mycelium overgrew the casing soil. When it appeared on the casing soil surface, the air temperature was lowered to 17–18 °C. The casing soil was watered to maintain constant moisture content. The growing chamber was aerated to keep CO_2 concentration below 1500 ppm. The fruit bodies were collected fully developed but still completely closed. Individuals from the control and treatment FT_1 were harvested simultaneously, whereas from the other experimental systems (FT_5, FT_{10}, $FT_{15,}$ and FT_{20}) 2, 3, 4, and 5 days later, respectively. The delay in the harvesting of fruiting bodies was due to the fact that increasing the amount of sludge (waste) addition delayed their setting.

Just two flushes of fruit bodies were produced and harvested. The interval between consecutive harvests was 8 days in each case. The yield included whole fruit bodies collected from 1 container, and the determined yield was a mean value calculated based on 8 containers belonging to the same treatment. None of the fruit bodies showed any signs of distortion or discoloration (Figure 1). The experiment was performed in May 2020.

Table 1. Concentration of elements [mg kg^{-1} dry weight] in flotation tailings used in experiment.

Elements Group	Element	Concentration	Elements Group	Element	Concentration
MEEs	Ca	12,700 ± 1510	REEs	Lu	0.197 ± 0.016
	K	7920 ± 231		Nd	95.6 ± 4.76
	Mg	2210 ± 187		Pr	0.814 ± 0.037
	Na	384 ± 24.6		Sc	0.980 ± 0.102
ETEs	B	27.9 ± 4.18		Sm	0.269 ± 0.035
	Co	4.87 ± 1.04		Tb	0.196 ± 0.021
	Cr	22.2 ± 2.29		Tm	0.446 ± 0.087
	Cu	238 ± 19.8		Y	1.08 ± 0.113
	Fe	884 ± 55.7		Yb	0.127 ± 0.036
	Mn	97.7 ± 10.1		ΣREEs	123 ± 9.94
	Ni	28.3 ± 1.95			
	Se	3.77 ± 0.96	PGEs	Ir	1.96 ± 0.224
	Zn	4200 ± 1670		Pd	0.893 ± 0.055
TEWDHE	Ag	5.47 ± 1.02		Pt	11.7 ± 0.978
	As	818 ± 27.9		Rh	0.617 ± 0.042
	Ba	77.9 ± 6.64		Ru	0.135 ± 0.012
	Cd	188 ± 13.7	NNEs	Al	476 ± 35.2
	Pb	163 ± 11.0		Au	2.87 ± 0.138
	Tl	10.6 ± 1.14		Bi	1.43 ± 0.117
REEs	Ce	16.5 ± 2.01		Ga	0.201 ± 0.013
	Dy	1.23 ± 0.11		Ge	0.228 ± 0.024
	Er	2.64 ± 0.676		In	1.16 ± 0.112
	Eu	0.654 ± 0.024		Li	1.65 ± 0.097
	Gd	0.375 ± 0.016		Sb	7.65 ± 0.921
	Ho	0.127 ± 0.09		Sr	80.7 ± 5.34
	La	1.26 ± 0.076		Te	0.836 ± 0.101

MEEs—major essential elements; ETEs—essential trace elements; TEWDHE—trace elements with detrimental health effects; REEs—rare earth elements; PGEs—platinum group elements; NNEs—nutritionally non-essential elements.

Figure 1. Preparation of experimental systems.

2.2. Analytical Procedure

All collected fruit bodies were carefully washed with distilled water from a Milli-Q Academic System (non-TOC) (Merck Millipore, Darmstadt, Germany) to remove substrate particles, and subsequently dried with paper towels and weighed. The mushrooms were then dried at 40 ± 1 °C to a constant weight in an electric oven (SLW 53 STD, Pol-Eko, Poland) and ground in a laboratory Cutting Boll Mill PM 200 (Retsch, Haan, Germany).

An accurately weighed 0.200–0.500 (±0.001) g of a sample was digested with 10 mL of concentrated nitric acid (HNO_3; 65%; Sigma-Aldrich, Darmstadt, Germany) in closed Teflon containers in a microwave digestion system (Mars 6 Xpress, CEM, Matthews, NC, USA). Finally, the samples were filtered (Qualitative Filter Papers Whatman) and diluted to a volume of 15.0 mL with demineralized water (Direct-Q system, Millipore, USA). The inductively coupled plasma mass spectrometry system PlasmaQuant MS Q (Analytik Jena, Germany) was used to determine the following conditions: plasma gas flow 9.0 L min^{-1}, nebulizer gas flow 1.05 L min^{-1}, auxiliary gas flow 1.5 L min^{-1}, radio frequency (RF) power 1.35 kW. The interferences were reduced using the integrated collision reaction cell (iCRC) working sequentially in three modes: with helium (He) as the collision gas, hydrogen (H) as the reaction gas, and without gas addition. The uncertainty for the analytical procedure, including sample preparation, was at the level of 20%. The detection limits were determined at the level of 0.001–0.010 mg kg^{-1} dry weight (DW) for all elements determined (3 times the standard deviation of the blank analysis (n = 10)). The accuracy was checked by analysis of the reference materials CRM 2709—soil; CRM S-1—loess soil; CRM 667—estuarine sediments; CRM 405—estuarine sediments; CRM NCSDC (73349)—bush branches and leaves, and the recovery (80–120%) was acceptable for most of the elements determined. For uncertified elements, recovery was defined using the standard addition method.

All the determined major and trace elements were divided into 6 groups, according to Kalač (2019):

(a) Major essential elements (MEEs): calcium (Ca), potassium (K), magnesium (Mg) and sodium (Na);
(b) Essential trace elements (ETEs): boron (B), cobalt (Co), copper (Cu), chromium (Cr), iron (Fe), manganese (Mn), nickel (Ni), selenium (Se) and zinc (Zn);
(c) Trace elements with detrimental health effects (TEWDHE): silver (Ag), arsenic (As), barium (Ba), cadmium (Cd), lead (Pb) and thallium (Tl);
(d) Rare earth elements (REEs): cerium (Ce), dysprosium (Dy), erbium (Er), europium (Eu), gadolinium (Gd), holmium (Ho), lanthanum (La), lutetium (Lu), neodymium (Nd), praseodymium (Pr), scandium (Sc), samarium (Sm), terbium (Tb), thulium (Tm), yttrium (Y) and ytterbium (Yb);
(e) Platinum group elements (PGEs): iridium (Ir), palladium (Pd), platinum (Pt), rhodium (Rh), ruthenium (Ru);
(f) Nutritionally non-essential elements (NNEs): aluminium (Al), gold (Au), bismuth (Bi), gallium (Ga), germanium (Ge), indium (In), lithium (Li), rhenium (Re), antimony (Sb), strontium (Sr), and tellurium (Te).

2.3. Statistical Analysis and Calculation

All statistical analyses were performed using the Agricole package (R). The analyses were performed in accordance with the procedure implemented in the R 3.6.1 environment [41]. To compare the content of determined elements in compost with different proportions of flotation tailings, a one-way ANOVA with Tukey's HSD (statistically significant difference) post hoc: test was used. The same analysis was performed to compare the content of elements in fruit bodies growing in particular experimental treatments, separately for both flushes. This analysis used the Stat and Agricole package. A heatmap with a cluster analysis (implemented in the package Heatmaply) was performed to visualize multidimensional data separately for particular groups of elements and all elements jointly. An empirical normalization transformation brings data to the 0 to 1 scale and it allows comparison of variables of different scales, but it also keeps the shape of the distribution. A dendrogram was computed and reordered based on row and columns means. Additionally, to compare Ca, K, Mg, and Na contents together in compost with flotation tailings and fruit bodies from both flushes produced from particular experimental treatments, the rank-sum test was performed [42].

To estimate the efficiency of the element accumulation by mushrooms growing under particular treatments, the bioaccumulation factor (BAF) was calculated as a ratio of metal content in the whole fruit body dry matter to its concentration in substrate dry matter.

3. Results

3.1. Fructification and Biomass Yield of *A. bisporus*

The fastest dynamic growth of *A. bisporus* fruit bodies was observed in the control treatment in both flushes. In contrast, the slowest growth was apparent for mushrooms growing under the FT_{15} and FT_{20} treatments. Generally, the size of fruit bodies growing under FT_5 and FT_{15} treatments was greater than for the rest (including the control), which is partially visible in Figure 2. The same color characterized all the collected fruit bodies from both flushes. No negative symptoms as a result of the presence of toxic elements in flotation tailings were noted.

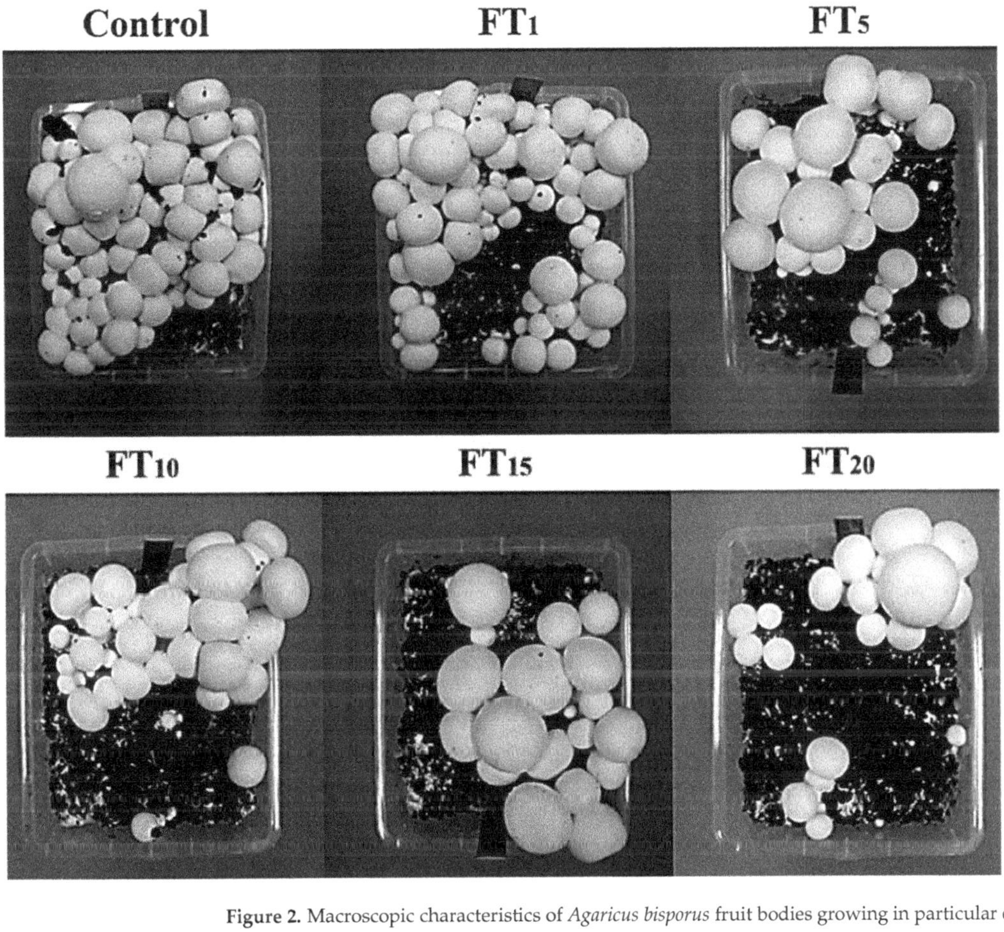

Figure 2. Macroscopic characteristics of *Agaricus bisporus* fruit bodies growing in particular experimental systems.

Within the first flush, decreasing biomass yields of 339, 268, 267, 265, 178, and 111 g in the control, FT_5, FT_{15}, FT_1, FT_{10}, and FT_{20}, respectively, were recorded (Figure 3). The determined quadratic equation ($y = -1.24x^2 - 26.3x + 349$; $R^2 = 0.6628$) indicates a clear decrease in the biomass yield with an increasing proportion of flotation tailings in the substrate. The biomass of *A. bisporus* collected from the second flush increased from the control (134 g) to treatments FT_1, FT_5, and FT_{10} with similar biomasses of 183, 194, and 191 g, respectively. The growth of fruit bodies under the FT_{15} system was related to the same biomass (134 g) as for the control, whereas the lowest biomass was observed for the FT_{20} (84.1 g). These changes are described by a quadratic equation ($y = -13.7x^2 + 84.3x + 65.8$), which reflects ($R^2 = 0.9777$) the fruit body response in relation to the increasing proportion of flotation tailings.

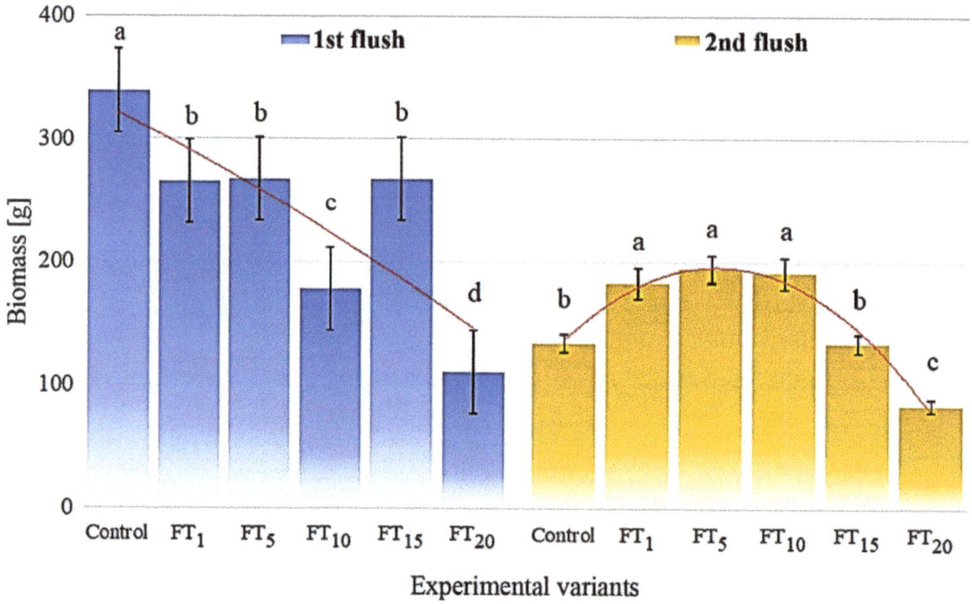

Figure 3. Biomass of *Agaricus bisporus* treated with particular flotation tailing addition; identical superscripts (a, b, c . . .) denote non-significant differences between means in columns determined in compost with flotation tailings and fruit bodies separately according to the post hoc Tukey's HSD test.

3.2. Content of Elements in Substrates

The addition of flotation tailings to the compost led to a lower mean content of MEEs from 6380 to 2550 mg kg^{-1} for Ca, from 3030 to 1560 mg kg^{-1} for K, from 726 to 282 mg kg^{-1} for Mg, and from 144 to 68.5 mg kg^{-1} for Na, for the control and FT_{20} (Table 2). Regarding the content of all MMEs in substrates, the highest similarities were observed between FT_1 and FT_5 or FT_{10} and FT_{15} (Figure 4a).

The content of ETEs in compost with flotation tailings significantly increased for all the elements included in this group except Mn, where the highest content was determined in the control and the lowest in the FT_{20} system (55.6 and 17.0 mg kg^{-1}, respectively) (Table 3). This observation confirms the heatmap, where the content of Mn in the control and FT1 belong to a separate group of objects (Figure 4b). The lowest and the highest mean contents of elements in the control and FT_{20} system were, in mg kg^{-1}, as follows: 1.32 and 4.62 (B); 0.057 and 0.623 (Co); 0.287 and 3.66 (Cr); 6.35 and 14.7 (Cu); 29.5 and 158 (Fe); 0.405 and 2.12 (Ni); 0.027 and 0.239 (Se); and also 45.2 and 781 (Zn). The highest mean contents determined in the substrate from the FT_{20} system were: 359, 1093, 1275, 231, 536, 885, and 1728% of the content for the control, respectively, for B, Co, Cr, Cu, Fe, Ni, Se, and Zn.

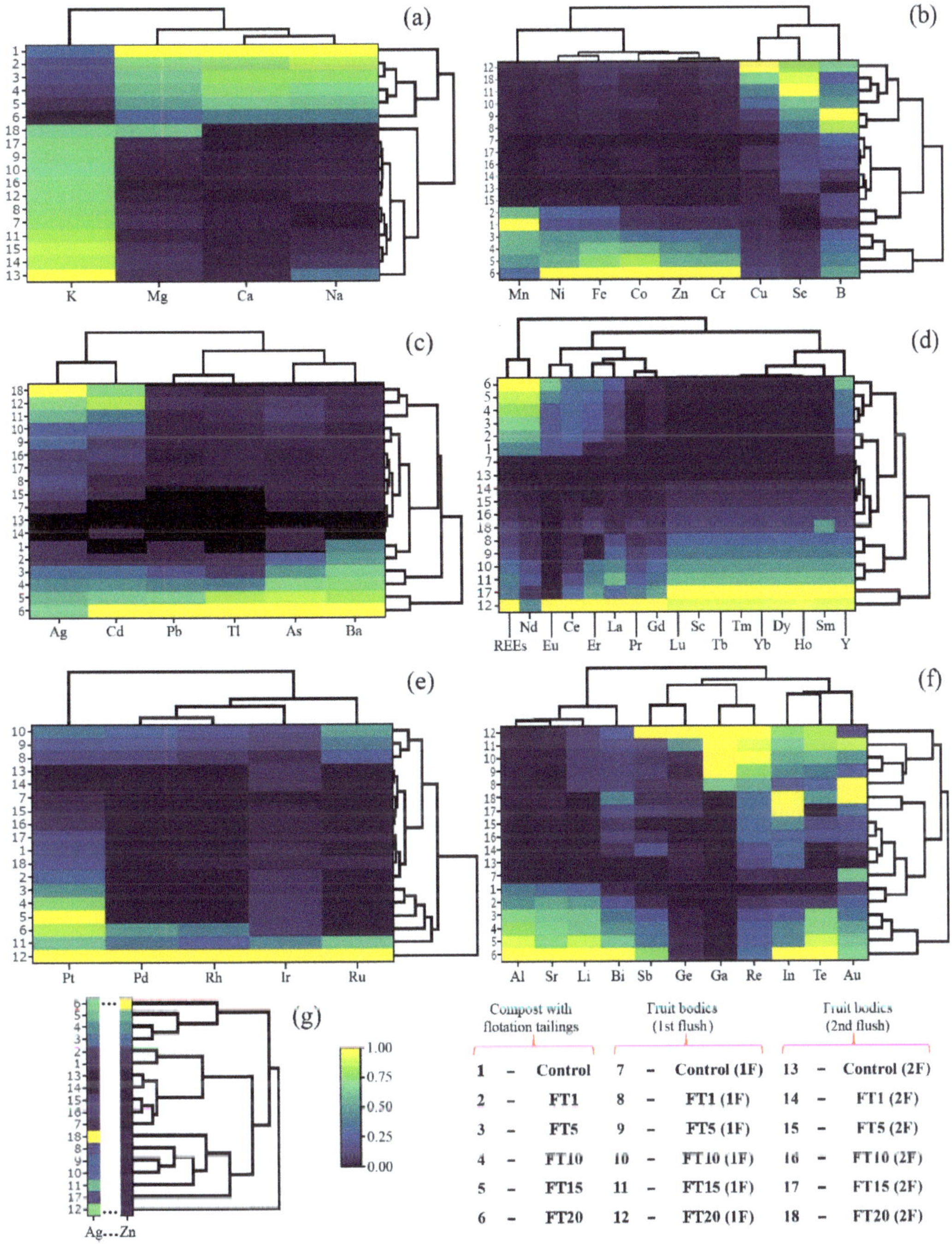

Figure 4. Correlation for compost with flotation tailings (1–6) and *Agaricus bisporus* collected from the 1st (7–12) and the 2nd flush (13–18) concerning the content of MEEs (**a**), ETEs (**b**), TEWDHE (**c**), REEs (**d**), PGEs (**e**), NNEs (**f**), and all elements jointly (**g**).

Table 2. Content of major essential elements [mg kg^{-1} dry weight] in compost with flotation tailings before the experiment and fruit bodies from particular flushes.

System	Ca	K	Mg	Na
	Compost without/with Flotation Tailings			
Control	6380 ^{a}	3030 ^{a}	726 ^{a}	144 ^{a}
FT_1	4720 ^{b}	2420 ^{ab}	536 ^{b}	124 ^{ab}
FT_5	4930 ^{b}	2350 ^{ab}	495 ^{b}	114 ^{ab}
FT_{10}	4700 ^{b}	2130 ^{bc}	474 ^{b}	101 ^{b}
FT_{15}	4020 ^{b}	1820 ^{bc}	428 ^{bc}	94.0 ^{bc}
FT_{20}	2550 ^{c}	1560 ^{c}	282 ^{c}	68.5 ^{bc}
System	1st flush			
Control	109 ^{a}	5280 ^{b}	148 ^{a}	17.5 ^{c}
FT_1	92.6 ^{b}	5150 ^{bc}	140 ^{ab}	18.0 ^{bc}
FT_5	83.1 ^{b}	4900 ^{d}	133 ^{ab}	18.3 ^{abc}
FT_{10}	51.9 ^{c}	4800 ^{e}	122 ^{b}	19.2 ^{abc}
FT_{15}	43.5 ^{c}	5540 ^{a}	120 ^{b}	20.5 ^{ab}
FT_{20}	24.6 ^{d}	5010 ^{c}	119 ^{b}	20.9 ^{a}
System	2nd flush			
Control	100 ^{a}	6420 ^{a}	148 ^{b}	62.8 ^{a}
FT_1	82.0 ^{b}	5660 ^{b}	165 ^{b}	26.3 ^{b}
FT_5	63.9 ^{c}	5770 ^{ab}	150 ^{b}	22.7 ^{b}
FT_{10}	48.2 ^{d}	4990 ^{c}	111 ^{b}	19.2 ^{b}
FT_{15}	40.6 ^{e}	4970 ^{c}	142 ^{b}	18.6 ^{b}
FT_{20}	10.4 ^{f}	4710 ^{c}	494 ^{a}	17.9 ^{b}

$n = 3$; identical superscripts (a, b, c . . .) denote non-significant differences between means in columns determined in compost with flotation tailings and fruit bodies separately according to the post hoc Tukey's HSD test.

Table 3. Content of essential trace elements [mg kg^{-1} dry weight] in compost with flotation tailings before the experiment and fruit bodies from particular flushes.

System	B	Co	Cr	Cu	Fe	Mn	Ni	Se	Zn
	Compost without/with Flotation Tailings								
Control	1.32 ^{e}	0.057 ^{e}	0.287 ^{d}	6.35 ^{d}	29.5 ^{e}	55.6 ^{a}	0.405 ^{c}	0.027 ^{c}	45.2 ^{d}
FT_1	2.03 ^{d}	0.060 ^{e}	0.351 ^{d}	6.66 ^{d}	42.8 ^{e}	33.3 ^{b}	0.611 ^{c}	0.033 ^{c}	54.0 ^{d}
FT_5	2.73 ^{c}	0.217 ^{d}	1.31 ^{c}	7.72 ^{cd}	76.6 ^{d}	32.3 ^{b}	1.01 ^{b}	0.108 ^{b}	255 ^{c}
FT_{10}	3.34 ^{bc}	0.367 ^{c}	1.50 ^{c}	9.26 ^{bc}	99.0 ^{c}	31.7 ^{b}	1.08 ^{b}	0.185 ^{a}	364 ^{b}
FT_{15}	3.91 ^{b}	0.513 ^{b}	2.07 ^{b}	11.0 ^{b}	117 ^{b}	26.6 ^{cb}	1.19 ^{b}	0.216 ^{a}	441 ^{b}
FT_{20}	4.62 ^{a}	0.623 ^{a}	3.66 ^{a}	14.7 ^{a}	158 ^{a}	17.0 ^{c}	2.12 ^{a}	0.239 ^{a}	781 ^{a}
System	1st flush								
Control	3.05 ^{e}	0.010 ^{c}	0.061 ^{c}	1.64 ^{e}	2.04 ^{d}	0.40 ^{d}	0.017 ^{c}	0.367 ^{f}	4.17 ^{d}
FT_1	7.19 ^{de}	0.019 ^{bc}	0.180 ^{b}	8.05 ^{de}	8.56 ^{c}	1.26 ^{c}	0.048 ^{bc}	1.05 ^{e}	14.6 ^{c}
FT_5	8.30 ^{d}	0.019 ^{bc}	0.176 ^{b}	14.1 ^{d}	9.00 ^{bc}	1.31 ^{bc}	0.080 ^{ab}	1.35 ^{d}	14.8 ^{c}
FT_{10}	4.43 ^{c}	0.050 ^{bc}	0.175 ^{b}	32.9 ^{c}	9.98 ^{bc}	1.40 ^{abc}	0.087 ^{ab}	1.65 ^{c}	16.1 ^{bc}
FT_{15}	3.81 ^{b}	0.025 ^{ab}	0.184 ^{b}	44.2 ^{b}	10.3 ^{b}	1.46 ^{ab}	0.105 ^{ab}	2.62 ^{b}	17.9 ^{b}
FT_{20}	5.73 ^{a}	0.062 ^{a}	0.254 ^{a}	103 ^{a}	17.1 ^{a}	1.52 ^{a}	0.124 ^{a}	2.11 ^{a}	26.4 ^{a}
System	2nd flush								
Control	0.892 ^{e}	0.010 ^{b}	0.064 ^{b}	3.21 ^{b}	2.48 ^{b}	0.37 ^{b}	0.014 ^{b}	0.318 ^{b}	6.63 ^{d}
FT_1	1.53 ^{d}	0.013 ^{ab}	0.064 ^{b}	3.35 ^{b}	2.84 ^{b}	0.49 ^{b}	0.026 ^{ab}	0.511 ^{b}	6.95 ^{cd}
FT_5	1.96 ^{c}	0.013 ^{ab}	0.061 ^{b}	4.31 ^{b}	2.20 ^{b}	0.52 ^{b}	0.037 ^{ab}	0.092 ^{b}	7.30 ^{cd}
FT_{10}	2.27 ^{b}	0.023 ^{ab}	0.062 ^{b}	8.94 ^{b}	2.30 ^{b}	0.67 ^{b}	0.053 ^{ab}	0.466 ^{b}	9.07 ^{bc}
FT_{15}	2.36 ^{ab}	0.023 ^{ab}	0.058 ^{b}	7.52 ^{b}	3.54 ^{b}	0.64 ^{b}	0.053 ^{ab}	0.432 ^{b}	9.86 ^{b}
FT_{20}	2.62 ^{a}	0.030 ^{a}	0.273 ^{a}	68.8 ^{a}	16.9 ^{a}	1.79 ^{a}	0.075 ^{a}	2.38 ^{a}	20.8 ^{a}

$n = 3$; identical superscripts (a, b, c . . .) denote non-significant differences between means in columns determined in compost with flotation tailings and fruit bodies separately according to the post hoc Tukey's HSD test

A significant increase in the mean TEWDHE content in the next treatment was also observed with the lowest value for the control (0.017, 1.55, 5.03, 0.036, 0.359, and 0.017 mg kg^{-1}, respectively, for Ag, As, Ba, Cd, Pb, and Tl) and the highest for the FT_{20} system (0.107, 145, 11.6, 39.7, 15.8, and 1.87 mg kg^{-1}, respectively) (Table 4). It is worth underlining that a high similarity was also observed between Ag and Cd, Pb and Tl, and As and Ba (Figure 4c).

Table 4. Content of trace elements with detrimental health effects [mg kg^{-1} dry weight] in compost with flotation tailings before the experiment and fruit bodies from particular flushes.

System	Ag	As	Ba	Cd	Pb	Tl
	Compost without/with Flotation Tailings					
Control	0.017 ^{d}	1.55 ^{f}	5.03 ^{e}	0.036 ^{e}	0.359 ^{e}	0.017 ^{d}
FT_1	0.023 ^{d}	35.2 ^{e}	5.76 ^{de}	1.51 ^{e}	0.675 ^{e}	0.108 ^{d}
FT_5	0.060 ^{c}	71.2 ^{d}	7.52 ^{cd}	13.1 ^{d}	3.07 ^{d}	0.189 ^{d}
FT_{10}	0.077 ^{bc}	100 ^{c}	8.42 ^{bc}	18.7 ^{c}	6.55 ^{c}	0.743 ^{c}
FT_{15}	0.100 ^{ab}	126 ^{b}	9.48 ^{b}	25.0 ^{b}	10.2 ^{b}	1.58 ^{b}
FT_{20}	0.107 ^{a}	145 ^{a}	11.6 ^{a}	39.7 ^{a}	15.8 ^{a}	1.87 ^{a}
System	1st flush					
Control	0.017 ^{b}	0.330 ^{d}	0.141 ^{d}	0.019 ^{f}	0.047 ^{d}	0.010 ^{c}
FT_1	0.035 ^{b}	4.22 ^{c}	0.285 ^{c}	1.14 ^{e}	0.171 ^{c}	0.064 ^{b}
FT_5	0.055 ^{b}	5.76 ^{c}	0.337 ^{c}	3.87 ^{d}	0.264 ^{bc}	0.054 ^{b}
FT_{10}	0.049 ^{b}	9.85 ^{b}	0.273 ^{c}	9.46 ^{c}	0.272 ^{bc}	0.057 ^{b}
FT_{15}	0.100 ^{a}	16.4 ^{a}	0.707 ^{a}	15.2 ^{b}	0.328 ^{b}	0.105 ^{a}
FT_{20}	0.118 ^{a}	14.6 ^{a}	0.511 ^{b}	33.3 ^{a}	1.03 ^{a}	0.118 ^{a}
System	2nd flush					
Control	0.010 ^{b}	0.137 ^{d}	0.128 ^{d}	0.021 ^{c}	0.032 ^{b}	0.011 ^{b}
FT_1	0.010 ^{b}	1.19 ^{c}	0.155 ^{d}	0.320 ^{c}	0.032 ^{b}	0.011 ^{b}
FT_5	0.027 ^{b}	2.04 ^{b}	0.214 ^{c}	1.22 ^{bc}	0.092 ^{b}	0.010 ^{b}
FT_{10}	0.027 ^{b}	1.98 ^{b}	0.288 ^{b}	5.09 ^{b}	0.062 ^{b}	0.062 ^{a}
FT_{15}	0.030 ^{b}	2.22 ^{b}	0.318 ^{b}	3.02 ^{bc}	0.173 ^{b}	0.058 ^{a}
FT_{20}	0.163 ^{a}	3.02 ^{a}	0.390 ^{a}	27.9 ^{a}	0.546 ^{a}	0.013 ^{b}

$n = 3$; identical superscripts (a, b, c . . .) denote non-significant differences between means in columns determined in compost with flotation tailings and fruit bodies separately according to the post hoc Tukey's HSD test.

Mean ∑REEs ranged from 9.12 to 21.5 mg kg^{-1} for the control and FT_{20}, respectively. The dominant REEs were Nd, from 7.92 to 18.7 mg kg^{-1} and Ce from 0.983 to 1.67 mg kg^{-1} (Table 5). This observation is confirmed by a heatmap, where, based on all experimental systems, the highest similarity is between ∑REEs and Nd (creating a separate group). At the same time, Ce, especially with Eu and Er, La, Pr, and Gd, create another group (Figure 4d). Moreover, the control and FT_1 were similar, whereas the remaining systems created a new group of objects.

Platinum was a dominant PGE with a mean content ranging from 0.323 to 1.42 mg kg^{-1}, respectively, for the control and FT_{20} system (Table 6, Figure 4e). The addition of flotation tailings caused an increase in Pt content in the substrate. An increase in Ir content with the addition of flotation tailings was also observed from 0.150 to 0.323 mg kg^{-1}, respectively, for the control and FT_{20} system. In the case of Pd and Rh, a significantly higher content of these elements was only observed in substrate FT_{20} compared with the other experimental systems.

For the NNEs, a significantly higher mean content of Al, Au, Bi, In, Li, Re, Sb, Sr, and Te was observed for the substrate in the FT_{15} and FT_{20} systems than for the control (Table 7). There were no significant differences in Ga and Ge content between the substrates in the control and particular treatments. All these observations are confirmed by a heatmap, where the control and FT_1 systems create a separate group of objects and the highest contents, especially of Al, Bi, In, Li, Sr, and Te, are visible (Figure 4f). According to Figure 4g, where a heatmap for all detectable elements was prepared, the control and FT_1 systems create a separate group, the same as the rest of the treatments used in this study. In contrast, there is generally an increased content of the studied elements in substrates with increased flotation tailings.

Table 5. Content of rare earth elements [mg kg^{-1} dry weight] in compost with flotation tailings before the experiment and fruit bodies from particular flushes.

System	Ce	Dy	Er	Eu	Gd	Ho	La	Lu	Nd	Pr	Sc	Sm	Tb	Tm	Y	Yb	Σ_{REEs}
	Compost without/with Flotation Tailings																
Control	0.983 ^{b}	0.010 ^{a}	0.041 ^{d}	0.010 ^{a}	0.013 ^{b}	0.010 ^{a}	0.027 ^{c}	0.010 ^{a}	7.92 ^{d}	0.010 ^{b}	0.010 ^{a}	0.010 ^{a}	0.010 ^{a}	0.010 ^{a}	0.040 ^{c}	0.010 ^{a}	9.12 ^{d}
FT_1	1.38 ^{ab}	0.010 ^{a}	0.293 ^{c}	0.010 ^{a}	0.027 ^{ab}	0.010 ^{a}	0.093 ^{b}	0.010 ^{a}	10.6 ^{c}	0.010 ^{b}	0.010 ^{a}	0.010 ^{a}	0.010 ^{a}	0.010 ^{a}	0.053 ^{bc}	0.010 ^{a}	12.6 ^{c}
FT_5	1.42 ^{ab}	0.010 ^{a}	0.337 ^{cb}	0.013 ^{a}	0.030 ^{ab}	0.010 ^{a}	0.123 ^{b}	0.010 ^{a}	12.5 ^{c}	0.010 ^{b}	0.010 ^{a}	0.010 ^{a}	0.010 ^{a}	0.010 ^{a}	0.090 ^{bc}	0.010 ^{a}	14.6 ^{c}
FT_{10}	1.46 ^{a}	0.010 ^{a}	0.340 ^{cb}	0.010 ^{a}	0.033 ^{ab}	0.010 ^{a}	0.117 ^{b}	0.010 ^{a}	14.8 ^{b}	0.010 ^{b}	0.010 ^{a}	0.010 ^{a}	0.010 ^{a}	0.010 ^{a}	0.097 ^{bc}	0.010 ^{a}	17.0 ^{b}
FT_{15}	1.53 ^{a}	0.010 ^{a}	0.423 ^{ab}	0.020 ^{a}	0.033 ^{ab}	0.010 ^{a}	0.143 ^{b}	0.010 ^{a}	16.5 ^{ab}	0.010 ^{b}	0.010 ^{a}	0.010 ^{a}	0.010 ^{a}	0.010 ^{a}	0.107 ^{b}	0.010 ^{a}	18.8 ^{b}
FT_{20}	1.67 ^{a}	0.010 ^{a}	0.537 ^{a}	0.027 ^{a}	0.043 ^{a}	0.010 ^{a}	0.213 ^{a}	0.010 ^{a}	18.7 ^{a}	0.093 ^{a}	0.010 ^{a}	0.010 ^{a}	0.010 ^{a}	0.010 ^{a}	0.173 ^{a}	0.010 ^{a}	21.5 ^{a}
System	1st flush																
Control	0.010 ^{c}	0.010 ^{f}	bDL	bDL	0.013 ^{c}	0.010 ^{f}	0.013 ^{f}	0.010 ^{f}	0.077 ^{c}	0.013 ^{b}	0.010 ^{f}	0.010 ^{f}	0.011 ^{f}	0.010 ^{f}	0.010 ^{f}	0.013 ^{f}	0.221 ^{f}
FT_1	0.142 ^{bc}	0.074 ^{e}	bDL	bDL	0.142 ^{bc}	0.088 ^{e}	0.170 ^{e}	0.084 ^{e}	1.14 ^{c}	0.142 ^{b}	0.085 ^{e}	0.096 ^{e}	0.089 ^{e}	0.086 ^{e}	0.081 ^{e}	0.082 ^{e}	2.50 ^{e}
FT_5	0.265 ^{bc}	0.119 ^{d}	bDL	bDL	0.265 ^{bc}	0.114 ^{d}	0.265 ^{d}	0.109 ^{d}	1.59 ^{c}	0.303 ^{b}	0.108 ^{d}	0.120 ^{d}	0.123 ^{d}	0.105 ^{d}	0.117 ^{d}	0.111 ^{d}	3.71 ^{d}
FT_{10}	0.378 ^{bc}	0.142 ^{c}	0.568 ^{b}	bDL	0.426 ^{ba}	0.142 ^{c}	0.426 ^{c}	0.142 ^{c}	3.31 ^{b}	0.426 ^{b}	0.142 ^{c}	0.142 ^{c}	0.142 ^{c}	0.142 ^{c}	0.142 ^{c}	0.142 ^{c}	6.81 ^{c}
FT_{15}	0.851 ^{b}	0.180 ^{b}	0.624 ^{a}	bDL	0.511 ^{b}	0.175 ^{b}	0.568 ^{b}	0.160 ^{b}	4.20 ^{b}	0.568 ^{b}	0.164 ^{b}	0.170 ^{b}	0.165 ^{b}	0.176 ^{b}	0.170 ^{b}	0.172 ^{b}	8.85 ^{b}
FT_{20}	4.94 ^{a}	0.255 ^{a}	1.46 ^{b}	0.047	1.45 ^{a}	0.252 ^{a}	0.937 ^{a}	0.255 ^{a}	8.12 ^{a}	2.384 ^{a}	0.275 ^{a}	0.256 ^{a}	0.265 ^{a}	0.238 ^{a}	0.255 ^{a}	0.244 ^{a}	21.6 ^{a}
System	2n d flush																
Control	0.010 ^{b}	0.010 ^{e}	bDL ^{c}	bDL	0.010 ^{b}	0.010 ^{b}	0.010 ^{b}	0.010 ^{b}	0.013 ^{d}	0.010 ^{b}	0.010 ^{b}	0.010 ^{c}	0.010 ^{b}	0.010 ^{b}	0.010 ^{b}	0.010 ^{b}	0.144 ^{e}
FT_1	0.022 ^{b}	0.022 ^{d}	bDL ^{c}	bDL	0.021 ^{b}	0.023 ^{b}	0.026 ^{b}	0.023 ^{b}	0.137 ^{cd}	0.023 ^{bc}	0.023 ^{b}	0.027 ^{c}	0.020 ^{b}	0.025 ^{b}	0.019 ^{b}	0.024 ^{b}	0.433 ^{de}
FT_5	0.064 ^{b}	0.025 ^{c}	bDL ^{c}	bDL	0.023 ^{b}	0.031 ^{b}	0.028 ^{b}	0.029 ^{b}	0.292 ^{c}	0.024 ^{bc}	0.022 ^{b}	0.032 ^{c}	0.037 ^{b}	0.018 ^{b}	0.027 ^{b}	0.030 ^{b}	0.685 ^{d}
FT_{10}	0.098 ^{b}	0.042 ^{b}	0.112 ^{b}	bDL	0.098 ^{b}	0.044 ^{b}	0.043 ^{b}	0.042 ^{b}	0.619 ^{b}	0.127 ^{bc}	0.046 ^{b}	0.042 ^{c}	0.040 ^{b}	0.043 ^{b}	0.044 ^{b}	0.043 ^{b}	1.49 ^{c}
FT_{15}	1.41 ^{a}	0.273 ^{a}	0.718 ^{a}	bDL	0.736 ^{a}	0.272 ^{a}	0.279 ^{a}	0.276 ^{a}	1.35 ^{a}	0.915 ^{a}	0.277 ^{a}	0.279 ^{a}	0.276 ^{a}	0.278 ^{a}	0.274 ^{a}	0.276 ^{a}	7.88 ^{a}
FT_{20}	0.289 ^{b}	0.046 ^{b}	0.159 ^{b}	bDL	0.116 ^{b}	0.043 ^{b}	0.040 ^{b}	0.033 ^{b}	1.31 ^{a}	0.231 ^{b}	0.048 ^{b}	0.144 ^{b}	0.050 ^{b}	0.043 ^{b}	0.049 ^{b}	0.045 ^{b}	2.65 ^{b}

$n = 3$; identical superscripts (a, b, c …) denote non-significant differences between means in columns determined in compost with flotation tailings and fruit bodies separately according to the post hoc Tukey's HSD test; bDL—below the detection limit.

Table 6. Content of platinum group elements [mg kg^{-1} dry weight] in compost with flotation tailings before the experiment and fruit bodies from particular flushes.

System	Ir	Pd	Pt	Rh	Ru
	Compost without/with Flotation Tailings				
Control	0.150 ^{c}	0.010 ^{b}	0.323 ^{e}	0.010 ^{b}	0.010 ^{a}
FT_1	0.237 ^{b}	0.010 ^{b}	0.486 ^{e}	0.010 ^{b}	0.010 ^{a}
FT_5	0.247 ^{b}	0.010 ^{b}	0.803 ^{d}	0.010 ^{b}	0.010 ^{a}
FT_{10}	0.283 ^{ab}	0.010 ^{b}	1.22 ^{c}	0.010 ^{b}	0.010 ^{a}
FT_{15}	0.277 ^{ab}	0.010 ^{b}	1.65 ^{a}	0.010 ^{b}	0.010 ^{a}
FT_{20}	0.323 ^{a}	0.077 ^{a}	1.42 ^{b}	0.090 ^{a}	0.010 ^{a}
System			1st flush		
Control	0.023 ^{d}	0.010 ^{b}	0.087 ^{e}	0.010 ^{b}	0.010 ^{f}
FT_1	0.160 ^{cd}	0.030 ^{b}	0.278 ^{d}	0.030 ^{b}	0.030 ^{e}
FT_5	0.360 ^{cd}	0.040 ^{b}	0.366 ^{d}	0.040 ^{b}	0.040 ^{d}
FT_{10}	0.407 ^{c}	0.050 ^{b}	0.612 ^{c}	0.050 ^{b}	0.050 ^{c}
FT_{15}	1.18 ^{b}	0.060 ^{b}	1.02 ^{b}	0.060 ^{b}	0.060 ^{b}
FT_{20}	2.79 ^{a}	0.150 ^{a}	1.67 ^{a}	0.210 ^{a}	0.090 ^{a}
System			2nd flush		
Control	0.227 ^{a}	0.010 ^{a}	0.032 ^{c}	0.010 ^{a}	0.010 ^{a}
FT_1	0.197 ^{a}	0.010 ^{a}	0.032 ^{c}	0.010 ^{a}	0.010 ^{a}
FT_5	0.073 ^{bc}	0.010 ^{a}	0.122 ^{bc}	0.010 ^{a}	0.010 ^{a}
FT_{10}	0.073 ^{bc}	0.010 ^{a}	0.155 ^{b}	0.010 ^{a}	0.010 ^{a}
FT_{15}	0.050 ^{c}	0.010 ^{a}	0.288 ^{a}	0.010 ^{a}	0.011 ^{a}
FT_{20}	0.110 ^{b}	0.010 ^{a}	0.351 ^{a}	0.010 ^{a}	0.010 ^{a}

$n = 3$; identical superscripts (a, b, c …) denote non-significant differences between means in columns determined in compost with flotation tailings and fruit bodies separately according to the post hoc Tukey's HSD test.

Table 7. Content of nutritionally non-essential elements [mg kg^{-1} dry weight] in compost with flotation tailings before the experiment and fruit bodies from particular flushes.

System	Al	Au	Bi	Ga	Ge	In	Li	Re	Sb	Sr	Te
	Compost without/with Flotation Tailings										
Control	26.7 ^{f}	0.187 ^{d}	0.024 ^{e}	0.010 ^{a}	0.010 ^{a}	0.013 ^{d}	0.033 ^{c}	0.013 ^{c}	0.012 ^{e}	3.43 ^{c}	0.010 ^{b}
FT_1	32.0 ^{e}	0.357 ^{c}	0.045 ^{e}	0.010 ^{a}	0.013 ^{a}	0.027 ^{cd}	0.047 ^{bc}	0.027 ^{c}	0.072 ^{d}	5.90 ^{b}	0.023 ^{b}
FT_5	47.0 ^{d}	0.427 ^{c}	0.079 ^{d}	0.010 ^{a}	0.020 ^{a}	0.048 ^{c}	0.060 abc	0.064 ^{b}	0.267 ^{c}	6.80 ^{b}	0.077 ^{a}
FT_{10}	52.5 ^{c}	0.633 ^{b}	0.104 ^{c}	0.010 ^{a}	0.023 ^{a}	0.089 ^{b}	0.063 abc	0.082 ^{ab}	0.433 ^{b}	7.10 ^{b}	0.077 ^{a}
FT_{15}	62.5 ^{b}	0.727 ^{ab}	0.134 ^{b}	0.010 ^{a}	0.027 ^{a}	0.093 ^{b}	0.077 ^{ab}	0.084 ^{ab}	0.452 ^{b}	7.59 ^{b}	0.100 ^{a}
FT_{20}	68.5 ^{a}	0.777 ^{a}	0.198 ^{a}	0.011 ^{a}	0.030 ^{a}	0.198 ^{a}	0.087 ^{a}	0.099 ^{a}	0.936 ^{a}	10.9 ^{a}	0.110 ^{a}
System	1st flush										
Control	0.203 ^{d}	0.757 ^{c}	0.013 ^{c}	0.010 ^{c}	0.017 ^{c}	0.016 ^{c}	0.010 ^{a}	0.026 ^{d}	0.013 ^{b}	0.013 ^{d}	0.013 ^{c}
FT_1	0.642 ^{c}	1.14 ^{a}	0.037 ^{b}	0.016 ^{b}	0.037 ^{c}	0.049 ^{bc}	0.019 ^{a}	0.161 ^{c}	0.105 ^{b}	0.037 ^{cd}	0.037 ^{bc}
FT_5	0.779 ^{bc}	0.343 ^{e}	0.036 ^{b}	0.019 ^{a}	0.025 ^{c}	0.095 ^{ab}	0.019 ^{a}	0.218 ^{b}	0.161 ^{b}	0.057 ^{c}	0.050 abc
FT_{10}	0.779 ^{bc}	0.580 ^{d}	0.036 ^{b}	0.019 ^{a}	0.037 ^{c}	0.116 ^{a}	0.019 ^{a}	0.273 ^{a}	0.125 ^{b}	0.109 ^{b}	0.069 ^{ab}
FT_{15}	0.931 ^{a}	1.01 ^{b}	0.044 ^{a}	0.019 ^{a}	0.243 ^{b}	0.133 ^{a}	0.019 ^{a}	0.264 ^{a}	0.175 ^{b}	0.228 ^{a}	0.094 ^{a}
FT_{20}	1.035 ^{a}	0.231 ^{f}	0.057 ^{a}	0.019 ^{a}	0.511 ^{a}	0.155 ^{a}	0.020 ^{a}	0.285 ^{a}	1.240 ^{a}	0.264 ^{a}	0.094 ^{a}
System	2n d flush										
Control	0.313 ^{b}	0.073 ^{d}	0.013 ^{c}	0.010 ^{a}	0.023 ^{b}	0.064 ^{b}	0.010 ^{a}	0.058 ^{b}	0.011 ^{d}	0.011 ^{c}	0.010 ^{a}
FT_1	0.343 ^{b}	0.147 ^{d}	0.024 ^{bc}	0.010 ^{a}	0.020 ^{b}	0.070 ^{b}	0.010 ^{a}	0.061 ^{b}	0.256 ^{a}	0.032 ^{c}	0.010 ^{a}
FT_5	0.363 ^{b}	0.277 ^{c}	0.030 ^{bc}	0.010 ^{a}	0.010 ^{c}	0.096 ^{b}	0.010 ^{a}	0.095 ^{a}	0.010 ^{d}	0.061 ^{bc}	0.027 ^{a}
FT_{10}	0.407 ^{b}	0.293 ^{c}	0.037 ^{b}	0.010 ^{a}	0.010 ^{c}	0.062 ^{b}	0.010 ^{a}	0.096 ^{a}	0.093 ^{c}	0.031 ^{c}	0.030 ^{a}
FT_{15}	0.417 ^{b}	0.420 ^{b}	0.043 ^{b}	0.012 ^{a}	0.053 ^{a}	0.202 ^{a}	0.011 ^{a}	0.099 ^{a}	0.144 ^{b}	0.086 ^{b}	0.010 ^{a}
FT_{20}	1.32 ^{a}	1.15 ^{a}	0.092 ^{a}	0.012 ^{a}	0.047 ^{a}	0.195 ^{a}	0.013 ^{a}	0.098 ^{a}	0.078 ^{c}	0.195 ^{a}	0.059 ^{a}

$n = 3$; identical superscripts (a, b, c …) denote non-significant differences between means in columns determined in compost with flotation tailings and fruit bodies separately according to the post hoc Tukey's HSD test.

3.3. *Content of Elements in A. bisporus Fruit Bodies*

The content of Ca significantly decreased in fruit bodies collected from the first (from 109 to 24.6 mg kg^{-1}) and the second yield (from 100 to 24.6 mg kg^{-1}) with an increase of flotation tailings in addition to the compost (Table 2). A similar trend was also observed for K and Mg (from the first yield only). The increased addition of flotation tailings to the compost did not generally cause significant changes in Mg content in fruit bodies from the second yield or Na in mushrooms from both yields. In this experiment, fruit bodies growing under a particular system were characterized by a high similarity between Ca and Na (Figure 4a). It is worth underlining that a heatmap with a separate object created for fruit bodies growing under the FT_{20} system from the second yield shows the effect of the lowest content of Ca, K, and Na (10.4, 4710, and 17.9 mg kg^{-1}, respectively) and the highest content of Mg (194 mg kg^{-1}). Only effective accumulation (BAF > 1) of K in *A. bisporus* from both yields and Mg in fruit bodies from the second yield under FT_{20} was observed (Figure 5).

The efficiency of ETE accumulation by mushroom species was highly diverse, which confirms a clear limited accumulation of Co, Cr, Fe, or Zn in mushrooms, especially from the second yield, and the opposite situation for Cu (Table 3). The highest mean content of ETEs was determined for mushrooms growing in substrates with higher additions of flotation tailings collected from the first (FT_{10} and/or FT_{15} and/or FT_{20}) and the second yield (FT_{20}), which confirms the heatmap where a separate group of objects was indicated for these treatments (Figure 4b). An apparent decrease of Mn content in the fruit bodies of *A. bisporus* was also confirmed by the heatmap, where this metal (similarly to B) is separated from the others (Figure 4b). BAF values higher than 1 were only observed for B, Cu, and Se in fruit bodies, mainly from the first yield (Figure 5).

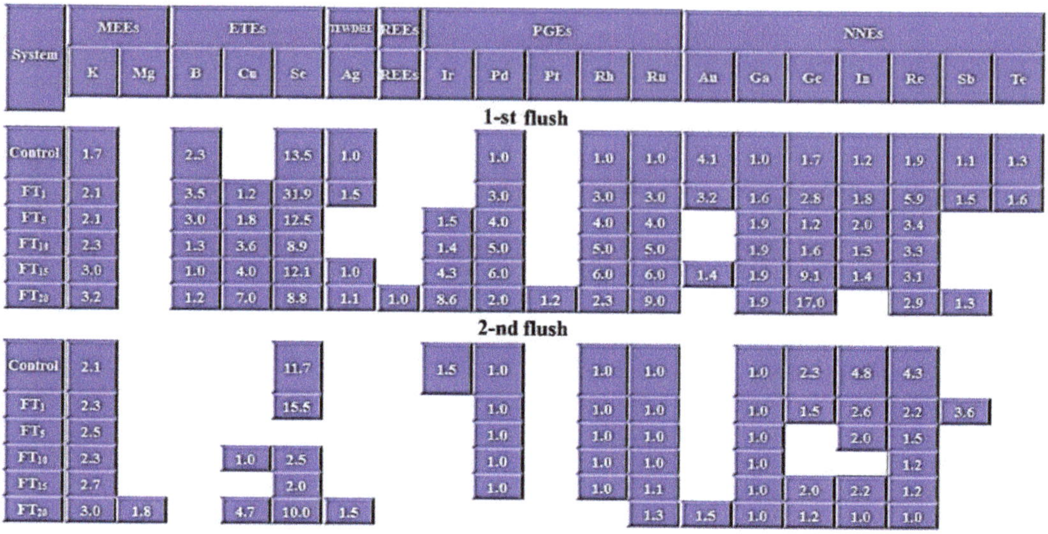

Figure 5. Bioaccumulation factor values for elements in fruit bodies collected from particular treatments.

The content of TEWDHE in fruit bodies collected from the first yield was the highest under the FT_{15} and/or FT_{20} experimental systems, whereas from the second yield it was only under the FT_{20} system, except for Tl, where the highest content of this metal was recorded under FT_{10} and FT_{15} (0.062 and 0.058 mg kg^{-1}, respectively) (Table 4). Despite this result, all mushrooms growing under the mentioned experimental systems created a separate group of objects, which undoubtedly shows similarities and differences concerning the other systems (Figure 4c). BAF > 1 was calculated only for Ag in *A. bisporus* bodies from the first (control, FT_1, FT_{15}, and FT_{20}) and the second (FT_{20}) yield (Figure 5).

∑REE content was the highest for *A. bisporus* growing under the FT_{20} system obtained from the first yield and FT_4 from the second yield (21.6 and 7.88 mg kg^{-1}, respectively) (Table 5), which confirms the graphical interpretation of the obtained results (a heatmap), where mushrooms growing on both these substrates are included in the same separate group of objects (Figure 4d). The dominant REEs were Nd and Ce in fruit bodies collected from both the first (1.14–8.12 and 0.142–4.94 mg kg^{-1}, respectively) and the second yield (0.137–1.35 and 0.022–1.41 mg kg^{-1}, respectively). The lower content of REEs in mushrooms growing under FT_{20} compared with FT_{15} from the second yield was probably the effect of the lower content of these elements in the substrate after the first yield and their generally limited accumulation in the second yield. The BAF calculated for ∑REEs in the first yield of *A. bisporus* growing only under FT_{20} was higher than 1 (Figure 5).

Effective accumulation of PGEs was determined, especially for Ir and Pt, whose highest mean content in fruit bodies under FT_{20} was 2.79 and 1.67 mg kg^{-1}, respectively (first yield). In this case, the heatmap shows that mushrooms growing under this treatment create absolutely separate objects (Figure 4e). The efficiency of PGE accumulation by *A. bisporus* after the second yield was lower (Table 6). The BAF calculated for particular PGEs shows that values higher than 1 were observed, especially for the fruit bodies collected in the first yield growing under all the experimental systems (Pd, Rh, and Ru), FT_5-FT_{20} (Ir), and FT_{20} (Pt) (Figure 5). In the second yield, BAF > 1 was also calculated, although it was usually an effect of a similar and low concentration of elements in the substrate and the content of these elements in mushrooms.

The efficiency of NNE accumulation in fruit bodies was diverse, with generally the highest content found in mushrooms under the highest addition of flotation tailings, except for Au (1.14 mg kg^{-1} under FT_1) and Sb (0.256 mg kg^{-1} under FT_1) in bodies from the first and the second yield, respectively (Table 7). According to the heatmap, the high similarity concerning the content of all NNEs between mushrooms from the first yield was the same as between mushrooms from the second yield (Figure 4f). Effective accumulation (BAF > 1) of Ga, Ge, In, and Re was recorded for mushrooms growing under almost all experimental systems from both yields, whereas for the other NNEs only for mushrooms under the addition of selected flotation tailings (Figure 5).

Based on all determined elements in the collected fruit bodies growing under the control (both yields) and the FT_1, FT_5, and FT_{10} systems from the second yield, a similarity was noted between them and their content in the substrates of the control FT_1 (Figure 4g). On the other hand, *A. bisporus* growing under the remaining experimental systems enriched with flotation tailings (except for mushrooms under FT_{20} from the first yield) were characterized by a similar content of all determined elements, which are clearly shown as a separate group. The above-mentioned mushrooms growing under the FT_{20} system and collected from the first yield created a separate object characterized by the highest content of all studied elements jointly.

4. Discussion

Recently, there have been more and more reports on the possibility of using *A. bisporus* in remediation. In our experiment, five different quantities of flotation tailings were used to study the mycoremediation potential of this species. In the first flushes, a reduction in the yield was confirmed with the addition of flotation tailings. The literature confirms that supplementation of the substrate with metals/metalloids may reduce the growth dynamics of this species. Rzymski et al. [43] showed supplementation with Cu, Se, and Zn resulted in the biomass of fruiting bodies decreasing significantly at higher element addition (0.8 mM). Also, the addition of higher Hg concentrations (0.4 and 0.5 mM) to the growing medium reduced the growth of *A. bisporus* biomass [44].

The substrate in our experiment was richer in MEEs than mixtures of substrate and flotation tailings. The same tendency was confirmed in fruit bodies growing in experimental systems with their use. A different trend was observed concerning other groups of elements. The control substrate (compost) contained a significantly lower content of other elements than the substrate with flotation tailings. Because of the ability of *A. bisporus* to accumulate selected elements, the composition of waste affects the composition of its fruit (different amounts of waste addition resulting in different levels of elements in the substrate). Differences in the accumulation of heavy metals (Cd, Cr, and Pb) in *A. bisporus* fruiting bodies depending on the concentration of these elements in the growing medium were confirmed by Zhou et al. [45]. Nagy et al. [46] also confirmed that the maximum removal efficiencies from monocomponent aqueous solutions by *A. bisporus* for Cd and Zn took place at the highest concentrations of the substrate. The ability of *A. bisporus* to effectively accumulate selected elements shown in our study, is also well demonstrated by the results of our previous studies on supplementation during the cultivation of these mushrooms. Effective uptake of Cu, Se, and Zn from the enriched medium was confirmed by Rzymski et al. [43]. Supplementation with 0.6 mmol L^{-1} of Cu, Se, and Zn resulted in an over 3-fold, 2.5-fold, and 10-fold increase in their concentrations in fruiting bodies, respectively, whereas Rzymski et al. [44] demonstrated that Hg uptake increased in a concentration-dependent manner and exceeded 116 mg kg^{-1} in *A. bisporus* caps after 0.5 mM was added to the substrate.

BAF values may measure the mycoremediation efficiency of different mushroom species growing in polluted substrates and provide direct information about the potential of particular mushroom species to accumulate elements. This is crucial because this process has numerous dependent factors (environmental and genetic) [47], and is also the case for mushroom cultivation using different substrates as previously described [48]. In this study, effective accumulation was observed for selected elements only, which reflected the chosen additions of flotation tailings added to the compost. It suggests that *A. bisporus* application can be limited and/or used for the accumulation of selective elements only [13,49].

In general, mushroom fruit bodies collected after the mycoremediation process are waste that can be a substrate for the recovery of elements in the case of very high contents of especially precious elements [50]. The risk of contaminated food is genuine [51]. The problem of assessing the quality of fruit bodies (especially for significant and toxic trace elements) that could be consumed has been closely associated with a lack of appropriate legal regulations for many years [52,53]. However, even in countries where legislation exists, the regulations are limited to specific and usually toxic trace elements only [54,55]. The analysis and the further possibility of using the "product" from the mycoremediation of post-industrial wastes seems to be unlikely. Simultaneously, the risk of consuming contaminated fruit bodies is high, as may be shown in the study of Pająk et al. [56], who collected 10 mushroom species from polluted forest ecosystems.

5. Conclusions

The possibility of using mushroom fruiting bodies to decontaminate contaminated substrates effectively is an essential aspect of bioremediation due to their ability to accumulate elements effectively. The *A. bisporus* strain tested in this study effectively accumulated selected elements (Ag, Au, B, Cu, Ga, Ge, In, Ir, K, Mg, Pd, Pt, Re, Rh, Ru, Sb, Se, and Te), as evidenced by values of BAF > 1. Although this efficiency was not spectacular, to be able to recover elements in pure form, fundamentally enriched the fruiting bodies. The results of these studies indicate the potential for using *A. bisporus* fruiting bodies after the mycoremediation process in industry, although this must be preceded by larger-scale tests. This application seems to be the most favorable for media contaminated with selected elements, the absorption of which by fruiting bodies is the most efficient. Our study confirmed that the key to mycoremediation is determining the right fungal species to target a specific pollutant. However, due to the particularly effective accumulation of As and Cd, the post-flotation sediment subjected to this process should contain the lowest possible concentrations of both of these elements. Further research is necessary to determine the long-term potency of such a method.

Author Contributions: Conceptualization, S.B., M.S. and M.M.; methodology, M.S., M.M. and P.N.; software, P.N. and A.B.; validation, S.B., M.S. and M.M; formal analysis, S.B. and P.N.; investigation, S.B., M.M., P.N.; resources, M.M. and A.B.; data curation, S.B., M.M. and M.G.; writing—original draft preparation, S.B., M.M. and M.G.; writing—review and editing, S.B., M.M. and P.K.; visualization, S.B. and M.M.; supervision, M.M. and P.K.; project administration, S.B., M.S. and M.M. All authors have read and agreed to the published version of the manuscript.

Funding: This research received no external funding.

Institutional Review Board Statement: Not applicable.

Informed Consent Statement: Not applicable.

Data Availability Statement: Not applicable.

Conflicts of Interest: The authors declare no conflict of interest.

References

1. Gupta, S.; Wali, A.; Gupta, M.; Annepu, S.K. Fungi: An Effective Tool for Bioremediation. *Plant-Microbe Interact. Agro-Ecol. Perspect.* **2017**, *2*, 593–606. [CrossRef]
2. Kaur, P.; Balomajumder, C. Effective Mycoremediation Coupled with Bioaugmentation Studies: An Advanced Study on Newly Isolated *Aspergillus* sp. in Type-II Pyrethroid-Contaminated Soil. *Environ. Pollut.* **2020**, *261*, 114073. [CrossRef] [PubMed]
3. Ngadin, A.A.; Taghavi, E.; Eaton, T. The Development of White-Rot Fungi as a Mycoremediation Product. In *Applied Mycology*; Springer: Cham, Switzerland, 2022; pp. 75–94. [CrossRef]
4. Hegde, G.M.; Aditya, S.; Wangdi, D.; Chetri, B.K. Mycoremediation: A Natural Solution for Unnatural Problems. In *Fungal Diversity, Ecology and Control Management*; Springer: Singapore, 2022; pp. 363–386. [CrossRef]
5. Shourie, A.; Vijayalakshmi, U. Fungal Diversity and Its Role in Mycoremediation. *Geomicrobiol. J.* **2022**, *39*, 426–444. [CrossRef]
6. Dey, P.; Malik, A.; Singh, D.K.; Haange, S.-B.; von Bergen, M.; Jehmlich, N. Insight into the Molecular Mechanisms Underpinning the Mycoremediation of Multiple Metals by Proteomic Technique. *Front. Microbiol.* **2022**, *13*, 872576. [CrossRef]

7. Kumar, V.; Dwivedi, S.K. Mycoremediation of Heavy Metals: Processes, Mechanisms, and Affecting Factors. *Environ. Sci. Pollut. Res.* **2021**, *28*, 10375–10412. [CrossRef]
8. Purohit, J.; Chattopadhyay, A.; Biswas, M.K.; Singh, N.K. Mycoremediation of Agricultural Soil: Bioprospection for Sustainable Development. In *Mycoremediation and Environmental Sustainability*; Springer: Cham, Switzerland, 2018; pp. 91–120. [CrossRef]
9. Akhtar, N.; Mannan, M.A.U. Mycoremediation: Expunging Environmental Pollutants. *Biotechnol. Rep.* **2020**, *26*, e00452. [CrossRef]
10. Koul, B.; Ahmad, W.; Singh, J. Mycoremediation: A Novel Approach for Sustainable Development. In *Microbe Mediated Remediation of Environmental Contaminants*; Woodhead Publishing: Cambridge, UK, 2021; pp. 409–420. [CrossRef]
11. Dey, P.; Malik, A.; Mishra, A.; Singh, D.K.; von Bergen, M.; Jehmlich, N. Mechanistic Insight to Mycoremediation Potential of a Metal Resistant Fungal Strain for Removal of Hazardous Metals from Multimetal Pesticide Matrix. *Environ. Pollut.* **2020**, *262*, 114255. [CrossRef]
12. Bosco, F.; Mollea, C. Mycoremediation in Soil. In *Environmental Chemistry and Recent Pollution Control Approaches*; IntechOpen: London, UK, 2019. [CrossRef]
13. Prasad, Y.; Sachin, D. Biosorption of Cu, Zn, Fe, Cd, Pb and Ni by Non-Treated Biomass of Some Edible Mushrooms. *Asian J. Exp. Biol. Sci.* **2013**, *4*, 190–195.
14. Akinyele, J.B.; Fakoya, S.; Adetuyi, C.F. Anti-Growth Factors Associated with Pleurotus Ostreatus in a Submerged Liquid Fermentation. *Malays. J. Microbiol.* **2012**, *8*, 135–140. [CrossRef]
15. Corral-Bobadilla, M.; González-Marcos, A.; Alba-Elías, F.; Diez de Santo Domingo, E. Valorization of Bio-Waste for the Removal of Aluminum from Industrial Wastewater. *J. Clean. Prod.* **2020**, *264*, 121608. [CrossRef]
16. Corral-Bobadilla, M.; González-Marcos, A.; Vergara-González, E.P.; Alba-Elías, F. Bioremediation of Waste Water to Remove Heavy Metals Using the Spent Mushroom Substrate of *Agaricus bisporus*. *Water* **2019**, *11*, 454. [CrossRef]
17. Hanif, M.A.; Bhatti, H.N. Remediation of Heavy Metals Using Easily Cultivable, Fast Growing, and Highly Accumulating White Rot Fungi from Hazardous Aqueous Streams. *Desalination Water Treat.* **2015**, *53*, 238–248. [CrossRef]
18. Vaseem, H.; Singh, V.K.; Singh, M.P. Heavy Metal Pollution Due to Coal Washery Effluent and Its Decontamination Using a Macrofungus, *Pleurotus ostreatus*. *Ecotoxicol. Environ. Saf.* **2017**, *145*, 42–49. [CrossRef]
19. Kapahi, M.; Sachdeva, S. Mycoremediation Potential of Pleurotus Species for Heavy Metals: A Review. *Bioresour. Bioprocess.* **2017**, *4*, 32. [CrossRef]
20. Boopathy, R. Factors Limiting Bioremediation Technologies. *Bioresour. Technol.* **2000**, *74*, 63–67. [CrossRef]
21. Puglisi, I.; Faedda, R.; Sanzaro, V.; lo Piero, A.R.; Petrone, G.; Cacciola, S.O. Identification of Differentially Expressed Genes in Response to Mercury I and II Stress in Trichoderma Harzianum. *Gene* **2012**, *506*, 325–330. [CrossRef]
22. Pihurov, M.; Vlăduț, V.; Bordean, D.; Ferdes, M. Mycoremediation of Heavy Metal-Contaminated Soil. In Proceedings of the International Symposium, ISB-INMA-TEH, Agricultural and Mechanical Engineering, Bucharest, Romania, 31 October–1 November 2019; INMA: Bucharest, Romania, 2019; pp. 423–428.
23. Bhatt, P.; Chen, S. Optimization of Mycoremediation Process for the Isolated Fungi. In *Mycoremediation Protocols*; Humana: New York, NY, USA, 2022; pp. 101–107. [CrossRef]
24. Robichaud, K.; Girard, C.; Dagher, D.; Stewart, K.; Labrecque, M.; Hijri, M.; Amyot, M. Local Fungi, Willow and Municipal Compost Effectively Remediate Petroleum-Contaminated Soil in the Canadian North. *Chemosphere* **2019**, *220*, 47–55. [CrossRef]
25. Damodaran, D.; Suresh, G.; Mohan, R. Bioremediation of soil by removing heavy metals using Saccharomyces Cerevisiae. In *Proceedings of the 2011 2nd International Conference on Environmental Science and Technology IPCBEE*; IACSIT Press: Singapore, 2011; Volume 6, pp. 22–27.
26. Gray, S.N. Fungi as Potential Bioremediation Agents in Soil Contaminated with Heavy or Radioactive Metals. *Biochem. Soc. Trans.* **1998**, *26*, 666–670. [CrossRef]
27. Damodaran, D.; Vidya Shetty, K.; Raj Mohan, B. Uptake of Certain Heavy Metals from Contaminated Soil by Mushroom—*Galerina vittiformis*. *Ecotoxicol. Environ. Saf.* **2014**, *104*, 414–422. [CrossRef]
28. Sithole, S.C.; Mugivhisa, L.L.; Amoo, S.O.; Olowoyo, J.O. Pattern and Concentrations of Trace Metals in Mushrooms Harvested from Trace Metal-Polluted Soils in Pretoria, South Africa. *S. Afr. J. Bot.* **2017**, *108*, 315–320. [CrossRef]
29. Rhodes, C.J. Mycoremediation (Bioremediation with Fungi)—Growing Mushrooms to Clean the Earth. *Chem. Speciat. Bioavailab.* **2015**, *26*, 196–198. [CrossRef]
30. Bilal, M.; Park, I.; Hornn, V.; Ito, M.; Hassan, F.U.; Jeon, S.; Hiroyoshi, N. The Challenges and Prospects of Recovering Fine Copper Sulfides from Tailings Using Different Flotation Techniques: A Review. *Minerals* **2022**, *12*, 586. [CrossRef]
31. Qu, G.; Chen, B.; Zhang, D.; Wu, F.; Jin, C.; Li, H.; Liu, S.; Li, Y.; Qin, J. Pollutants' Migration and Transformation Behavior in Phosphorus Ore Flotation Tailings Treated with Different Additives. *Appl. Geochem.* **2022**, *143*, 105358. [CrossRef]
32. Antonijević, M.M.; Dimitrijević, M.D.; Stevanović, Z.O.; Serbula, S.M.; Bogdanovic, G.D. Investigation of the Possibility of Copper Recovery from the Flotation Tailings by Acid Leaching. *J. Hazard. Mater.* **2008**, *158*, 23–34. [CrossRef]
33. Galjak, J.; Đokić, J.; Dervišević, I.; Milentijević, G.; Mojsić, M.; Živković, B. Assessment of Pollution and Distribution of Heavy Metals in the Soil Near the Flotation Gornje Polje. *Pol. J. Environ. Stud.* **2022**, *31*, 1–10. [CrossRef]
34. Zhai, Q.; Liu, R.; Wang, C.; Wen, X.; Li, X.; Sun, W. A Novel Scheme for the Utilization of Cu Slag Flotation Tailings in Preparing Internal Electrolysis Materials to Degrade Printing and Dyeing Wastewater. *J. Hazard. Mater.* **2022**, *424*, 127537. [CrossRef]
35. Kinnunen, P.; Karhu, M.; Yli-Rantala, E.; Kivikytö-Reponen, P.; Mäkinen, J. A Review of Circular Economy Strategies for Mine Tailings. *Clean. Eng. Technol.* **2022**, *8*, 100499. [CrossRef]

36. Zhao, K.; Zhu, X.; Ma, H.; Ji, J.; Jin, X.; Sun, J. Design and Experiment of the Environment Control System for the Industrialized Production of Agaricus Bisporus. *Int. J. Agric. Biol. Eng.* **2021**, *14*, 97–107. [CrossRef]
37. Shahnaz, M.; Sharma, S.; Dev, D.; Prasad, D.N. Cultivation Technology and Antibacterial Activity of *Agaricus bisporus* (U-03). *Int. J. Pharm. Chem. Anal.* **2020**, *7*, 135–144. [CrossRef]
38. Kryczyk, A.; Piotrowska, J.; Sito, M.; Sulkowska-Ziaja, K.; Dobosz, K.; Opoka, W.; Muszynska, B. Remediation Capacity of Cd and Pb Ions by Mycelia of *Imleria badia*, *Laetiporus sulphureus*, and *Agaricus bisporus* in Vitro Cultures. *J. Environ. Sci. Health Part B* **2017**, *52*, 617–622. [CrossRef]
39. Kumar, P.; Kumar, V.; Goala, M.; Singh, J.; Kumar, P. Integrated Use of Treated Dairy Wastewater and Agro-Residue for *Agaricus bisporus* Mushroom Cultivation: Experimental and Kinetics Studies. *Biocatal. Agric. Biotechnol.* **2021**, *32*, 101940. [CrossRef]
40. Ugya, A.Y.; Imam, T.S. Efficiency of the decomposition process of agaricus bisporus in the mycoremediation of refinery wastewater: Romi stream case study. *World J. Pharm. Res.* **2017**, *6*, 200–211. [CrossRef]
41. *The R Foundation for Statistical Computing Platform, 2019*, version 3.6.1, The R Foundation for Statistical Computing Platform: x86_64-w64-mingw32/x64 (64-bit); R Core Team: Vienna, Austria, 2019.
42. Morrison, D.F. *Multivariate Statistical Methods*; McGraw-Hill Book Company: New York, NY, USA, 1990; p. 495.
43. Rzymski, P.; Mleczek, M.; Niedzielski, P.; Siwulski, M.; Gąsecka, M. Cultivation of *Agaricus bisporus* Enriched with Selenium, Zinc and Copper. *J. Sci. Food Agric.* **2017**, *97*, 923–928. [CrossRef]
44. Rzymski, P.; Mleczek, M.; Siwulski, M.; Gąsecka, M.; Niedzielski, P. The Risk of High Mercury Accumulation in Edible Mushrooms Cultivated on Contaminated Substrates. *J. Food Compos. Anal.* **2016**, *51*, 55–60. [CrossRef]
45. Zhou, L.; Guo, S.; Shu, M.; Liu, X.; Liu, X.; Guo, X. Accumulation Characteristics and Resistance to Heavy Metal Contamination of *Agaricus bisporus* Varieties in Shanxi Province, China. *Agric. Biotechnol.* **2018**, *7*, 220–224.
46. Nagy, B.; Mânzatu, C.; Măicăneanu, A.; Indolean, C.; Barbu-Tudoran, L.; Majdik, C. Linear and Nonlinear Regression Analysis for Heavy Metals Removal Using *Agaricus bisporus* Macrofungus. *Arab. J. Chem.* **2017**, *10*, S3569–S3579. [CrossRef]
47. Ali, A.; Guo, D.; Mahar, A.; Wang, P.; Shen, F.; Li, R.; Zhang, Z. Mycoremediation of Potentially Toxic Trace Elements—A Biological Tool for Soil Cleanup: A Review. *Pedosphere* **2017**, *27*, 205–222. [CrossRef]
48. Koutrotsios, G.; Danezis, G.; Georgiou, C.; Zervakis, G.I. Elemental Content in *Pleurotus ostreatus* and *Cyclocybe cylindracea* Mushrooms: Correlations with Concentrations in Cultivation Substrates and Effects on the Production Process. *Molecules* **2020**, *25*, 2179. [CrossRef]
49. Nagy, B.; Măicăneanu, A.; Indolean, C.; Mânzatu, C.; Silaghi-Dumitrescu, L.; Majdik, C. Comparative Study of Cd(II) Biosorption on Cultivated *Agaricus bisporus* and Wild *Lactarius piperatus* Based Biocomposites. Linear and Nonlinear Equilibrium Modelling and Kinetics. *J. Taiwan Inst. Chem. Eng.* **2014**, *45*, 921–929. [CrossRef]
50. Liang, X.; Gadd, G.M. Metal and Metalloid Biorecovery Using Fungi. *Microb. Biotechnol.* **2017**, *10*, 1199–1205. [CrossRef]
51. Chatterjee, S.; Sarma, M.K.; Deb, U.; Steinhauser, G.; Walther, C.; Gupta, D.K. Mushrooms: From Nutrition to Mycoremediation. *Environ. Sci. Pollut. Res.* **2017**, *24*, 19480–19493. [CrossRef] [PubMed]
52. Peintner, U.; Schwarz, S.; Mešić, A.; Moreau, P.A.; Moreno, G.; Saviuc, P. Mycophilic or Mycophobic? Legislation and Guidelines on Wild Mushroom Commerce Reveal Different Consumption Behaviour in European Countries. *PLoS ONE* **2013**, *8*, e63926. [CrossRef] [PubMed]
53. European Union. *Commission Regulation (EC) No 629/2008 of 2 July 2008 Amending Regulation (EC) No 1881/2006 Setting Maximum Levels for Certain Contaminants in Foodstuffs*; European Union: Brussels, Belgium, 2008.
54. EFSA (European Food Safety Authority). Panel on Dietetic Products Nutrition & Allergies. In *Dietary Reference Values for Nutrients*; Summary Report, Technical Report; EFSA Supporting Publication: Parma, Italy, 2017. [CrossRef]
55. EFSA Panel on Contaminants in the Food Chain (CONTAM). Scientific Opinion on Arsenic in Food. *EFSA J.* **2009**, *7*, 1351. [CrossRef]
56. Pająk, M.; Gąsiorek, M.; Jasik, M.; Halecki, W.; Otremba, K.; Pietrzykowski, M. Risk Assessment of Potential Food Chain Threats from Edible Wild Mushrooms Collected in Forest Ecosystems with Heavy Metal Pollution in Upper Silesia, Poland. *Forests* **2020**, *11*, 1240. [CrossRef]

Journal of *Fungi*

Article

Comparative Copper Resistance Strategies of *Rhodonia placenta* and *Phanerochaete chrysosporium* in a Copper/Azole-Treated Wood Microcosm

Gaurav Pandharikar [1,†], Kévin Claudien [1,†], Christophe Rose [2], David Billet [3,4], Benoit Pollier [5], Aurélie Deveau [1,*], Arnaud Besserer [6] and Mélanie Morel-Rouhier [1,*]

[1] Université de Lorraine, INRAE, IAM, 54000 Nancy, France; gaurav-girish.pandharikar@inrae.fr (G.P.); kevin.claudien@univ-lorraine.fr (K.C.)
[2] Université de Lorraine, AgroParisTech, INRAE, Silva, 54000 Nancy, France; christophe.rose@inrae.fr
[3] Université de Lorraine, CNRS, LIEC, 54000 Nancy, France; david.billet@univ-lorraine.fr
[4] Pôle de Compétences Chimie Analytique Environnementale, ANATELo, Université de Lorraine, 54000 Nancy, France
[5] Unit INRAE, UR 1138, BEF, 54280 Champenoux, France; benoit.pollier@inrae.fr
[6] Université de Lorraine, INRAE, LERMAB, F-54000 Nancy, France; arnaud.besserer@univ-lorraine.fr
* Correspondence: aurelie.deveau@inrae.fr (A.D.); melanie.morel@univ-lorraine.fr (M.M.-R.)
† These authors contributed equally to this work.

Abstract: Copper-based formulations of wood preservatives are widely used in industry to protect wood materials from degradation caused by fungi. Wood treated with preservatives generate toxic waste that currently cannot be properly recycled. Despite copper being very efficient as an antifungal agent against most fungi, some species are able to cope with these high metal concentrations. This is the case for the brown-rot fungus *Rhodonia placenta* and the white-rot fungus *Phanerochaete chrysosporium*, which are able to grow efficiently in pine wood treated with Tanalith E3474. Here, we aimed to test the abilities of the two fungi to cope with copper in this toxic environment and to decontaminate Tanalith E-treated wood. A microcosm allowing the growth of the fungi on industrially treated pine wood was designed, and the distribution of copper between mycelium and wood was analysed within the embedded hyphae and wood particles using coupled X-ray fluorescence spectroscopy and Scanning Electron Microscopy (SEM)/Electron Dispersive Spectroscopy (EDS). The results demonstrate the copper biosorption capacities of *P. chrysosporium* and the production of copper-oxalate crystals by *R. placenta*. These data coupled to genomic analysis suggest the involvement of additional mechanisms for copper tolerance in these rot fungi that are likely related to copper transport (import, export, or vacuolar sequestration).

Keywords: copper; wood; detoxification; ligninolytic fungi; oxalate; biosorption

Citation: Pandharikar, G.; Claudien, K.; Rose, C.; Billet, D.; Pollier, B.; Deveau, A.; Besserer, A.; Morel-Rouhier, M. Comparative Copper Resistance Strategies of *Rhodonia placenta* and *Phanerochaete chrysosporium* in a Copper/Azole-Treated Wood Microcosm. *J. Fungi* **2022**, *8*, 706. https://doi.org/10.3390/jof8070706

Academic Editors: Miha Humar and Ivan Širić

Received: 10 June 2022
Accepted: 29 June 2022
Published: 4 July 2022

1. Introduction

Wood is an organic composite that is mainly characterized by its hygroscopic behaviour, orthotropic composition, and variable natural durability [1]. Wood can be considered as an ecological construction material because it is renewable, sustainable, and provides a solution to CO_2 sequestration [2]. However, the susceptibility of timber to both biotic (fungi, bacteria, insects) and abiotic degradation (UV, erosion) when used in outdoor terrestrial environments may be a limitation [3]. Direct soil contact is one of the most severe exposure situations that wood can be subjected to. Indeed, in-ground wood is permanently wet and stays in direct contact with wood-degrading organisms such as fungi and bacteria, which, in turn, become well established and can proliferate through the readily available nutrient sources in the wood [4]. To protect timber against microbes at the industrial scale, wood is commonly impregnated with copper (Cu)-based preservatives [2,5]. The most widely used are chromium copper salts, which protect timber for 20 years if properly impregnated by vacuum pressure treatments.

Wood preservative formulations that are currently found on the market include ACQ (alkaline Cu quaternary), CA (Cu-azole), Cu-citrate, and Cu-ethanolamine [6,7]. Among these, Tanalith E®, a Cu-Azole-based compound, is the most widely used wood preservative in Europe [8]. The antifungal activities of CA compounds rely on the complementary effects of azoles that inhibit the fungal ergosterol biosynthesis pathway [6,9,10] and the toxicity of high doses of Cu. The main effects of an excess of free cytosolic Cu ions are the inactivation of metalloenzymes by metal displacement, the perturbation of Fe-S cluster assembly, and the generation of reactive oxygen species (ROS) through Fenton chemistry, causing biological damage to the fungal cell [11,12].

However, some wood-decay fungi exhibit high detoxification and tolerance properties against Cu-based preservatives and are able to cope with high amounts of Cu [13,14]. This could lead to resistance to Cu-based antifungal compounds, thus making them inefficient to protect wood against fungal decay. On the other hand, these abilities to resist high Cu concentrations may be exploited for the detoxification of treated wood waste. Indeed, while being essential for wood preservation, these harmful wood preservatives pose a significant threat to the ecosystem and human health when treated wood becomes waste. Currently, it is not possible to recycle this biomass due to the toxicity of the compounds used for its preservation. Nowadays, wood biomass is treated by specific incineration, pyrolysis, and gasification techniques. Although these methods of chemical disposal are low-cost and efficient, they still release toxic pollutants and other harmful components into the atmosphere that potentially damage the environment and public health [15]. To overcome this problem, alternative sustainable solutions to treat wood waste need to be developed. In recent years, some studies have shown that fungi may be used as biocatalysts for copper decontamination in wood waste [16]. These fungi rely on their ability to both leach Cu from the wood and to resist its toxicity.

The cellular and molecular mechanisms responsible for Cu tolerance have been described in ascomycetes. Two major mechanisms have been proposed for Cu detoxification. One of them, which has been well described in *Saccharomyces cerevisiae,* relies on Cu sequestration by metallothioneins [17,18]. The second mechanism relies on Cu extrusion by Cu-transporting ATPases [19]. Few data are available in fungi, but in human, Cu-transporting ATPases change their location from the trans-Golgi compartment to the cell membrane in response to Cu toxicity to act as export pumps conferring copper resistance [20]. Other mechanisms such as the down regulation of Cu importers, ion sequestration into the vacuole, the production of extracellular chelators, or cell wall biosorption could be involved [11].

In contrast to ascomycetes, the molecular mechanisms of Cu homeostasis and resistance mechanisms to preservatives are unexplored in Basidiomycetes. As some of them can cope with high amounts of Cu through the detoxification of copper via the production of copper oxalate, bioleaching, and/or biosorption activities [21,22], they may thus be exploited for the detoxification process of treated wood waste. In the present work, we aimed to decipher the strategies used by brown-rot *Rhodonia placenta* and white-rot *Phanerochaete chrysosporium* fungi, which are known to have high resistance to Cu, to colonize pine wood treated with Tanalith E. We combined experimental and in silico approaches to (i.) identify how the two fungi cope with toxic levels of Cu in the presence of azoles and (ii.) identify the potential molecular players involved in Cu homeostasis and resistance. For this purpose, we have developed a fungal microcosm to work directly on treated wood waste as it is released after industrial use. We used it to test the growth of the two fungi on industrial wood-treated sawdust. We then measured the levels of Cu in the wood, fungal hyphae, and liquid media by coupling X-ray fluorescence spectroscopy (XRF), Inductively Coupled Plasma Atomic Emission Spectrometry (ICP-OES), and Scanning Electron Microscopy (SEM)/Electron Dispersive Spectroscopy (EDS) to determine the fungal ability to detoxify treated wood. In parallel, comparative genomics were used to explore the potential mechanisms used by the two fungi to cope with Cu toxicity at the cellular level. Our results indicate that the two fungi use different strategies to cope with Tanalith E toxicity.

2. Material and Methods

2.1. Microcosm Set-Up

A biological microcosm was developed to grow *Rhodonia placenta* (Fr.) Niemelä, K.H. Larss. & Schigel (*Rp*) (older names: *Oligoporus placenta*, *Postia placenta*) and *Phanerochaete chrysosporium* RP78 (Burds.) Hjortstam & Ryvarden (*Pc*) in submerged cultures with treated wood. The wood samples were collected from pine wood planks that were industrially impregnated (autoclave class IV impregnation) with Tanalith E3474 preservative product, raising the concentration within the wood to 16.7 kg/m^3. Tanalith E3474 (Arch Timber Protection Ltd., Castleford, UK) is a commercial formulation composed of 16.4% *w*/*w* copper (copper (II) carbonate–copper (II) hydroxide 1:1), 0.18 % *w*/*w* tebuconazole, and 0.18% *w*/*w* propiconazole. After treatment, the wood was dried, and samples were milled with a cutting mill SM 100 (Retch) to obtain particles with a size between 0.5 and 2 mm. To remove residual unbound products within the sawdust, three mechanical leaching steps were performed over 24 h in ultrapure water by continuous magnetic stirring in 1 L flasks. Subsequently, sawdust was dried overnight at 80 °C in an oven. Parallelly fungal preculture was prepared in 1% malt broth by adding one 7 mm fungal plug per flask and kept at 28 °C in an incubator for 4 days. Biological microcosm prepared in Erlenmeyer flasks containing 25 mL of 1 % malt medium and 2% sawdust (treated sawdust (TSD) or non-treated sawdust (NTSD)) were autoclaved (120 °C—20 min).The fungal precultures that had been grown for 4 days were inoculated in the autoclaved microcosm containing sawdust. Subsequently, the biological microcosm was incubated at 28 °C for ten days and with shaking at 80 rpm shaking for *P. chrysosporium* and without shaking for *R. placenta*.

2.2. Respiration Tests

To evaluate the biological activity of *R. placenta* and *P. chrysosporium* in the presence of NTSD and TSD, a fungal respiration assay was conducted. Sealed flasks containing 1% malt medium and 2% leached sawdust were inoculated with four days grown fungal precultures. Fungal activity was followed using a carbometer (LAMBDA laboratory instruments), allowing for the non-destructive quantification of the CO_2 released during a ten-day growth period. A 120 mL amount of air was harvested every two days and replaced with new 0.2 µm filtered atmospheric air to avoid anoxia in the system. Five biological replicates were used.

2.3. X-ray Fluorescence Spectroscopy and Inductively Coupled Plasma Atomic Emission Spectrometry for Copper Quantification in the Liquid Phase

To quantify the precise amount of Cu leached by the fungi in the liquid phase of the microbial microcosms, X-ray Fluorescence Spectrometry (XRF) and Inductively Coupled Plasma Atomic Emission Spectrometry (ICP-OES) were coupled. At 10 days post inoculation, the liquid phase was separated from the sawdust colonized by the fungus by centrifugation at 8350× *g* for 15 min. For the quantification of Cu by XRF (Thermo Scientific, Waltham, MA, USA), 500 µL of the liquid phase was diluted with 1% malt in 2 mL polypropylene cups that were transparent to X-rays. Spectra were analysed using the UniQuant (Thermo Scientific) program after calibration according to the manufacturer's instructions using different elements. For Cu quantification by ICP-OES (Agilent 720/725 ICP-OES), the liquid phase was filtered through 0.45 µm polytetrafluoroethylene filters, and 10 mL samples were directly used for Cu quantification. A standard curve was prepared using commercially available Cu standard solutions (Merck). Prior to analysing the main experimental samples, semiquantitative analyses were performed to determine the accurate quantity of Cu at highest and lowest precision levels in the biological samples. The quantification results were generated through ICP Expert II (Agilent) software and were further analysed. Three and six biological repeats were performed for the XRF and ICP analyses, respectively.

2.4. Scanning Electron Microscopy (SEM) and Electron Dispersive Spectroscopy (EDS) Microanalyses for Copper Quantification in Sawdust and Fungal Hyphae

Coupling electron microscopy with microanalysis (SEM-EDS) allowed for the amount of Cu inside the wood and inside the fungal hyphae to be determined independently. Following 10 days of incubation in the microcosm, the sawdust colonized by the fungus was separated from the liquid phase by centrifugation at 8350× *g* for 15 min. Wood chips corresponding to the solid sawdust/fungus samples were slowly desiccated overnight in a freeze dryer (FreeZone6literbenchtop, Labconco, Kansas City, MO, USA). Subsequently, wood chips were press-stuck on stubs using a conductive carbon cement (LEIT_C, agar scientific) and dried at 50 °C for one hour in oven. A canned air duster was used to ensure the steadiness of the samples on the stubs. Samples were coated with two layers of carbon (10 nm final carbon layer thickness) using four high-current pulses on carbon threads (ACE 600, Leica microsystems, Wetzlar, Germany). Samples were initially observed using a Field Emission Gun SEM (FEGSEM-SIGMA HD-VP; Zeiss, Oberkochen, Germany) placed in high-vacuum mode (10^{-4} Pa), at a 20 kV high-acceleration voltage, a 1 nA current beam, and a 9 mm working distance (requisite analytical distance) in the way to identify and select Sites of Interest (SI). The numeric Back Scattered Detector signal (QBSD; Zeiss) was used to acquire all SI images, and further EDS analyses were performed with a spectrometer (EDS-SDD 80 mm^2 detector; Oxford instruments, Abingdon, UK) for each SI (three to nine SI per sample; ten microanalyses per SI). Each obtained spectrum was deconvoluted by the software algorithm (INCA software, Oxford Instruments, Abingdon, UK), and the results produce the estimation of the semi-quantitative mass fraction of Cu in the sample. To distinguish between the copper and calcium inside different-coloured crystals, INCA mapping software was used. The reliability of the data relies on the repeatability of the measurements: up to 120 spectra were obtained for each sample condition, and 3 biological samples were prepared for each condition (Refer to Supplemental Data S1 for more details on sampling).

2.5. High Performance Liquid Chromatography for Soluble Oxalate Quantification

For the quantification of soluble oxalate, samples were harvested ten days post inoculation and centrifuged at 8346× *g* for 15 min. The liquid phase of the samples was harvested and filtered with 0.2 µm PTFE (polytetrafluoroethylene) filters. Filtered samples were processed via high-pressure ion chromatography (HPIC) ICS-5000+ (thermo) coupled with ab ISQ EM single-quadrupole mass spectrometer (MS). Oxalate was retained on the Thermo IonPac AS11-HC 2 × 250 mm column set and was detected with a conductimetric detector. The separation was enhanced with mass spectrometric detection in SIM mode. The hydroxide eluent started with a low concentration (1 mM KOH) to separate the weakly retained anions. After maintaining this concentration for 8 min, the eluent concentration was gradually increased to elute the retained anions more strongly. The KOH concentration was increased to 30 mM at 28 min, during which time the oxalate eluted (retention time 22 min). The oxalate's limit of quantification was 50 µg/L. A 10 millilitre amount of 0.2 µm filtered liquid phase was used for quantification. Three biological replicates from each treatment were performed.

2.6. Statistical Analyses

All statistical analyses performed with R studio and GraphPad Prism version 7.04. All figures were generated through GraphPad Prism 7.04. All experimental data were expressed as mean ± s.e. To test the normality of the samples, the Shapiro–Wilk (W) test was performed. For the fungal respiration assay, two-way ANOVA was performed. Differences between the conditions over the time period were tested using Tukey's post hoc multiple comparison test. Data generated on the ICP, XRF, and SEM-EDS were analysed using one-way ANOVA. Then, Tukey's post hoc multiple comparison test was performed to identify possible statistical differences between the different conditions.

2.7. Genome Mining to Compare Copper Related Genes in R. placenta and P. chrysosporium

Gene sequences related to Cu homeostasis were first searched for within the *Saccharomyces* genome database (Available online: https://www.yeastgenome.org/ (accessed on 9 February 2022) using "copper" as a keyword. The list was then manually curated to the keep genes coding for proteins involved in Cu transport, chelation, and reduction and the regulation of gene expression. The amino acid sequences of these candidates were used as a template for a Blast search using *R. placenta* MAD-698-R-SB12 v1.0 and the *P. chrysosporium* RP78 v2.2 genomes from the Mycocosm of the Joint Genome Institute database (Available online: https://mycocosm.jgi.doe.gov/mycocosm/home (accessed on 9 February 2022)). To complete the analysis, a search using "copper" as keyword was also performed on both genomes. The obtained sequences were compared to the ones retrieved using the Blast search tool. Additional sequences were manually checked for annotation and were included in the dataset. Finally, each sequence was blast back to the corresponding genome to retrieve all isoforms for a specific gene family. Evolutionary analyses were conducted in MEGA X [23]. The evolutionary history was inferred using the neighbor-joining method [24]. The trees were drawn to scale, with branch lengths in the same units as those of the evolutionary distances used to infer the phylogenetic tree. The evolutionary distances were computed using the p-distance method [25] and are in the units of the number of amino acid differences per site. All ambiguous positions were removed for each sequence pair (pairwise deletion option).

3. Results

3.1. Efficient Growth of Both the Brown-Rot Rhodonia Placenta and the White-Rot Phanerochaete Chrysosporium on Tanalith E3474-Treated Wood

Phanerochaete chrysosporium and *R. placenta* were cultivated in the presence of non-treated or treated pine sawdust supplemented with a small volume of malt medium. The supplementation with malt aimed to provide the fungi with a carbon source and other nutrients to remove the efficiency of the wood-degradative systems and to only focus on the detoxification of the treated wood. The system was set-up to allow the whole analysis of both the liquid phase and the solid phase composed of a complex embedded structure of hyphae and wood particles (Figure S1). Scanning Electron Microscopy images of the solid phase clearly showed that both fungi were able to colonize treated wood particles, without obvious modifications to the hyphae morphology compared to the non-treated sawdust condition (Figure 1A). To assess the metabolic activity of the fungi on the treated sawdust, a respiration assay was performed from day 2 to 10 after fungal inoculation on sawdust (Figure 1B). The cumulative amount of CO_2 released at day 10 was lower in TSD compared to the control without sawdust and NTSD samples for *P. chrysosporium*. The overall amount of CO_2 released per day was relatively constant all along the kinetics (around 10 mg for *R. placenta* and 17 mg for *P. chrysosporium*). Globally, the presence of Tanalith E within the wood substrates had little to no impact on the primary metabolism of *R. placenta* and *P. chrysosporium* or on their ability to colonize wood particles.

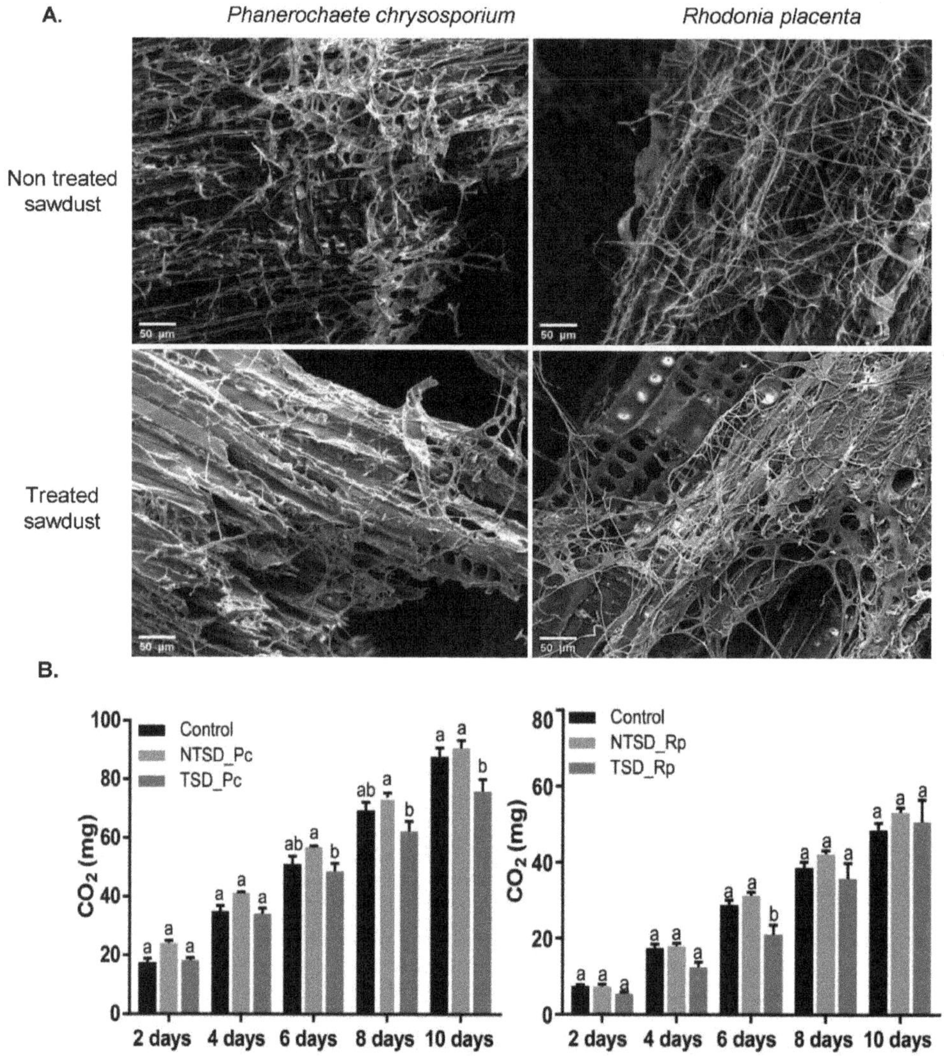

Figure 1. Fungal colonization of non-treated and treated sawdust: (**A**) Scanning Electron Microscopy images of *P. chrysosporium* and *R. placenta* colonization of treated (TSD) and non-treated (NTSD) sawdust after 10 days; (**B**) cumulative CO_2 production for *P. chrysosporium* and *R. placenta* grown on malt (control), non-treated (NTSD_Pc/NTSD_Rp), and treated (TSD_Pc/TSD_Rp) sawdust over 10 days; mean ± s.e., $n = 5$; two-way ANOVA and Tukey's post hoc test (p-value ≤ 0.05). Identical letters indicate no significant differences in the respiration between treatments at a given time point.

3.2. Comparative Copper Bioleaching Ability of *R. placenta* and *P. chrysosporium* from Tanalith E3474 Treated Wood

First, total Cu was quantified in NTSD and TSD by X-Ray fluorescence before and after leaching. A total of 2.78 g of copper/kg of dry wood (2780 ppm) was measured in the starting TSD material. The water leaching steps released 0.7 % of the Cu measured in TSD. In NTSD, almost no Cu was detected. Cu levels in the liquid phase of the microcosm were quantified at 10 days of culture by both XRF and ICP-OES (Figure 2A,B). As expected, negligible amounts of Cu were detected in the liquid phase of the non-treated sawdust setup (<2 ppm, data not shown). For TSD, a control without fungi was analysed to estimate the amount of Cu leached by the malt medium and the sterilization step. Less than 20 ppm was released in the culture medium without fungi (Figure 2A,B).

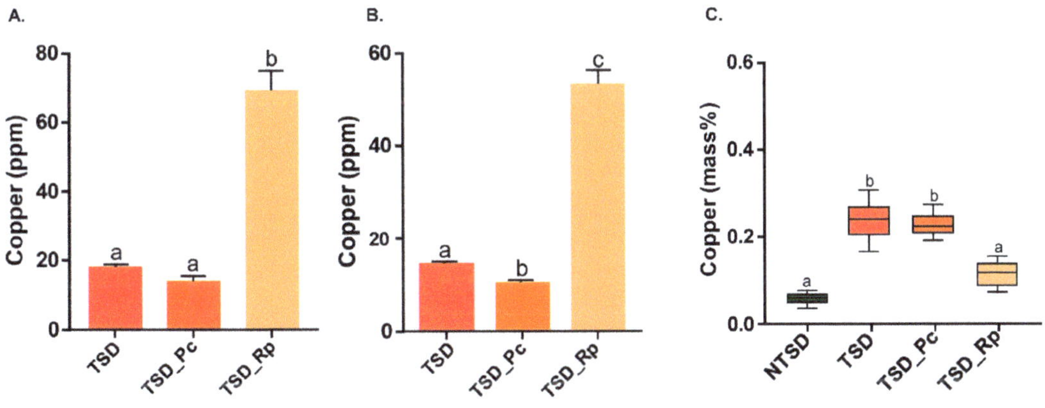

Figure 2. Copper quantification in liquid and solid phases of the microcosm after 10 days of *P. chrysosporium* and *R. placenta* growth with TSD (TSD_Pc/TSD_Rp). Amount of copper retrieved in the liquid phase measured by XRF (**A**) and ICP-OES (**B**). TSD represents the control without fungus. XRF = mean ± s.e., $n = 3$; ICP-OES = mean ± s.e., $n = 6$. (**C**) Copper quantified in wood by SEM-EDS microanalyses. The amount of copper was expressed as the estimated mass fraction of copper in the sample (mean ± s.e., $n = 9$; one-way ANOVA and Tukey's post hoc test (p-value ≤ 0.05), different letters to show statistically significant differences).

In the presence of *P. chrysosporium* (TSD_Pc), a decrease of about 25% in the amount of Cu in the liquid phase was measured compared to the TSD condition without fungi. This difference was observed using both methods; however, it was only statistically significant for the ICP-OES analysis. Additionally, Cu was quantified in the sawdust by Scanning Electron Microscopy with Energy Dispersive Spectroscopy (SEM-EDS). The Cu levels in the TSD_Pc samples were similar to the control without mycelium (TSD) (Figure 2C). Taken together, these results suggest that *P. chrysosporium* has little ability to bioleach Cu from treated sawdust in liquid medium. In contrast, a strong release (about 38 to 51 ppm) of Cu was measured in the liquid phase for *R. placenta* (Figure 2A,B), which was correlated with a significant decrease (about 54 %) in Cu in the treated sawdust particles (TSD_Rp) (Figure 2C). The Cu level in TSD_Rp was relatively similar to that of NTSD, showing successful Cu leaching. It is well-documented that metal bioleaching in *R. placenta* is mainly ensured by organic acid secretion by fungi [26,27]. Thus, soluble oxalate was quantified in the liquid phase of the culture system at 10 days (Figure 3). The amount of soluble oxalate reached 300 mg/L for *R. placenta* after 10 days of culture in treated sawdust, suggesting that the high Cu leaching efficiency of *R. placenta* relied on its high oxalate production. In contrast, low levels (6 mg/L) of soluble oxalate were secreted by *P. chrysosporium*.

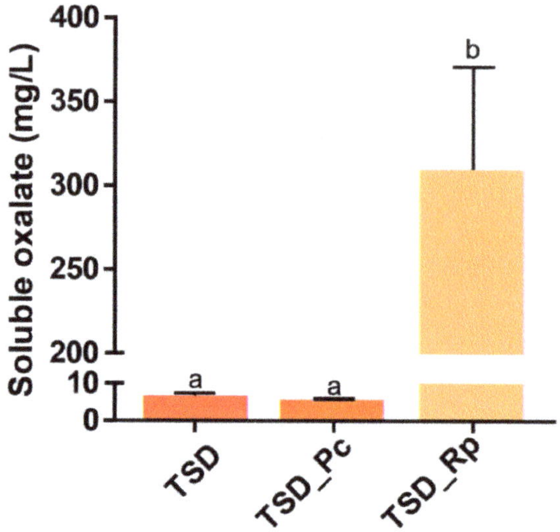

Figure 3. Soluble oxalate quantification in the liquid phase of the microcosm after 10-day culture of *P. chrysosporium* and *R. placenta* with TSD (TSD_Pc/TSD_Rp). TSD represents the control without fungus (mean ± s.e., $n = 3$, one-way ANOVA and Tukey's post hoc test (p-value ≤ 0.05), different letters to show statistically significant differences).

3.3. SEM-EDS Based Comparative Analysis of Copper Detoxification Strategies between R. placenta and P. chrysosporium

To better understand how *R. placenta* and *P. chrysosporium* resist the toxicity of the high amounts of Cu within the treated sawdust, Cu was quantified in the hyphae of both fungi. Since the solid phase of the setup is a complex matrix of embedded wood particles and hyphae (Figure S1), SEM coupled with EDS microanalyses were used to quantify the Cu in the hyphae independently from the wood particles (Figure 4A). Cu accumulation was measured in the *P. chrysosporium* (TSD_Pc) hyphae, suggesting the mycelial biosorption of the metal within the cell wall and/or the internalization of Cu within *P. chrysosporium* cells (Figure 4B). Contrary to classical Cu quantification performed on the whole fungal sample, the SEM-EDS approach allowed us to discriminate the Cu biosorbed onto the fungal hyphae from the Cu immobilized at the surface in the form of copper-acid crystals. The Cu levels of *R. placenta* (TSD_Rp) hyphae remained very low; however crystals of two different shapes associated with the *R. placenta* fungal hyphae: ball and diamond shape, were detected (Figure 5A). Microanalyses identified the ball shape (green) as copper oxalate crystals and the diamond shape (red) as calcium oxalate crystals. The same analyses on *P. chrysosporium* showed only a few oxalate crystals. However, no Cu-oxalate crystals were observed, and the few crystals that were detected were Ca-oxalate crystals (Figure 5B). Overall, these results confirm that in the case of Tanalith E-treated wood, *P. chrysosporium* likely uses biosorption at the cell wall as one mechanism to ensure resistance to copper. In contrast, *R. placenta* uses oxalate for extracellular Cu immobilization. However, it is unlikely that the immobilization Cu-oxalate crystals was used as a single mechanism by *R. placenta* for Cu tolerance. Indeed, the strong oxalate-dependent bioleaching released high amounts of soluble Cu in the liquid phase of the microcosm with TSD that the fungus has to cope with.

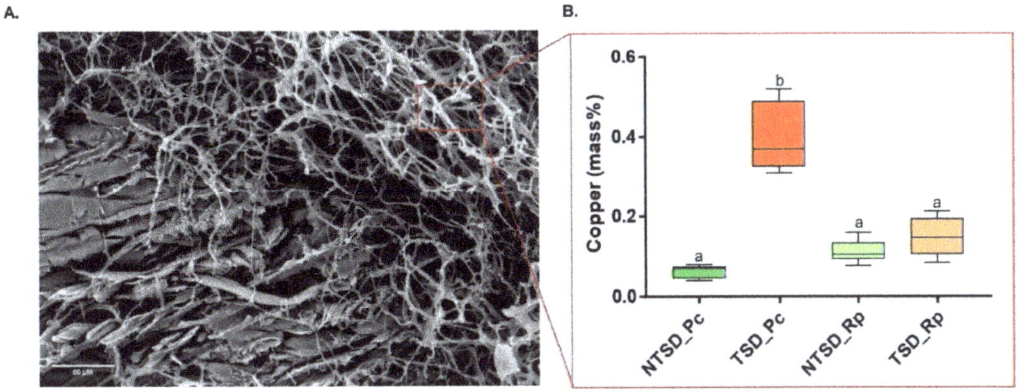

Figure 4. SEM-EDS-based copper quantification in fungal hyphae of *P. chrysosporium* and *R. placenta* after 10 days in the microcosm with TSD or NTSD. (**A**) SEM image of *P. chrysosporium* hyphae. Zones only focusing on mycelium and avoiding detecting copper in the wood substrate were selected for analysis. (**B**) Copper amount in fungal hyphae (mean ± s.e., $n = 9$, one-way ANOVA and Tukey's post hoc test (p-value ≤ 0.05), different letters to show statistically significant differences).

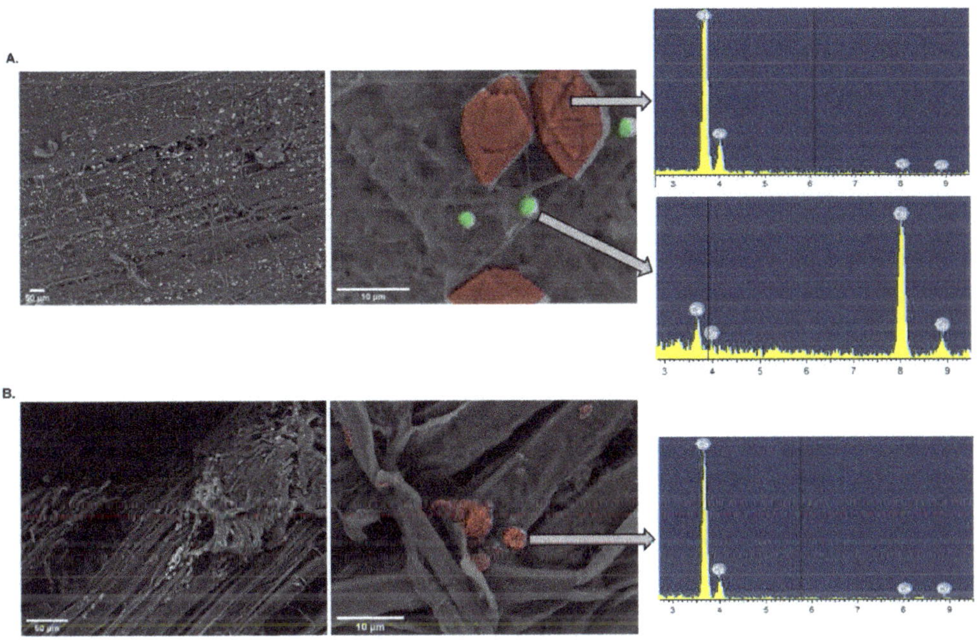

Figure 5. SEM images of oxalate crystals after 10 day culture of *P. chrysosporium* and *R. placenta* with TSD. (**A**) Oxalate crystals produced by *R. placenta*. Red crystals (diamond shape) and green crystals (ball shape) represent calcium and copper oxalate crystals, respectively. (**B**) Oxalate crystals produced by *P. chrysosporium*. Only calcium oxalate crystals were detected.

3.4. Comparative Genomics of Copper-Related Genes in R. placenta and P. chrysosporium

To obtain an overview of the putative mechanisms involved in the cell response of *R. placenta* and *P. chrysosporium* to copper, we performed a comparative genomic analysis focused on copper-related genes (Figure 6). Since Cu homeostasis is well-studied in

S. cerevisiae, a search of orthologues was conducted in the genomes using *S. cerevisiae* sequences as a template. *R. placenta* and *P. chrysosporium* contain a higher number of genes related to Cu transport compared to *S. cerevisiae*, and interestingly, *P. chrysosporium* exhibits more genes coding for the Cu transporters from the CTR, FET, and ATPases families, such as the ferric reductase and cupric reductase from the FRE family compared to *R. placenta* (Figure 6A). ScCtr2 is a low-affinity Cu transporter of the vacuolar membrane [28]. Only one orthologue of ScCtr2 was found in *R. placenta* and *P. chrysosporium*. *P. chrysosporium* exhibits four sequences, (compared to one in *R. placenta*) related to the high-affinity Cu transporters ScCtr1 and ScCtr3 in the plasma membrane (Figure 6B). FET proteins participate in Cu transport. In yeast, ScFET3 and ScFET5 have been characterized as multicopper oxidases that oxidize ferrous iron (Fe^{2+}) to ferric iron (Fe^{3+}) for subsequent cellular uptake by transmembrane permease Ftr1p [29]. These proteins are required for high-affinity iron uptake and are involved in mediating resistance to Cu ion toxicity. ScFET4 is described as a low-affinity Fe^{2+} transporter of the plasma membrane [30]. Four and five FET-related sequences were detected in the *R. placenta* and *P. chrysosporium* genomes, respectively, compared to three in *S. cerevisiae* (Figure 6C). Ion-transporting P-type ATPases belong to an extended family of transporters. In *S. cerevisiae*, ScCCC2 was identified as being specific to Cu transport and PCA1, which is defined as a cadmium-transporting P-type ATPase, and may also have a role in Cu and iron homeostasis [31]. Figure 6D shows the phylogenetic relationship between the P-ATPase genes. The ones related to Cu transport are underlined in pale yellow. Two orthologues of ScCCC2 were identified in both the *R. placenta and P. chrysosporium* genomes, while one and two orthologues of ScPCA1 were identified in these genomes, respectively (Figure 6D). FRE proteins have been described as ferric reductases and cupric reductases that reduce siderophore-bound iron and oxidized Cu prior to uptake by transporters. In *S. cerevisiae*, FRE1 and FRE2 have been described as being able to oxidize Cu prior to uptake [32]. With few exceptions, sequences from yeast cluster together, and sequences for both basidiomycetes cluster separately, suggesting divergence among this protein family (Figure 6E). Another interesting difference between *S. cerevisiae* and the two basidiomycetes is that no gene coding for metallothionein was detected in the latter genomes. Metallothioneins are small proteins that are rich in cysteines that act in intracellular Cu scavenging. They have been proposed to be one of the two main mechanisms responsible for Cu tolerance in *S. cerevisiae* [19]. This is obviously not the case in basidiomycetes. Globally, this genomic analysis suggests that the difference in gene numbers between *S. cerevisiae* and the two basidiomycetes but also between *R. placenta* and *P. chrysosporium* could reflect the divergence in the cellular response for regulating Cu transport between these fungi.

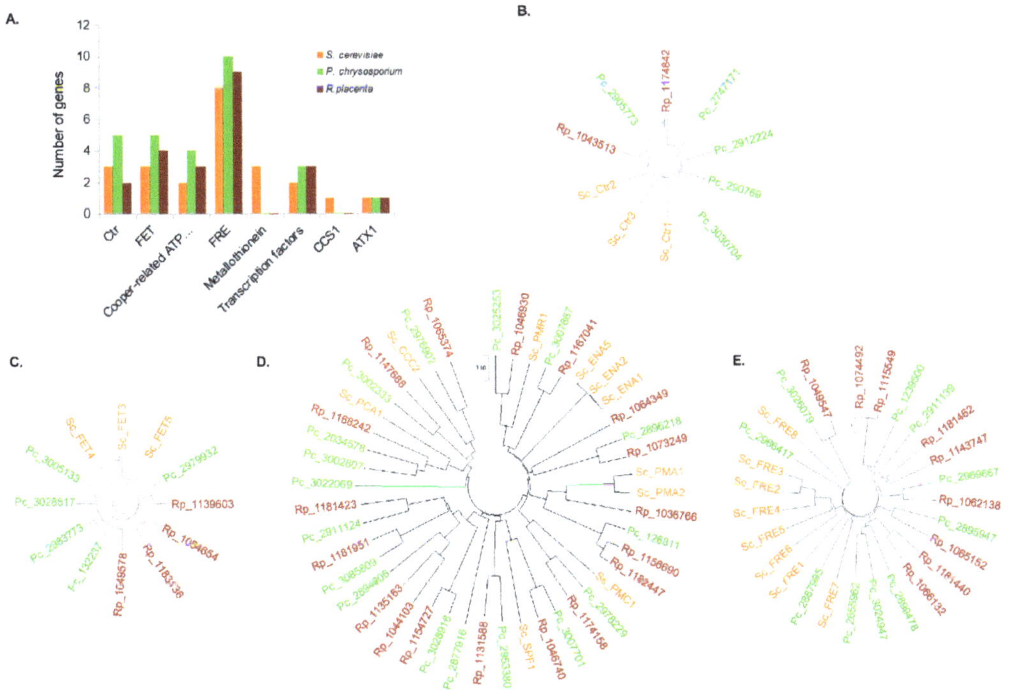

Figure 6. Comparative genomic analysis of copper-related genes in *P. chrysosporium*, *R. placenta*, and *S. cerevisiae*. (**A**) Number of genes coding for the well-known copper-related proteins in *S. cerevisiae*, related to copper transporters (Ctr, FET, ATPases and FRE), intracellular chelating proteins (Metallothionein), transcription factors, the copper chaperone for superoxide dismutase CCS1, and the cytosolic copper metallochaperone ATX1. Phylogenetic relationship between *S. cerevisiae* (in orange), *R. placenta* (in red), and *P. chrysosporium* (in green) genes coding for Ctr (**B**), FET (**C**), P-ATPases (**D**), and FRE (**E**) orthologues.

4. Discussion

Fungicides such as Tanalith E used for wood preservation contains a combination of Cu and azoles that inhibit the development of wood-degrading microorganisms by disrupting different basic metabolic processes [5,8,33]. Yet, as illustrated by our study and previous ones [21,34,35], some wood-degrading fungi are able to cope with these toxic compounds. In the present case, SEM microscopic images and respiration measurements indicated the successful colonization of the treated sawdust in the microcosm by both the fungi without impairing fungal growth and primary metabolism when using the preservative treatment. However, while both fungi were able to grow well in the presence of Tanalith E, the resistance mechanisms to Cu toxicity deployed by the two fungi differed drastically. Our data suggest that in the presence of Tanalith E, the resistance of the brown-rot fungus *R. placenta mainly* relied on the immobilization of Cu by oxalic acid, while the white-rot fungus *P. chrysosporium* used biosorption to resist Cu toxicity. Such behaviors are consistent with previous studies since biosorption and metal precipitation are the two main mechanisms used by fungi to tolerate or survive against different toxic metal stresses [11,22,36–39].

White- and brown-rot fungi differ in their ability to degrade wood. White-rot fungi are unique in terms of lignin degradation, while brown-rot fungi circumvent lignin to degrade holocellulose via iron-dependent oxidative chemistry [12,40,41]. Both groups of fungi can produce oxalate during wood degradation, although at variable levels depending on the species, and oxalic acid secretion may promote wood decay by reducing pH and mobilizing iron [36,42]. In addition, some fungi, especially brown rots, have been shown to be Cu tolerant due to high oxalate production [43,44]. Oxalic acid produced by fungi accumulates as oxalate salt crystals on the outside of fungal hyphae and work as metal chelators rendering the Cu ion inert [40,45]. Likewise, SEM coupled to EDS microanalyses on the Cu-oxalate crystals associated with the fungal hyphae of *R. placenta* identified higher levels of Cu incorporated inside the Cu-oxalate crystals and very low levels of Cu present on the wood. Similar observations were made with *Fomitopsis palustris*, for which microscopic observations of the mycelial mat scraped off from wood blocks highlighted Cu-oxalate complexes at the interface between both fungal mats and the wood surfaces [46]. If oxalic acid was constitutively produced by the fungus in malt extract in our experimental conditions, the levels of oxalate produced by *R. placenta* were two times higher in the presence of TSD compared to NTSD (data not shown), suggesting that the oxalate production by *R. placenta* not only occurred during primary metabolic process but was also stimulated in the presence of copper, likely for copper-based fungicide tolerance.

Interestingly, the ICP and XRF results showed high levels of Cu in the liquid phase, indicating that not all the Cu was immobilized in the Cu-oxalate crystals or was alternatively released from the crystals due to physicochemical modifications in the medium (pH or metabolite secretion for example). These results suggest that the oxalate-mediated chelation of Cu by *R. placenta* was able to mobilize Cu from the wood surface to the liquid phase. These results confirm the strong bioleaching of Cu by *R. placenta*, leading to Cu removal from the wood. Overall, the levels of soluble oxalate were directly correlated with the ability to cleanse Cu-treated wood. However, since the oxalate-dependent bioleaching activity of *R. placenta* released Cu in the liquid phase, it is likely that the complexation of Cu in Cu-oxalate crystals was not sufficient to explain the resistance phenotype of the fungus towards copper, since under these conditions, Cu is still present in high amounts in the liquid phase that is in direct contact with the hyphae. Therefore, it is thus unlikely that a single mechanism was employed by the fungus for Cu tolerance when grown on treated wood.

The ability to produce oxalic acid varies according to the fungal species. Likewise, we found very low levels of oxalate in the *P. chrysosporium* microcosm after 10 days of culture on TSD. Multiple studies on white-rot fungi, especially *P. chrysosporium*, have showed their ability to degrade oxalate rather than to produce and secrete this acid [47,48], and this could be dependent on the culture conditions. The factors affecting oxalate production are primarily the carbon and nitrogen sources in the culture medium and the pH of the environment [49]. When nutrients are plentiful, such as in cultures grown under high-carbon and -nitrogen conditions, no oxalate can be detected in cultures of *P. chrysosporium* [50]. Since the microcosm contained 1 % malt, this could be the reason for the low level of oxalate production by *P. chrysosporium* in this condition. However, the amount of oxalic acid produced did not necessarily indicate the degree of Cu tolerance exhibited by the fungi. For example, *R. placenta* produced higher levels of oxalic acid than *Fibroporia (Poria) vaillantii* but was less Cu tolerant, suggesting that other mechanisms were also involved [51]. Accordingly, we demonstrated that *P. chrysosporium* is tolerant to Tanalith E-treated sawdust despite the lack of oxalate production. One of the probable mechanisms of resistance could be the biosorption ability of this species, which was highlighted by SEM coupled to EDS microanalyses. The high capacity for biosorption was already demonstrated for *P. chrysosporium*, with most studies describing the biosorption of metals, and water contaminants in particular, from aqueous systems [22,52–54]. There is limited research on the mechanisms and performance for Cu biosorption in fungi. Indeed, biosorption is a complex mechanism influenced by many physico-chemical and biological factors. The fungal cell wall can act as a cation exchanger, and Cu^{2+} ions can bind or be complexed by carboxylic, phosphate, amine, or sulfhydryl groups of proteins at the cell surface [55]. A high fungal biomass is thus correlated with a high capacity for biosorption [56]. However, no relationship between mycelial growth among different fungal species and the removal of the metal elements present in chromated Cu arsenate-treated wood was found [57]. This points out that this process is part of the Cu resistance mechanism but not of the Cu remediation process. This is consistent with our results since we quantified Cu accumulation in *P. chrysosporium* hyphae but no Cu removal from the treated wood. The biosorbed Cu was thus likely the soluble Cu from the solution due to leaching from TSD after autoclave sterilization in the malt medium. The existence of mucilaginous sheaths around hyphae can also provide a matrix for fungus–mineral interactions and metal transformations harbouring mineral-weathering and metal-chelating agents. Accordingly, the extracellular polymeric substances (EPS) of *P. chrysosporium* were shown to contribute to Pb stress resistance by Pb immobilization [58,59]. Metal diffusion and the precipitation of metal oxalates that occur in this well-hydrated mucilaginous microenvironment were successfully observed using the wet-mode environmental SEM (ESEM) technique [60]. However, this structure is not easily observed by conventional SEM in high-vacuum mode.

Cu can also accumulate inside the cells. The transport of the metal across the cell membrane yields intracellular accumulation, which is dependent on the cell metabolism. This means that this kind of intracellular biosorption may only take place with viable cells. It is often associated with an active defence system of the microorganism, which reacts in the presence of toxic metal [61]. Interestingly, the analysis of *R. placenta* and *P. chrysosporium* genomes highlighted additional Cu transporter genes compared to what has been described in *S. cerevisiae*. Metallothioneins (MT) have a crucial role in the intracellular sequestration of copper. In basidiomycetes, such proteins have been functionally characterized in *Hebeloma cylindrosporum* [62], *Cystoderma carcharias* [63] or *Laccaria bicolor* [64], and sequences have been detected in other Basidiomycete species such as *Pisolithus arhizus*, *Paxillus involutus*, or *Agaricus bisporus* [62]. Surprisingly, these sequences, which were identified in mycorrhizal species, were not found in the *R. placenta* and *P. chrysosporium* genomes, suggesting that either such proteins are absent in these wood decay fungi or, alternatively, are too divergent to be detected by our approach. The only MT detected in *R. placenta* and *P. chrysosporium* genomes is a Cu-thionein that is functionally characterized in the ectomycorrhizal fungus *Suillus luteus* [65]. Because this MT is ubiquitous in the subphylum Agaricomycotina and displays highly conserved features, it has been proposed to be important for basic cellular functions. Its role in Cu tolerance has been demonstrated in *S. luteus*; however, nothing is known concerning the *R. placenta* and *P. chrysosporium* isoforms. Further analysis will be necessary to test whether these proteins are involved in Cu resistance in the two fungi and act as a complement to the oxalate and biosorption protection mechanism.

5. Conclusions and Perspectives

In our study, the extracellular production of oxalic acid by *R. placenta* could help fungi to precipitate Cu from treated wood waste, reducing its toxicity. This ability to precipitate strong amounts of Cu from treated solid wood in a short time is quite commendable for copper extraction and thus for the bioremediation of the treated wood. In contrast, *P. chrysosporium* used copper biosorption from the liquid phase to protect itself. Despite it not being as efficient as bioleaching, the ability of *P. chrysosporium* to resist and degrade several organic pollutants is of great interest in the context of wood preservation. Indeed, in the current work, we focused on the fungal ability to detoxify or tolerate Cu levels while Tanalith E also contains azoles that also need to be degraded to obtain full bioremediation. The combination of fungi with different abilities may be a powerful tool to valorize copper azole-containing wood wastes. Finally, both the fungi have the capability to produce a large number of enzymes, transporters, and siderophores that could be further studied to understand the role of these pathways in the detoxification process. The identification of new biocatalysts from these fungi could possibly help us to successfully remove the toxic compounds from wood waste.

Supplementary Materials: The following supporting information can be downloaded at https://www.mdpi.com/article/10.3390/jof8070706/s1, Figure S1: Picture of the microcosm containing *P. chrysosporium* and treated sawdust. Supplemental Data S1: Distribution of sites of interest and microanalysis procedure.

Author Contributions: Conceptualization, A.D., A.B. and M.M.-R.; data curation, G.P. and K.C.; formal analysis, G.P., K.C., D.B. and B.P.; funding acquisition, M.M.-R.; investigation, A.D., A.B. and M.M.-R.; methodology, G.P., K.C. and A.B.; project administration, M.M.-R.; supervision, C.R., A.D., A.B. and M.M.-R.; validation, M.M.-R.; writing—original draft, G.P. and K.C.; writing—review and editing, A.D., A.B. and M.M.-R. All authors have read and agreed to the published version of the manuscript.

Funding: This work was supported by two grants overseen by the French National Research Agency (ANR) (ANR-18-CE04-0012 and ANR-11-LABX-0002-01, Lab of Excellence ARBRE).

Institutional Review Board Statement: Not applicable.

Informed Consent Statement: Not applicable.

Data Availability Statement: Not applicable.

Acknowledgments: We thank Carla Blot for her help in the genome analysis. This work was partly carried out in the Pôle de compétences en Chimie analytique environnementale, ANATELo, LIEC laboratory, UMR 7360 CNRS—Université de Lorraine. The SEM imaging was completed using the SILVATECH platform (Silvatech, INRAE, 2018. Structural and functional analysis of tree and wood Facility, doi: 10.15454/ 1.5572400113627854E12) from UMR 1434 SILVA, 1136 IAM, 1138 BEF and the 4370 EA LERMAB EEF research center INRAE Nancy-Lorraine. The SILVATECH facility is supported by the French National Research Agency through the Lab of Excellence ARBRE. We thank Mr Fenneteau (from Bois et Compagnie) for providing treated wood.

Conflicts of Interest: The authors declare no conflict of interest.

References

1. Verbist, M.; Nunes, L.; Jones, D.; Branco, J.M. *Service Life Design of Timber Structures*; Woodhead Publishing: Sawston, UK, 2019; Volume 11, pp. 311–336. [CrossRef]
2. Ramage, M.H.; Henry, B.; Marta, B.-W.; George, F.; Thomas, R.; Darshil, U.S.; Guanglu, W.; Li, Y.; Patrick, F.; Danielle, D.-T.; et al. The wood from the trees: The use of timber in construction. *Renew. Sustain. Energy Rev.* **2017**, *68*, 333–359. [CrossRef]
3. Rowell, R.M. (Ed.) *Handbook of Wood Chemistry and Wood Composites*, 2nd ed.; CRC Press: Boca Raton, FL, USA, 2012. [CrossRef]
4. Marais, B.N.; Brischke, C.; Militz, H. Wood durability in terrestrial and aquatic environments–A review of biotic and abiotic influence factors. *Wood Mater. Sci. Eng.* **2020**, *17*, 82–105. [CrossRef]
5. Civardi, C.; van den Bulcke, J.; Schubert, M.; Michel, E.; Butron, M.I.; Boone, M.N.; Dierick, M.; Van Acker, J.; Wick, P.; Schwarze, F.W.M.R. Penetration and Effectiveness of Micronized Copper in Refractory Wood Species. *PLoS ONE* **2016**, *11*, e0163124. [CrossRef]
6. Freeman, B.M.H.; Mcintyre, C.R. Copper-based wood preservatives. *For. Prod. J.* **2008**, *58*, 6–27.
7. Kartal, S.N.; Terzi, E.; Yilmaz, H.; Goodell, B. Bioremediation and decay of wood treated with ACQ, micronized ACQ, nano-CuO and CCA wood preservatives. *Int. Biodeterior. Biodegrad.* **2015**, *99*, 95–101. [CrossRef]
8. Helsen, L.; Hardy, A.; Van Bael, M.K.; Mullens, J. Tanalith E 3494 impregnated wood: Characterisation and thermal behaviour. *J. Anal. Appl. Pyrolysis* **2007**, *78*, 133–139. [CrossRef]
9. Woo, C.; Daniels, B.; Stirling, R.; Morris, P. Tebuconazole and propiconazole tolerance and possible degradation by Basidiomycetes: A wood-based bioassay. *Int. Biodeterior. Biodegrad.* **2010**, *64*, 403–408. [CrossRef]
10. Hitchcock, C.A.; Pye, G.W.; Troke, P.F.; Johnson, E.M.; Warnock, D.W. Fluconazole resistance in *Candida glabrata*. *Antimicrob. Agents Chemother.* **1993**, *37*, 1962–1965. [CrossRef] [PubMed]
11. Robinson, J.R.; Isikhuemhen, O.S.; Anike, F.N. Fungal–metal interactions: A review of toxicity and homeostasis. *J. Fungi* **2021**, *7*, 225. [CrossRef]
12. Macomber, L.; Imlay, J.A. The iron-sulfur clusters of dehydratases are primary intracellular targets of copper toxicity. *Proc. Natl. Acad. Sci. USA* **2009**, *106*, 8344–8349. [CrossRef]
13. Bridžiuvienė, D.; Levinskaitė, L. Fungal tolerance towards copper-based wood preservatives. *Biologija* **2007**, *53*, 54–61. [CrossRef]
14. Clausen, C.A.; Green, F. Oxalic acid overproduction by copper-tolerant brown-rot basidiomycetes on southern yellow pine treated with copper-based preservatives. *Int. Biodeterior. Biodegrad.* **2003**, *51*, 139–144. [CrossRef]
15. Xing, D.; Magdouli, S.; Zhang, J.; Koubaa, A. Microbial remediation for the removal of inorganic contaminants from treated wood: Recent trends and challenges. *Chemosphere* **2020**, *258*, 127429. [CrossRef]

16. Deshmukh, R.; Khardenavis, A.A.; Purohit, H.J. Diverse Metabolic Capacities of Fungi for Bioremediation. *Indian J. Microbiol.* **2016**, *56*, 247–264. [CrossRef]
17. Smith, A.D.; Logeman, B.L.; Thiele, D.J. Copper Acquisition and Utilization in Fungi. *Annu. Rev. Microbiol.* **2017**, *71*, 597–623. [CrossRef]
18. Yasokawa, D.; Murata, S.; Kitagawa, E.; Iwahashi, Y.; Nakagawa, R.; Hashido, T.; Iwahashi, H. Mechanisms of copper toxicity in *Saccharomyces cerevisiae* determined by microarray analysis. *Environ. Toxicol. Int. J.* **2008**, *23*, 599–606. [CrossRef] [PubMed]
19. Antsotegi-Uskola, M.; Markina-Iñarrairaegui, A.; Ugalde, U. Copper Homeostasis in Aspergillus nidulans Involves Coordinated Transporter Function, Expression and Cellular Dynamics. *Front. Microbiol.* **2020**, *11*, 555306. [CrossRef]
20. Suzuki, M.; Gitlin, G.D. Intracellular localization of the Menkes and Wilson's disease proteins and their role in intracellular copper transport. *Pediatr. Int.* **1999**, *41*, 436–442. [CrossRef]
21. Pohleven, F.; Humar, M.; Amartey, S.; Benedik, J. *Tolerance of Wood Decay Fungi to Commercial Copper based Wood Preservatives*; IRG/WP 02-30291; International Research Group on Wood Preservation: Stockholm, Sweden, 2002.
22. Say, R.; Denizli, A.; Yakup Arica, M. Biosorption of cadmium(II), lead(II) and copper(II) with the filamentous fungus *Phanerochaete chrysosporium*. *Bioresour. Technol.* **2001**, *76*, 67–70. [CrossRef]
23. Kumar, S.; Stecher, G.; Li, M.; Knyaz, C.; Tamura, K. MEGA X: Molecular Evolutionary Genetics Analysis across computing platforms. *Mol. Biol. Evol.* **2018**, *35*, 1547–1549. [CrossRef]
24. Saitou, N.; Nei, M. The neighbor-joining method: A new method for reconstructing phylogenetic trees. *Mol. Biol. Evol.* **1987**, *4*, 406–425. [CrossRef]
25. Nei, M.; Kumar, S. *Molecular Evolution and Phylogenetics*; Oxford University Press: New York, NY, USA, 2000.
26. Dutton, M.V.; Evans, C.S. Oxalate production by fungi: Its role in pathogenicity and ecology in the soil environment. *Can. J. Microbiol.* **1996**, *42*, 881–895. [CrossRef]
27. Gadd, G.M.; Bahri-Esfahani, J.; Li, Q.; Rhee, Y.J.; Wei, Z.; Fomina, M.; Liang, X. Oxalate production by fungi: Significance in geomycology, biodeterioration and bioremediation. *Fungal Biol. Rev.* **2014**, *28*, 36–55. [CrossRef]
28. Rees, E.M.; Lee, J.; Thiele, D.J. Mobilization of intracellular copper stores by the Ctr2 vacuolar copper transporter. *J. Biol. Chem.* **2004**, *279*, 54221–54229. [CrossRef]
29. Kwok, E.Y.; Severance, S.; Kosman, D.J. Evidence for iron channeling in the Fet3p-Ftr1p high-affinity iron uptake complex in the yeast plasma membrane. *Biochemistry* **2006**, *45*, 6317–6327. [CrossRef]
30. Hassett, R.; Dix, D.R.; Eide, D.J.; Kosman, D.J. The Fe(II) permease Fet4p functions as a low affinity copper transporter and supports normal copper trafficking in *Saccharomyces cerevisiae*. *Biochem. J.* **2000**, *351*, 477–484. [CrossRef]
31. Adle, D.J.; Sinani, D.; Kim, H.; Lee, J. A cadmium-transporting P1B-type ATPase in yeast Saccharomyces cerevisiae. *J. Biol. Chem.* **2007**, *282*, 947–955. [CrossRef]
32. Georgatsou, E.; Mavrogiannis, L.A.; Fragiadakis, G.S.; Alexandraki, D. The yeast Fre1p/Fre2p cupric reductases facilitate copper uptake and are regulated by the copper-modulated Mac1p activator. *J. Biol. Chem.* **1997**, *272*, 13786–13792. [CrossRef]
33. Tascioglu, C.; Tsunoda, K. Biological performance of copper azole-treated wood and wood-based composites. *Holzforschung* **2010**, *64*, 399–406. [CrossRef]
34. Bak, J.S. Lignocellulose depolymerization occurs via an environmentally adapted metabolic cascades in the wood-rotting basidiomycete *Phanerochaete chrysosporium*. *Microbiologyopen* **2015**, *4*, 151–166. [CrossRef]
35. Sierra-Alvarez, R. Removal of copper, chromium and arsenic from preservative-treated wood by chemical extraction-fungal bioleaching. *Waste Manag.* **2009**, *29*, 1885–1891. [CrossRef] [PubMed]
36. Schilling, J.S.; Jellison, J. Metal accumulation without enhanced oxalate secretion in wood degraded by brown rot fungi. *Appl. Environ. Microbiol.* **2006**, *72*, 5662–5665. [CrossRef] [PubMed]
37. Alluri, H.K.; Ronda, S.R.; Settalluri, V.S.; Jayakumar Singh, B.; Suryanarayana, V.; Venkateshwar, P. Biosorption: An eco-friendly alternative for heavy metal removal. *Afr. J. Biotechnol.* **2007**, *6*, 2924–2931. [CrossRef]
38. Ide, M.; Ichinose, H.; Wariishi, H. Molecular identiWcation and functional characterization of cytochrome P450 monooxygenases from the brown rot basidiomycete *Postia placenta*. *Arch. Microbiol.* **2012**, *194*, 243–253. [CrossRef]
39. Sing, C.; Yu, J. Copper adsorption and removal from water by living mycelium of white rot fungus *Phanerochaete chrysosporium*. *Water Res.* **1998**, *32*, 2746–2752. [CrossRef]
40. Alfredsen, G.; Fossdal, C.G.; Nagy, N.E.; Jellison, J.; Goodell, B. Furfurylated wood: Impact on *Postia placenta* gene expression and oxalate crystal formation. *Holzforschung* **2016**, *70*, 947–962. [CrossRef]
41. Cragg, S.M.; Beckham, G.T.; Bruce, N.C.; Bugg, T.D.H.; Distel, D.L.; Dupree, P.; Etxabe, A.G.; Goodell, B.S.; Jellison, J.; McGeehan, J.E.; et al. Lignocellulose degradation mechanisms across the Tree of Life. *Curr. Opin. Chem. Biol.* **2015**, *29*, 108–119. [CrossRef]
42. Schilling, J.S. Oxalate Production and Cation Translocation during Wood Biodegradation by Fungi. Electronic PhD Theses and Dissertations. Forestry, University of Maine, 2006; 336. Available online: https://digitalcommons.library.umaine.edu/etd/336 (accessed on 9 February 2022).
43. Clausen, C.A.; Green, F.; Woodward, B.M.; Evans, J.W.; DeGroot, R.C. Correlation between oxalic acid production and copper tolerance in *Wolfiporia cocos*. *Int. Biodeterior. Biodegrad.* **2000**, *46*, 69–76. [CrossRef]
44. Schilling, J.S.; Jellison, J. Oxalate regulation by two brown rot fungi decaying oxalate-amended and non-amended wood. *Holzforschung* **2005**, *59*, 681–688. [CrossRef]

45. Arantes, V.; Goodell, B. Current understanding of brown-rot fungal biodegradation mechanisms: A review. *ACS Symp. Ser.* **2014**, *1158*, 3–21. [CrossRef]
46. Hattori, T.; Hisamori, H.; Suzuki, S.; Umezawa, T.; Yoshimura, T.; Sakai, H. Rapid copper transfer and precipitation by wood-rotting fungi can effect copper removal from copper sulfate-treated wood blocks during solid-state fungal treatment. *Int. Biodeterior. Biodegrad.* **2015**, *97*, 195–201. [CrossRef]
47. Mäkelä, M.R.; Hildén, K.; Lundell, T.K. Oxalate decarboxylase: Biotechnological update and prevalence of the enzyme in filamentous fungi. *Appl. Microbiol. Biotechnol.* **2010**, *87*, 801–814. [CrossRef] [PubMed]
48. Mäkelä, M.R.; Sietiö, O.M.; De Vries, R.P.; Timonen, S.; Hildén, K. Oxalate-metabolising genes of the white-rot fungus Dichomitus squalens are differentially induced on wood and at high proton concentration. *PLoS ONE* **2014**, *9*, e87959. [CrossRef] [PubMed]
49. Dutton, M.V.; Evans, C.S.; Atkey, P.T.; Wood, D.A. Oxalate production by Basidiomycetes, including the white-rot species *Coriolus versicolor* and *Phanerochaete chrysosporium*. *Appl. Microbiol. Biotechnol.* **1993**, *39*, 5–10. [CrossRef]
50. Kuan, I.C.; Tien, M. Stimulation of Mn peroxidase activity: A possible role for oxalate in lignin biodegradation. *Proc. Natl. Acad. Sci. USA* **1993**, *90*, 1242–1246. [CrossRef] [PubMed]
51. Köse, C.; Kartal, S.N. Tolerance of brown-rot and dry-rot fungi to CCA and ACQ wood preservatives. *Turkish J. Agric. For.* **2010**, *34*, 181–190. [CrossRef]
52. Iqbal, M.; Edyvean, R.G.J. Biosorption of lead, copper and zinc ions on loofa sponge immobilized biomass of *Phanerochaete chrysosporium*. *Miner. Eng.* **2004**, *17*, 217–223. [CrossRef]
53. Rudakiya, D.M.; Iyer, V.; Shah, D.; Gupte, A.; Nath, K. Biosorption Potential of *Phanerochaete chrysosporium* for Arsenic, Cadmium, and Chromium Removal from Aqueous Solutions. *Glob. Challenges* **2018**, *2*, 1800064. [CrossRef]
54. Noormohamadi, H.R.; Fat'hi, M.R.; Ghaedi, M.; Ghezelbash, G.R. Potentiality of white-rot fungi in biosorption of nickel and cadmium: Modeling optimization and kinetics study. *Chemosphere* **2019**, *216*, 124–130. [CrossRef]
55. Civardi, C.; Grolimund, D.; Schubert, M.; Wick, P.; Schwarze, F.W.M.R. Micronized copper-treated wood: Copper remobilization into spores from the copper-tolerant wood-destroying fungus *Rhodonia placenta*. *Environ. Sci. Nano* **2019**, *6*, 425–431. [CrossRef]
56. Dusengemungu, L.; Kasali, G.; Gwanama, C.; Ouma, K.O. Recent Advances in Biosorption of Copper and Cobalt by Filamentous Fungi. *Front. Microbiol.* **2020**, *11*, 3285. [CrossRef] [PubMed]
57. da Costa, L.G.; Brocco, V.F.; Paes, J.B.; Kirker, G.T.; Bishell, A.B. Biological and chemical remediation of CCA treated eucalypt poles after 30 years in service. *Chemosphere* **2022**, *286*, 131629. [CrossRef] [PubMed]
58. Li, N.; Zhang, X.; Wang, D.; Cheng, Y. Contribution characteristics of the in situ extracellular polymeric substances (EPS) in *Phanerochaete chrysosporium* to Pb immobilization. *Bioprocess Biosyst. Eng.* **2017**, *40*, 1447–1452. [CrossRef] [PubMed]
59. Li, N.; Liu, J.; Yang, R.; Wu, L. Distribution, characteristics of extracellular polymeric substances of *Phanerochaete chrysosporium* under lead ion stress and the influence on Pb removal. *Sci. Rep.* **2020**, *10*, 17633. [CrossRef] [PubMed]
60. Fomina, M.; Hillier, S.; Charnock, J.M.; Melville, K.; Alexander, I.J.; Gadd, G.M. Role of oxalic acid overexcretion in transformations of toxic metal minerals by *Beauveria caledonica*. *Appl. Environ. Microbiol.* **2005**, *71*, 371–381. [CrossRef] [PubMed]
61. Ahalya, N.; Ramachandra, T.V.; Kanamadi, R.D. Biosorption of heavy metals. *Res. J. Chem. Environ.* **2003**, *7*, 71–79. [CrossRef]
62. Ramesh, G.; Podila, G.K.; Gay, G.; Marmeisse, R.; Reddy, M.S. Different Patterns of Regulation for the Copper and Cadmium Metallothioneins of the Ectomycorrhizal Fungus *Hebeloma cylindrosporum*. *Appl. Environ. Microbiol.* **2009**, *75*, 2266–2274. [CrossRef]
63. Sácký, J.; Černý, J.; Šantrůček, J.; Borovička, J.; Leonhardt, T.; Kotrba, P. Cadmium hyperaccumulating mushroom *Cystoderma carcharias* has two metallothionein isoforms usable for cadmium and copper storage. *Fungal Genet. Biol.* **2021**, *153*, 103574. [CrossRef]
64. Liu, B.; Dong, P.; Zhang, X.; Feng, Z.; Wen, Z.; Shi, L.; Xia, Y.; Chen, C.; Shen, Z.; Lian, C.; et al. Identification and characterization of eight metallothionein genes involved in heavy metal tolerance from the ectomycorrhizal fungus *Laccaria bicolor*. *Environ. Sci. Pollut. Res.* **2022**, *29*, 14430–14442. [CrossRef]
65. Nguyen, H.; Rineau, F.; Vangronsveld, J.; Cuypers, A.; Colpaert, J.V.; Ruytinx, J. A novel, highly conserved metallothionein family in basidiomycete fungi and characterization of two representative SlMTa and SlMTb genes in the ectomycorrhizal fungus *Suillus luteus*. *Environ. Microbiol.* **2017**, *19*, 2577–2587. [CrossRef]

Journal of *Fungi*

MDPI

Article

Occurrence and Health Risk Assessment of Cadmium Accumulation in Three *Tricholoma* Mushroom Species Collected from Wild Habitats of Central and Coastal Croatia

Ivan Širić [1], Pankaj Kumar [2,*], Ebrahem M. Eid [3,4], Archana Bachheti [5], Ivica Kos [1], Dalibor Bedeković [1], Boro Mioč [1] and Miha Humar [6]

1 Faculty of Agriculture, University of Zagreb, Svetosimunska 25, 10000 Zagreb, Croatia; isiric@agr.hr (I.Š.); ikos@agr.hr (I.K.); dbedekovic@agr.hr (D.B.); bmioc@agr.hr (B.M.)
2 Agro-Ecology and Pollution Research Laboratory, Department of Zoology and Environmental Science, Gurukula Kangri (Deemed to Be University), Haridwar 249404, Uttarakhand, India
3 Biology Department, College of Science, King Khalid University, Abha 61321, Saudi Arabia; ebrahem.eid@sci.kfs.edu.eg
4 Botany Department, Faculty of Science, Kafrelsheikh University, Kafr El-Sheikh 33516, Egypt
5 Department of Environmental Science, Graphic Era (Deemed to be University), Dehradun 248002, Uttarakhand, India; bachheti.archana@gmail.com
6 Department of Wood Science and Technology, Biotechnical Faculty, University of Ljubljana, Jamnikarjeva 101, 1000 Ljubljana, Slovenia; miha.humar@bf.uni-lj.si
* Correspondence: rs.pankajkumar@gkv.ac.in

Abstract: This study deals with the biomonitoring of cadmium (Cd) heavy metal in the three selected *Tricholoma* mushroom species collected from wild habitats of central and coastal Croatia. For this, mushroom (*T. columbetta*: n = 38, *T. portentosum*: n = 35, and *T. terreum*: n = 34) and surface soil samples were collected from nine forest localities of Croatia and analyzed for Cd concentration using inductively coupled plasma–optical emission spectrometry (ICP–OES) through the acid digestion method. The findings revealed that Cd was present in *Tricholoma* spp. and surface soil. However, the maximum mean Cd concentration (mg/kg dry weight) was recorded in *T. portentosum* (cap: 0.98; stipe: 0.72), followed by *T. columbetta* (cap: 0.96; stipe: 0.73) and *T. terreum* (cap: 0.81; stipe: 0.63). The bioconcentration factor (BCF) value (>1) revealed that the selected *Tricholoma* spp. had the potential for Cd accumulation. Moreover, the principal component (PC) and hierarchical cluster (HC) analyses were used to derive the interactions and similarities between Cd levels *Tricholoma* spp. and sampling localities. The multivariate analysis suggested that central sampling localities had higher Cd levels as compared to coastal localities. However, the daily intake of metals (DIM < 0.426) and health risk index (HRI < 1) showed that there was no potential health risk associated with the consumption of selected *Tricholoma* spp. The findings of this study are helpful to understand the Cd accumulation behavior of wild edible *Tricholoma* spp. collected from Croatia.

Keywords: cadmium; health risk assessment; heavy metals; health hazard; mushrooms; *Tricholoma* spp.

Citation: Širić, I.; Kumar, P.; Eid, E.M.; Bachheti, A.; Kos, I.; Bedeković, D.; Mioč, B.; Humar, M. Occurrence and Health Risk Assessment of Cadmium Accumulation in Three *Tricholoma* Mushroom Species Collected from Wild Habitats of Central and Coastal Croatia. *J. Fungi* **2022**, *8*, 685. https://doi.org/10.3390/jof8070685

Academic Editor: Laurent Dufossé

Received: 30 May 2022
Accepted: 27 June 2022
Published: 29 June 2022

1. Introduction

Around the globe, population growth, intensive industrialization, and urbanization have led to environmental pollution, especially in soil and water. The anthropogenic disposal of pollutants in the environment, especially heavy metals, has become an unavoidable problem affecting different life forms. Cadmium (Cd) is a toxic metal found mostly in trace amounts in the Earth's crust, with an average concentration of 0.36 mg/kg in soils [1]. The presence of Cd in the soil is a consequence of natural processes and anthropogenic practices. In the natural pedogenetic processes, the soil takes up heavy metals from the parent substrate, whereas in the anthropogenic processes, various activities, such as urbanization, industrialization, trade, and agricultural production, lead to heavy metal mixing in environmental areas. The geogenic origin of Cd is usually associated with sulfur minerals, which oxidizes relatively quickly in the environment, and the metal cation separates from sulfur at an early stage of mineral degradation. In the later stages of pedogenesis, Cd is more common in the composition of Mn oxide [2]. Anthropogenic sources of Cd in the environment include atmospheric deposition, industrial and municipal waste discharge, phosphate fertilizers, pesticides, sewage sludge, ores, metal industry, mining, and incidents [1,3]. Cadmium is considered a very dangerous pollutant to the ecosystem, and, unlike other plants in the terrestrial environment, mushrooms effectively absorb it from the soil [4,5]. Cadmium is a carcinogenic element that adversely affects the kidneys, bones, cardiovascular system, and immune system and belongs to Group 1, according to the International Agency for Research on Cancer (IARC) classification [6]. Anthropogenic sources of Cd pollution have been very important causes of its deposition in forest soils in recent decades. Although everyday knowledge of the toxicity of Cd on the environment and human health has led to its reduction in use in some areas. Still, anthropogenic sources continue to increase in certain areas, which negatively affects the natural landscapes, including forests where mushrooms mainly grow.

Mushrooms belong to the kingdom of fungi and are classified as a distinct microbiological group of organisms of significant nutritional, pharmaceutical, and ecological value. Wild edible mushrooms are considered high-quality foods due to their natural nutritional benefits, including aromatic and antioxidant properties [7,8]. They are a good source of nutritionally important elements, such as K, Zn, Cu, and Mn [9] and B-group vitamins, vitamin D, proteins, and dietary fiber [10]. Additionally, many species of mushrooms are used as medicines to prevent diseases, such as hypertension [11] and hypercholesterolemia [12], and to improve the immune system [13]. Additionally, mushrooms play an important role in the ecosystem because they can degrade many complex molecules of plants and animals [9]. In a symbiotic relationship, the mushroom benefits from plants' easy access to food. Similarly, the plant benefits because the mushroom produces mycelia, which aids in the absorption of water and nutrients. However, it is known that mushrooms can accumulate certain heavy metals (essential and toxic) and develop their fruiting bodies under conditions that are toxic to most other organisms [14]. Mushrooms can absorb certain forms of heavy metals, such as Cd^{2+}, Cd^{6+} Hg^{2+}, As^{5+}, etc., in their fruiting bodies. In regard to this, intracellular speciation and uptake of metals are generally regulated by metallothioneins and GT-complexes that are directly connected to fungal physiology. Previous studies suggest that mushrooms have an effective system that allows them to absorb heavy metals in a form that does not affect their growth and development. Due to the extremely efficient system of absorption and storage of heavy metals, mushrooms have an extremely good bioaccumulation potential, which depends on many external (environmental) factors and internal mechanisms within the fungus [15]. Thus, various environmental factors, such as the type of soil, the content of organic matter in the soil, nitrogen, the pH value, and the concentration of metals in the soil, as well as the species of fungus or the morphological part of the fruiting body (cap and stalk), the fructification time, age of mycelium, and production of ligands, can influence the accumulation of heavy metals in mushrooms [16–21]. Biological and molecular mechanisms of heavy metal uptake in mushrooms have been presented by [22]. The authors state that mushrooms can accumulate metals by passive and active uptake mechanisms. Through a passive uptake mechanism, heavy metals are trapped in the cell structure and then adsorbed to binding sites. The active mechanism of metal uptake is carried out by the biological metabolic process of heavy metal transfer in the cell through the cell membrane [23]. Additionally, according to Mleczek et al. [24], the organic acids secreted by the mushrooms can chelate poorly soluble mineral components of the soil, facilitating and accelerating their uptake by the hyphae and their accumulation in the fruiting body of mushrooms.

Cadmium concentration and its distribution in different mushroom species has been studied by numerous authors around the world [9,14,15,17,25–30]. Additionally, the potential toxic effect of Cd from fungi was determined by Sarikurkcu et al. [31]. A wide range of Cd concentrations was found in the edible mushroom species, with concentrations in uncontaminated areas in the range of <0.5 to 2 mg/kg dm, while concentrations in contaminated areas were as high as 10 mg/kg dm [9,14,15,17,25–30,32]. Thus, high concentrations of Cd in mushroom edible parts may have adverse effects on human health. Some other studied species that can accumulate Cd are: *Agaricus bisporus* [33–35]; *A. campestris* [17,26,36]; *A. macrosporus* [37]; *Armillaria mellea* [30,38,39]; *Amanita muscaria* and *A. allies* [19]; *Boletus edulis* [32,36,40–44]; *Cantharellus cibarius* [40,45–48]; *Cystoderma carcharias* [49]; *Macrolepiota procera* [25,26,36,50,51]; and *Xerocomus badius* [24,30,36,41,52]. The concentration of Cd in mushroom species of the genus *Tricholoma* usually varies in the range of <0.5 to 1 mg/kg [17,26,30,36,53]. *Tricholoma* spp., belonging to the class Agaricomycetes and genus *Tricholoma*, are found throughout the world, but they are most common in temperate and subtropical climates in both the southern and northern hemispheres. *Tricholoma* spp. are distinguished by hyaline, subglobose to oblong spores, simple pileipellis structures, and a lack of well-differentiated sterile elements, including cystidia [54]. Some *Tricholoma* spp., such as *T. matsutake*, are characterized by a high accumulation of Cd (48.52 mg/kg dm) [29], therefore making it crucial to monitor the Cd content present in commonly consumed *Tricholoma* spp. in Croatia. Studies on the monitoring of Cd contamination of *Tricholoma* spp. are lacking, particularly in central and coastal Croatia. Therefore, keeping in mind the concerns regarding Cd occurrence, this paper aimed to (i) determine the Cd concentration in the *Tricholoma* spp. and its soil substrate; (ii) compare the distribution of the Cd in anatomical parts, i.e., cap and stipe; (iii) determine the suitability of the studied mushroom species as Cd bioaccumulators; and (iv) study the potential health risks associated with the consumption of Cd contaminated mushrooms of the genus *Tricholoma*.

2. Materials and Methods

2.1. Mushrooms and Forest soil Sampling

The sampling of mushrooms of the *Tricholoma* spp. and substrates (soil) was carried out in nine localities in Croatia, of which five localities were in the central zone and four in the coastal zone (Table 1 and Figure 1). The sampled localities are interspersed with mixed forests of deciduous and coniferous trees. A total of 107 samples of three mushroom species (*Tricholoma columbetta*: n = 38, *T. portentosum*: n = 35, and *T. terreum*: n = 34) were collected from July 2012 to November 2014. Fully developed and mature fruiting bodies of mushrooms were collected by random selection from two large regions in Croatia. At the same time, samples of topsoil (0–10 cm) were collected at mushroom sampling sites (n = 177; 10 samples for each site) using the quadrate sampling method [55]. After their collection, the mushroom bodies were thoroughly washed and cut into two anatomical parts, i.e., the cap (pileus) and the stipe (stipes), using a sterile knife, followed by drying at 60 °C to achieve a constant weight. The drying of the mushroom samples was performed in a food and plant dryer (MSG-01; MPM Product, Milanówek, Poland, and Ultra FD1000 dehydrator, Ezidri, Australia). After drying, the samples were ground in a laboratory mill (Retsch SM 200) and passed through a 1.0 mm diameter sieve, followed by storage in air-tight plastic bags until further Cd analysis. Similarly, the forest soil samples were also dried at room temperature, ground using a laboratory mill and passed through a 1.0 mm pore size sieve.

Table 1. Description of study sites of *Tricholoma* spp. mushroom sample collection.

Site Name and Sample Size	Code	Longitude (N)	Latitude €	Zone Type	*Tricholoma* spp.
Brezova Gora (n = 14)	BG	15.909140	46.281183	Central Croatia	
Medvednica, Stubaki (n = 14)	MS	15.969287	45.919902	Central Croatia	*T. columbetta* (n = 38)
Petrova Gora (n = 10)	PG	15.810489	45.239646	Central Croatia	

Table 1. *Cont.*

Site Name and Sample Size	Code	Longitude (N)	Latitude €	Zone Type	*Tricholoma* spp.
Ravna Gora (n = 11)	RG	14.940796	45.369653	Coastal Croatia	
Island Krk (n = 10)	IK	14.626754	45.090944	Coastal Croatia	*T. portentosum* (n = 34)
Labinština (n = 13)	L	14.135917	45.093335	Coastal Croatia	
Maksimir (n = 10)	M	16.052633	45.814058	Central Croatia	
Dugi Dol, Karlovac (n = 12)	DD	15.576698	45.354304	Central Croatia	*T. terreum* (n = 35)
Skrad (n = 13)	S	14.947011	45.425098	Coastal Croatia	

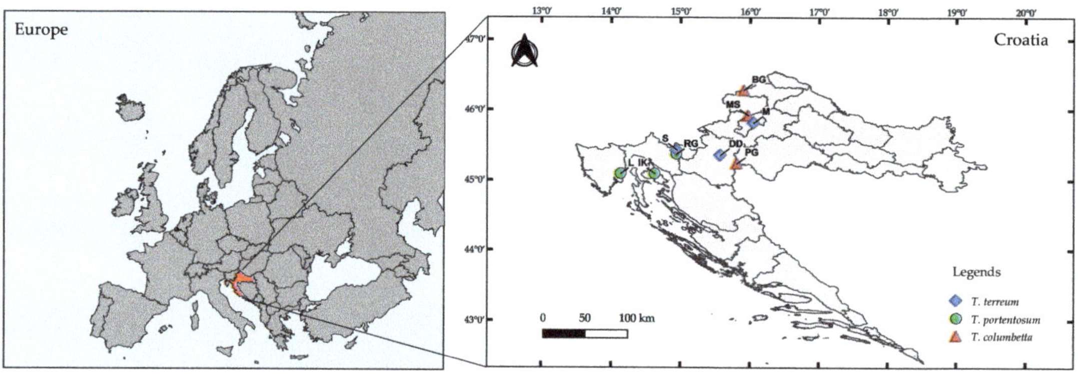

Figure 1. Map of the *Tricholoma* spp. mushroom sample collection sites across Croatia.

2.2. Analysis of Cadmium

The concentration of Cd was determined using inductively coupled plasma–optical emission spectroscopy (ICP–OES). Validation of the method for Cd content was performed using certified reference material (IAEA-336) lichens. The reported Cd concentration for the reference material was 0.117 mg/kg. The recovery result for Cd in this study was 0.120 ± 0.003 mg/kg (n = 3), which showed good agreement with the certified levels. Detection limits of Cd were 0.003 mg/kg. The laboratory glassware used to prepare samples for the determination of Cd was soaked in a solution of ethylenediaminetetraacetic acid (EDTA) at a concentration of 5% (w/v) for 24 h and then in 10% (v/v) HNO_3 for 24 h. A total of 0.5 g dried mushroom sample was digested with 5 mL of HNO_3 (65%, Suprapur, Merck, Darmstadt, Germany) in closed PTFE vessels inside a microwave destruction oven (Milestone Microwave Laboratory System, MLS 1200 mega, Shelton, CT, USA). The destruction program consisted of several steps, such as at a power of 100 W and duration of 5 min; at a power of 0 W and duration of 2 min ("standby time"); at a power of 250 W and duration of 5 min 20; at a power of 400 W and duration of 5 min; and at a power of 600 W and duration of 5 min. After destruction in the microwave, the samples were cooled in a water bath and transferred via a funnel into 25 mL plastic volumetric flasks. After that, the flasks were filled with distilled water. The soil samples were filtered through filter paper (Sigma-Aldrich, St. Louis, MO, USA). From the volumetric flasks, the samples were transferred to plastic tubes for measurement with ICP-OES (Optima 8000, Perkin Elmer, Waltham, MA, USA) equipped with an autosampler, by which Cd concentration was analyzed.

2.3. Bioconcentration and Health Risk Index (HRI) Calculation

Bioaccumulation by living organisms is expressed as the tendency of accumulating a specific quantity of heavy metals from their growing environment [35]. The bioconcentration factor (BCF) values were calculated as the ratio between the concentration of Cd in mushroom samples and the forest soil. Furthermore, the health risk index (HRI) was calculated according to Chui et al. [56] by using Equation (1):

$$HRI = DIC/RfD \quad (1)$$

According to the above equation, DIC represents daily Cd intake from the consumption of the analyzed mushroom species, while RfD represents the extent of exposure to oral contaminants during life and is mainly used in health assessments [57]. The following Equation (2) was used to calculate the daily intake of Cd [29,56]:

$$DIC = SM \times MCM/ABW \quad (2)$$

where SM—serving of mushroom (0.03 kg of dried mushrooms), MCM—Cd concentrations in mushrooms (mg/kg dry weight), and ABW—average body weight (70 kg for a regular consumer). In the calculations of the health risk index (HRI), the values of the daily consumption of dried mushrooms (30 g) and the bodyweight of a regular consumer (70 kg) were assumed [45].

2.4. Data Analysis and Statistics

All samples were analyzed in three replicates. Descriptive data analysis included minimum value, maximum value, median, and mean and standard deviation (SD) were calculated using the Statistica 10.0 (Statsoft, Tulsa, OK, USA). The map of the study area was generated using QGIS (Version 3.22.3-Białowieza, Open Source, Gispo Ltd., Helsinki, Finland) software, while the principal component and cluster analyses were performed using the OriginPro (Version 2022b, OriginLab, Northampton, MA, USA) software packages.

3. Results and Discussion

3.1. Cadium Contents in Tricholoma spp. Collected from Central and Coastal Croatia

In the current study, Cd contents in the analyzed *Tricholoma* spp. and forest topsoil are listed in Table 2. The results showed considerable differences in the content of Cd accumulated by three *Tricholoma* spp. across the sampling locations. Nevertheless, the differences in Cd content in forest soils from different sampling localities were also observed. The ICP–OES analysis revealed that Cd contents were present in both the cap and stipe regions of *Tricholoma* spp. at all locations of sample collection. Specifically, the samples of *T. columbetta* mushroom collected from Brezova Gora and *T. portentosum* from Labinstina showed identical Cd concentrations, i.e., 0.91 $\pm$ 0.13 and 0.89 $\pm$ 0.16 mg/kg, respectively. However, the concentration of Cd in the forest topsoil was relatively low, ranging between 0.07 and 0.57 mg/kg in Brezova Gora, with mean values lying within the range from 0.17 $\pm$ 0.03 mg/kg (Ravna Gora) to 0.28 $\pm$ 0.09 mg/kg (Stubaki). However, the Cd concentrations in *T. columbetta* and forest soil samples collected from the central region were almost identical to those in the species *T. portentosum* and associated soil, some of whose samples were collected at sites on the Croatian coastal locations. Regarding this, the highest Cd was found in the cap samples of *T. portentosum* at Ravna Gora (0.99 mg/kg), while the lowest average Cd content was found in the stem samples of *T. terreum* at Skrad (0.59 mg/kg). Moreover, the average Cd analysis in the full body of *Tricholoma* spp. showed that *T. columbetta* had the highest concentration, followed by *T. portentosum* and *T. terreum*. Overall, the analyzed samples of the *T. terreum* showed relatively lower Cd content as compared to *T. columbetta* and *T. potrentosum*. Figure 2a–c shows the correlation between Cd contents in soil and cap and stipe parts of three *Tricholoma* spp.

Table 2. Cadmium contents in *Tricholoma* spp. and related forest soil (mg/kg dry weight), $Q_{C/S}$ index, and BCF values (mean ± SD, median, and range).

Mushroom Species, Localities, and Number of Specimens	Cd Concentration				BCF_{cap}	BCF_{stipe}	$BCF_{fullbody}$	$Q_{C/S}$
	Cap	Stipe	Full Body	Soil				
T. columbetta, Brezova Gora, *n* = 14	0.91 ± 0.13 0.90 (0.75–1.11)	0.66 ± 0.06 0.65 (0.57–0.78)	0.78 ± 0.16 0.75 (0.57–1.10)	0.24 ± 0.14 0.23 (0.07–0.57)	5.44 ± 4.07 3.89 (1.37–14.84)	3.95 ± 2.93 2.84 (1.09–10.19)	5.04 ± 4.39 3.26 (1.93–8.14)	1.38 ± 0.12 1.40 (1.18–1.60)
T. columbetta Medvednica, Stubaki, *n* = 14	0.94 ± 0.11 0.96 (0.76–1.13)	0.73 ± 0.06 0.73 (0.61–0.83)	0.83 ± 0.14 0.80 (0.60–1.13)	0.38 ± 0.09 0.37 (0.19–0.52)	2.67 ± 0.86 2.43 (1.54–5.07)	2.08 ± 0.75 1.85 (1.32–4.19)	2.67 ± 0.70 2.16 (2.17–3.16)	1.29 ± 0.14 1.27 (1.10–1.60)
T. columbetta Petrova Gora, *n* = 10	0.92 ± 0.11 0.88 (0.81–1.17)	0.71 ± 0.05 0.72 (0.64–0.77)	0.81 ± 0.13 0.79 (0.63–1.17)	0.23 ± 0.08 0.21 (0.11–0.36)	4.54 ± 1.85 4.23 (2.45–9.06)	3.46 ± 1.06 3.40 (2.13–5.90)	4.49 ± 1.75 3.76 (3.25–5.73)	1.29 ± 0.15 1.23 (1.15–1.54)
T. portentosum Island Krk, *n* = 10	0.88 ± 0.06 0.90 (0.78–0.94)	0.66 ± 0.02 0.67 (0.63–0.69)	0.83 ± 0.17 0.80 (0.62–1.13)	0.27 ± 0.08 0.30 (0.12–0.39)	3.67 ± 1.85 3.00 (2.34–8.56)	2.78 ± 1.32 2.27 (1.61–6.19)	4.03 ± 1.60 2.67 (2.90–5.17)	1.33 ± 0.09 1.35 (1.18–1.45)
T. portentosum Ravna Gora, *n* = 11	0.99 ± 0.08 0.98 (0.88–1.13)	0.68 ± 0.03 0.68 (0.63–0.72)	0.77 ± 0.11 0.73 (0.63–0.94)	0.17 ± 0.04 0.17 (0.10–0.23)	6.17 ± 1.77 5.85 (4.35–9.36)	4.20 ± 1.04 3.99 (3.11–6.14)	5.19 ± 1.56 4.29 (4.09–6.30)	1.46 ± 0.11 1.46 (1.32–1.59)
T. portentosum Labinština, *n* = 13	0.89 ± 0.16 0.87 (0.69–1.19)	0.70 ± 0.08 0.72 (0.59–0.81)	0.79 ± 0.15 0.76 (0.58–1.18)	0.21 ± 0.05 0.20 (0.11–0.32)	4.59 ± 1.57 4.14 (2.63–7.92)	3.60 ± 1.11 3.49 (2.31–6.52)	4.48 ± 1.12 3.80 (3.69–5.27)	1.26 ± 0.14 1.23 (1.06–1.57)
T. terreum Maksimir, *n* = 10	0.80 ± 0.05 0.81 (0.73–0.87)	0.64 ± 0.01 0.63 (0.62–0.66)	0.71 ± 0.09 0.69 (0.61–0.87)	0.30 ± 0.09 0.33 (0.16–0.42)	2.78 ± 1.06 1.91 (1.86–5.43)	2.18 ± 0.76 1.91 (1.57–4.03)	2.94 ± 1.23 2.09 (2.07–3.81)	1.27 ± 0.06 1.28 (1.17–1.35)
T. terreum Dugi Dol, Karlovac, *n* = 12	0.77 ± 0.04 0.77 (0.68–0.83)	0.60 ± 0.02 0.59 (0.58–0.65)	0.68 ± 0.09 0.66 (0.57–0.83)	0.18 ± 0.07 0.17 (0.10–0.29)	4.64 ± 1.59 4.53 (2.82–8.25)	3.67 ± 1.26 3.69 (2.05–5.79)	4.28 ± 2.01 3.88 (2.86–5.70)	1.28 ± 0.09 1.28 (1.13–1.43)
T. terreum Skrad, *n* = 13	0.75 ± 0.03 0.74 (0.68–0.81)	0.59 ± 0.02 0.58 (0.57–0.65)	0.66 ± 0.08 0.66 (0.57–0.80)	0.23 ± 0.09 0.22 (0.12–0.34)	3.61 ± 1.36 3.43 (2.16–5.93)	2.85 ± 1.02 2.56 (1.79–4.54)	3.55 ± 1.69 3.00 (2.35–4.75)	1.26 ± 0.07 1.26 (1.14–1.39)

$Q_{C/S}$ (cap to stipe Cd content quotient); BCF_{cap}, BCF_{stipe}, and $BCF_{fullbody}$ (bioconcentration factor values for caps, stipes, and full body).

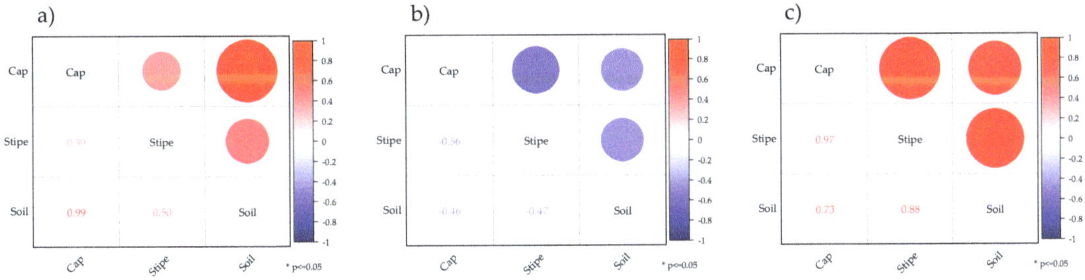

Figure 2. Correlation between Cd contents in the soil, cap, and stipe parts of *Tricholoma* spp. ((**a**) *T. columbetta*, (**b**) *T. portentosum*, and (**c**) *T. terreum*); * indicates level of significance ($p < 0.05$); refer to the color scale bar for interpretation of correlation coefficient values as text and circles.

The Cd levels in the investigated *Tricholoma* spp. mushrooms in this study are comparable to the results previously reported by other authors [26,27,36,58,59]. Comparatively, the average Cd concentrations determined in the forest topsoil of Ravna Gora (0.17 ± 0.07 mg/kg dm) region were analogous to those reported in the Szczecinek (0.17 ± 0.06 mg/kg) area of Poland [60] and Yunnan province (0.17 ± 0.03 mg/kg dm) of China [27]. In their study, Petkovšek and Pokorny [14] noted elevated levels of Cd in forest soils can be caused by pollution from nearby anthropogenic sources, such as industrial processes, smelters, and agricultural production. Similar levels of Cd (0.91 mg/kg) in *T. argyraceum* are also reported by Soylak et al. [58]. In addition, Saba et al. [61] reported similar results of Cd (0.91 mg/kg) in *Suillus gavillei* ectomycorrhizal mushroom species. Similarly, *T. terreum* samples collected from a Mediterranean region of Turkey showed average Cd values of 4.90 mg/kg [31], which is considerably higher than the results established in this study. On the other hand, Severoglu et al. [25] found very low Cd levels in the *T. terreum* samples collected in the central region of Turkey (0.05 mg/kg dm).

3.2. Bioconcentration Factor (BCF) of Cd Accumulation in Tricholoma spp.

In this study, the bioconcentration factor (BCF) values were calculated to estimate the Cd accumulation potential of selected *Tricholoma* spp. From the upper layer of forest soil (Table 2). The determined values of the BCF indicate whether Cd is actively bioaccumulated (BCF > 1) or not (BCF < 1) by selected *Tricholoma* spp. The BCF values for cap parts of *Tricholoma* spp. Were considerably higher as compared to those for stipes. The highest reported median BCF value was 5.85 for the *T. portentosum* in Ravna Gora with a mean value of 6.17 ± 1.77. Similarly, the specified BCF median value for *T. portentosum* was 3.16 times higher than in the case of *T. columbetta* in Medvednica, Stubaki, where medium BCF was only 1.85. Moreover, the BCF values established for *T. portentosum* in Ravna Gora, also indicated a potential for Cd accumulation. It is well known that mushrooms of the genus *Tricholoma* have a good Cd accumulation potential. Since Cd contamination in the upper layer of soil may be triggered by several anthropogenic activities, thus, higher BCF values were found for some sapling locations in this study. Therefore, the central Croatian sampling locations reported relatively higher BCF values as compared to those in coastal locations. Overall, the BCF values for *T. portentosum* in the Ravna Gora showed the highest Cd bioavailability of all the sampling locations. The concept of BCF is widely accepted by the scientific community for determining the hazardous metal accumulation by edible mushrooms. In a report by Širić et al. [21], the BCF values of Hg metal accumulated by four *Tricholoma* spp., such as *T. equestre*, *T. portentosum*, *T. columbetta*, and *T. terreum*, were observed between 18 to 37 in southern and northern regions of Europe. Similarly, Kojta et al. [60] also reported a BCF value > 40 for Cd accumulation by the *Macrolepiota procera* saprophytic mushroom in the Augustowska forest region of Poland, respectively.

3.3. PCA and HCA Results

Principal component analysis (PCA) is a widely accepted statistical tool for deriving the interactive effects of multiple variables based on their dominance [35]. In the current study, the data of Cd concentration in three *Tricholoma* spp. samples were analyzed using PCA based on their collection locations in central and coastal Croatia. In the case of *T. columbetta*, the data were orthogonally transformed onto two principal components, i.e., PC1 and PC2, with variances of 94.04 and 5.96%, respectively (Table 3). Regarding this, the highest concentration of Cd in *T. columbetta* was suggested in the stiped parts at the Petrova Gora (PG) site. However, the highest Cd contents in the cap parts of *T. columbetta* were observed at the Medvednica location as indicated by the vector length of the biplot axis (Figure 3a). Similarly, the PCA-based chemometric assessment of Cd contents in the *T. portentosum* mushroom collected from three coastal zones of Croatia revealed that the two extracted PCs had variances of 90.96% (PC1) and 9.04% (PC2). Contradictorily, maximum Cd levels were observed in the cap parts of *T. portentosum* mushroom collected from RG location as revealed by its vector length dominance in the PC1 data group (Figure 3b). Similarly, PCA results of Cd contents in *T. terreum* mushroom samples collected from two central zones and one coastal zone showed that the cap part indicated the highest concentration at central zone locations (Maksimir and Dugi Dol) (Figure 3c). The percentage of variance distribution among the two PCs was identified as 99.25% (PC1) and 0.75% (PC2). This theorizes that soils of central Croatian zones were more responsible for high Cd uptake by selected *Tricholoma* spp. Hence, the PCA tool was helpful to relate the effect of central and coastal Croatian sampling locations with Cd contents in *Tricholoma* spp. On the other hand, the similarities between sampling locations and Cd levels in *Tricholoma* spp. samples were evaluated using the hierarchical cluster analysis. As depicted in Figure 4a–c, it was observed that Brezova Gora and Medvednica locations showed the highest similarities in terms of Cd contents analyzed in *T. columbetta*; however, Petrova Gora showed a slight similarity, which might be because all three locations are within central Croatia. On the other hand, the Cd contents in T. portentosum mushroom showed no significant difference amongst the three sampling locations, viz., Ravna Gora, Island Krk, and Labinština. However, notable similarities were seen in the case of Cd levels in *T. terreum* mushrooms at the Skrad and Dugi Dol sampling sites. Regarding this, the Maksimir site showed high variation for Cd levels in *T. terreum* mushroom. Previously, Kumar et al. [34] used PCA and HCA approaches to derive the interrelationship between heavy metal levels in *Agaricus bisporus* and their sampling locations across the thirteen districts of Uttarakhand State in India. They revealed that PCA and HCA were useful to understanding the impact of sampling location on the availability of eight heavy metals, including Cd in *A. bisporus* samples. Similarly, Buruleanu et al. [62] also used the PCA tool to study the effect of heavy metal concentration on different biochemical constituents of wild and cultivated mushroom species in Romania. The results of these reports are in line with the current study and suggest that effective information can be derived from the multivariate analysis of Cd level data in *Tricholoma* spp. samples collected from central and coastal Croatian locations.

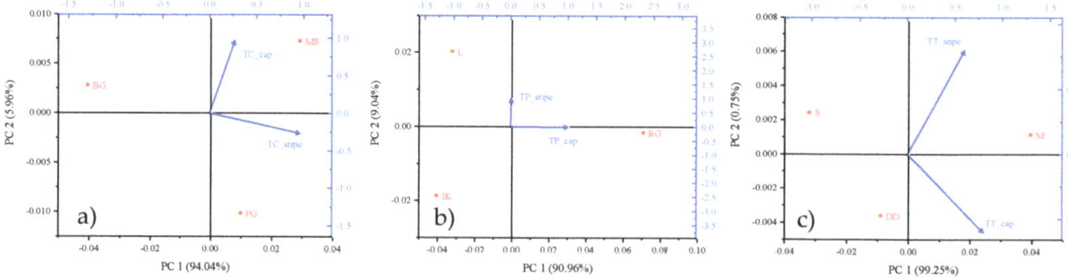

Figure 3. PCA biplot showing interactions between sampling locations and Cd contents in the cap and stipe parts of *Tricholoma* spp. ((**a**) *T. columbetta*, (**b**) *T. portentosum*, and (**c**) *T. terreum*).

Table 3. PCA matrix for Cd concentration in cap and stipe parts of *Tricholoma* spp.

Mushroom Species	Variables	Principal Component	
		PC 1	PC 2
T. columbetta (*n* = 38)	Variance (%)	94.04	5.95
	Eigenvalues	0.0013	0.0003
	Cd cap	0.99	−0.01
	Cd stipe	0.01	0.99
T. portentosum (*n* = 34)	Variance (%)	90.96	9.04
	Eigenvalues	0.0038	0.0001
	Cd cap	0.99	−0.01
	Cd stipe	0.01	0.99
T. terreum (*n* = 35)	Variance (%)	99.25	0.75
	Eigenvalues	0.0013	0.0005
	Cd cap	0.79	−0.60
	Cd stipe	0.60	0.80

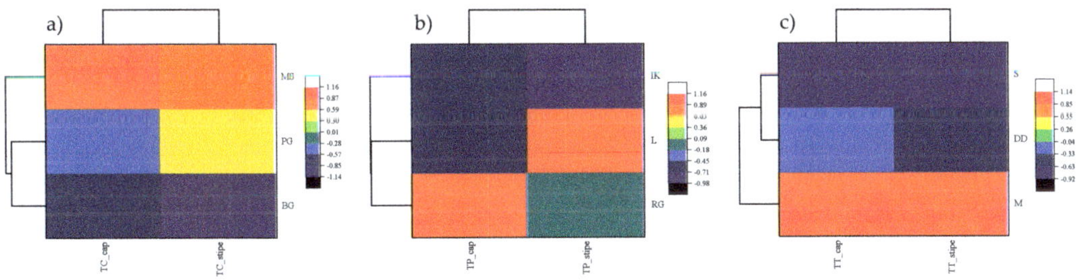

Figure 4. Clustered heatmap showing similarities between sampling locations and Cd contents in the cap and stipe of *Tricholoma* spp. ((**a**) and TC: *T. columbetta*, (**b**) and TP: *T. portentosum*, (**c**) and TT: *T. terreum*).

3.4. Health Risk Assessment of Cd Intake

In this study, the potential risk of Cd intake from the consumption of wild edible mushrooms *Tricholoma* spp. was established by using the provisional tolerable daily intake value PTDI (0.5 µg/kg bw/d) for a person of 70 kg body weight [63]. Based on the determined Cd concentrations in mushrooms and the assumed meal (300 g fresh or 30 dried mushrooms per day) [45], the daily intake of Cd (DIC) was calculated as given in Table 4. Here, the range of DIC values in the cap part of *Tricholoma* spp. was 0.329–0.426, while 0.254–0.312 for the stiped. The results showed that the highest DIC was determined in caps of *T. portentosum* at Ravna Gora (0.426 µg/kg body weight/serving). On the other hand, the highest DIC for Cd (0.312 µg/kg body weight/serving) was found in stipes of *T. columbetta*. However, the Skrad sampling location showed the lowest (0.254 g/kg body weight/serving) DIC values. In the case of the health risk index (HRI), the highest value was encountered in the case of *T. portentosum* (0.852) at the Ravna Gora location for the cap parts. For the stiped parts, the highest HRI values (0.624) were observed at the Medvednica, Stubaki, location in *T. columbetta*. Overall, the determined HRI values were below 1 for Cd levels in all analyzed samples of *Tricholoma* spp. in both central and coastal Croatia (Table 4). However, Leung et al. [64] stated that established health risk index values of 1 or less are considered safe for human health.

Table 4. Daily intakes of Cd and health risk index in wild edible *Tricholoma* spp. mushrooms.

Species	Locations	Daily Intakes of Cd (DIC, µg/kg Body Weight/Serving)		Health Risks Index (HRI)	
		Cap	Stipe	Cap	Stipe
T. columbetta	Brezova Gora	0.388	0.281	0.776	0.562
	Medvednica, Stubaki	0.403	0.312	0.806	0.624
	Petrova Gora,	0.393	0.306	0.787	0.612
T. portentosum	Island Krk	0.379	0.284	0.757	0.569
	Ravna Gora	0.426	0.292	0.852	0.585
	Labinština	0.382	0.301	0.764	0.602
T. terreum	Maksimir	0.334	0.272	0.689	0.544
	Dugi Dol, Karlovac	0.329	0.258	0.658	0.516
	Skrad	0.320	0.254	0.639	0.508

With the increasing number of wild mushrooms consumers, it has become a topic of great importance to biomonitoring the presence of toxic elements and their potential risks [61]. Regarding this, the elevated Cd levels in wild edible mushrooms can harm consumers' health, particularly as several species of *Tricholoma* mushroom are consumed in fresh or processed form. However, the practice of monitoring Cd levels before consuming these wild mushrooms is almost entirely lacking in Croatia. Being classified as a "probable" human carcinogen (IARC), the health risk assessment in the current study suggests an association between Cd exposure and the occurrence of cancer in humans [65]. There are several methods of preparing wild edible mushrooms, particularly *Tricholoma*, the most common of which are heat-treated, dried, or pickled. Maintaining regular physiological functions in the human body requires a diet with the optimal intake of essential elements (Fe, Zn, Cu, Mn, Mo, Se) [66], and their deficiency or excessive intake can cause health problems [67]. In addition to essential elements, there are also non-essential elements (Al, As, Ba, Cd, Hg, Ni, Pb) that have no biological functions in the body and are considered dangerous/toxic to consumers [66]. In regard to this, Cd is a well-known food contaminant possessing destructive health effects. Therefore, the toxicological effects of Cd associated with food consumption, in this case of mushrooms whose samples have HRI values > 1, may pose a health risk. The results presented in the current study are consistent with those reported in previous studies. Recently, Sarikurkcu et al. [31] and Chen et al. [27] found an HRI value for Cd greater than 1 in *T. terreum* and *T. matsutake* species. Similarly, Barea-Sepúlveda et al. [68] also calculated HRI values > 1 for Cd in the ectomycorrhizal mushroom species *A. caesarea*, whose samples were collected in Spain and Morocco. Thus, the HRI tool in the present study was helpful for biomonitoring the health risk associated with intake of Cd-contaminated *Tricholoma* spp.

4. Conclusions

This study investigated the occurrence of Cd metal in three *Tricholoma* mushroom species (*T. columbetta, T. portentosum*, and *T. terreum*) and their adjoining soil substrates across central and coastal Croatia. Results revealed that the highest Cd contents were observed in the *T. portentosum* mushroom followed by *T. columbetta* and *T. terreum*. However, the bioconcentration factor values revealed that selected *Tricholoma* spp. are good Cd accumulators and could uptake considerable amounts of Cd into their vegetative parts from soils. Overall, the health risk studies suggested no potential health risk associated with the consumption of *Tricholoma* spp. According to this study, exposure to Cd through the consumption of contaminated *Tricholoma* spp. is unlikely to cause adverse human health effects if the health risk index (HRI) value goes above 1. Furthermore, continuous monitoring of other toxic heavy metals in wild edible mushrooms in other regions of Croatia is highly recommended.

Author Contributions: Conceptualization, I.Š. and P.K.; Data curation, I.Š. and P.K.; Formal analysis, I.Š.; Funding acquisition, I.Š., I.K., D.B., B.M. and M.H.; Investigation, I.Š., I.K. and D.B.; Methodology, I.Š. and P.K.; Project administration, I.Š.; Resources, I.Š.; Software, I.Š. and P.K.; Supervision, I.Š.; Validation, E.M.E., A.B., I.K., D.B., B.M. and M.H.; Visualization, I.Š. and P.K.; Writing—original draft, I.Š. and P.K.; Writing—review and editing, E.M.E., A.B., I.K., D.B., B.M. and M.H. All authors have read and agreed to the published version of the manuscript.

Funding: This research was funded by the University Grants for Financing Scientific and Artistic Activities at the University of Zagreb, Faculty of Agriculture. All individuals included in this section have consented to the acknowledgment.

Institutional Review Board Statement: Not applicable.

Informed Consent Statement: Not applicable.

Data Availability Statement: Not applicable.

Acknowledgments: The authors are grateful to their host institutes for providing the necessary facilities to conduct this study.

Conflicts of Interest: The authors declare no conflict of interest. The study did not involve any animal or human experimentation/testing.

References

1. Kubier, A.; Wilkin, R.T.; Pichler, T. Cadmium in Soils and Groundwater: A Review. *Appl. Geochem.* **2019**, *108*, 104388. [CrossRef] [PubMed]
2. He, Z.L.; Yang, X.E.; Stoffella, P.J. Trace Elements in Agroecosystems and Impacts on the Environment. *J. Trace Elem. Med. Biol.* **2005**, *19*, 125–140. [CrossRef] [PubMed]
3. UNEP. *Final Review of Scientific Information on Lead*; Version of December; United Nations Environment Programme, Economy Division: Nairobi, Kenya, 2010.
4. Türkdoğan, M.K.; Kilicel, F.; Kara, K.; Tuncer, I.; Uygan, I. Heavy Metals in Soil, Vegetables and Fruits in the Endemic Upper Gastrointestinal Cancer Region of Turkey. *Environ. Toxicol. Pharmacol.* **2003**, *13*, 175–179. [CrossRef]
5. Cocchi, L.; Vescovi, L.; Petrini, L.E.; Petrini, O. Heavy Metals in Edible Mushrooms in Italy. *Food Chem.* **2006**, *98*, 277–284. [CrossRef]
6. International Agency for Research on Cancer. *Agents Classified by the IARC Monographs*; International Agency for Research on Cancer: Lyon, France, 2012; Volume 1–105.
7. Cardwell, G.; Bornman, J.F.; James, A.P.; Black, L.J. A Review of Mushrooms as a Potential Source of Dietary Vitamin D. *Nutrients* **2018**, *10*, 1498. [CrossRef] [PubMed]
8. Kalač, P. Mineral Composition and Radioactivity of Edible Mushrooms. In *Mineral Composition and Radioactivity of Edible Mushrooms*; Academic Press: Cambridge, MA, USA, 2019; pp. 1–392. ISBN 9780128175651.
9. Falandysz, J.; Borovička, J. Macro and Trace Mineral Constituents and Radionuclides in Mushrooms: Health Benefits and Risks. *Appl. Microbiol. Biotechnol.* **2013**, *97*, 477–501. [CrossRef] [PubMed]
10. Niazi, A.R.; Ghafoor, A. Different Ways to Exploit Mushrooms: A Review. *All Life* **2021**, *14*, 450–460. [CrossRef]
11. Talpur, N.A.; Echard, B.W.; Fan, A.Y.; Jaffari, O.; Bagchi, D.; Preuss, H.G. Antihypertensive and Metabolic Effects of Whole Maitake Mushroom Powder and Its Fractions in Two Rat Strains. *Mol. Cell. Biochem.* **2002**, *237*, 129–136. [CrossRef]
12. Jeong, S.C.; Jeong, Y.T.; Yang, B.K.; Islam, R.; Koyyalamudi, S.R.; Pang, G.; Cho, K.Y.; Song, C.H. White Button Mushroom (*Agaricus bisporus*) Lowers Blood Glucose and Cholesterol Levels in Diabetic and Hypercholesterolemic Rats. *Nutr. Res.* **2010**, *30*, 49–56. [CrossRef]
13. Zhao, S.; Gao, Q.; Rong, C.; Wang, S.; Zhao, Z.; Liu, Y.; Xu, J. Immunomodulatory Effects of Edible and Medicinal Mushrooms and Their Bioactive Immunoregulatory Products. *J. Fungi* **2020**, *6*, 269. [CrossRef]
14. Petkovšek, S.A.S.; Pokorny, B. Lead and Cadmium in Mushrooms from the Vicinity of Two Large Emission Sources in Slovenia. *Sci. Total Environ.* **2013**, *443*, 944–954. [CrossRef] [PubMed]
15. Alonso, J.; Salgado, M.J.; García, M.A.; Melgar, M.J. Accumulation of Mercury in Edible Macrofungi: Influence of Some Factors. *Arch. Environ. Contam. Toxicol.* **2000**, *38*, 158–162. [CrossRef] [PubMed]
16. Aloupi, M.; Koutrotsios, G.; Koulousaris, M.; Kalogeropoulos, N. Trace Metal Contents in Wild Edible Mushrooms Growing on Serpentine and Volcanic Soils on the Island of Lesvos, Greece. *Ecotoxicol. Environ. Saf.* **2012**, *78*, 184–194. [CrossRef]
17. Širić, I.; Humar, M.; Kasap, A.; Kos, I.; Mioč, B.; Pohleven, F. Heavy Metal Bioaccumulation by Wild Edible Saprophytic and Ectomycorrhizal Mushrooms. *Environ. Sci. Pollut. Res.* **2016**, *23*, 18239–18252. [CrossRef] [PubMed]
18. Borovička, J.; Konvalinková, T.; Žigová, A.; Ďurišová, J.; Gryndler, M.; Hršelová, H.; Kameník, J.; Leonhardt, T.; Sácký, J. Disentangling the Factors of Contrasting Silver and Copper Accumulation in Sporocarps of the Ectomycorrhizal Fungus *Amanita strobiliformis* from Two Sites. *Sci. Total Environ.* **2019**, *694*, 133679. [CrossRef] [PubMed]

19. Falandysz, J.; Mędyk, M.; Treu, R. Bio-Concentration Potential and Associations of Heavy Metals in *Amanita muscaria* (L.) Lam. from Northern Regions of Poland. *Environ. Sci. Pollut. Res.* **2018**, *25*, 25190–25206. [CrossRef]
20. Kavčič, A.; Mikuš, K.; Debeljak, M.; Teun van Elteren, J.; Arčon, I.; Kodre, A.; Kump, P.; Karydas, A.G.; Migliori, A.; Czyzycki, M.; et al. Localization, Ligand Environment, Bioavailability and Toxicity of Mercury in *Boletus* spp. and Scutiger Pes-Caprae Mushrooms. *Ecotoxicol. Environ. Saf.* **2019**, *184*, 109623. [CrossRef]
21. Širić, I.; Falandysz, J. Contamination, Bioconcentration and Distribution of Mercury in *Tricholoma* spp. Mushrooms from Southern and Northern Regions of Europe. *Chemosphere* **2020**, *251*, 126614. [CrossRef]
22. Choma, A.; Nowak, K.; Komaniecka, I.; Waśko, A.; Pleszczyńska, M.; Siwulski, M.; Wiater, A. Chemical Characterization of Alkali-Soluble Polysaccharides Isolated from a *Boletus edulis* (Bull.) Fruiting Body and Their Potential for Heavy Metal Biosorption. *Food Chem.* **2018**, *266*, 329–334. [CrossRef]
23. Malik, A. Metal Bioremediation through Growing Cells. *Environ. Int.* **2004**, *30*, 261–278. [CrossRef]
24. Mleczek, M.; Magdziak, Z.; Gąsecka, M.; Niedzielski, P.; Kalač, P.; Siwulski, M.; Rzymski, P.; Zalicka, S.; Sobieralski, K. Content of Selected Elements and Low-Molecular-Weight Organic Acids in Fruiting Bodies of Edible Mushroom *Boletus badius* (Fr.) Fr. from Unpolluted and Polluted Areas. *Environ. Sci. Pollut. Res.* **2016**, *23*, 20609–20618. [CrossRef]
25. Severoglu, Z.; Sumer, S.; Yalcin, B.; Leblebici, Z.; Aksoy, A. Trace Metal Levels in Edible Wild Fungi. *Int. J. Environ. Sci. Technol.* **2013**, *10*, 295–304. [CrossRef]
26. Širić, I.; Kasap, A.; Bedeković, D.; Falandysz, J. Lead, Cadmium and Mercury Contents and Bioaccumulation Potential of Wild Edible Saprophytic and Ectomycorrhizal Mushrooms, Croatia. *J. Environ. Sci. Health Part B Pestic. Food Contam. Agric. Wastes* **2017**, *52*, 156–165. [CrossRef] [PubMed]
27. Chen, H.X.; Chen, Y.; Li, S.; Zhang, W.; Zhang, Y.; Gao, S.; Li, N.; Tao, L.; Wang, Y. Trace Elements Determination and Health Risk Assessment of *Tricholoma matsutake* from Yunnan Province, China. *J. Consum. Prot. Food Saf.* **2020**, *15*, 153–162. [CrossRef]
28. Gałgowska, M.; Pietrzak-fiećko, R. Cadmium and Lead Content in Selected Fungi from Poland and Their Edible Safety Assessment. *Molecules* **2021**, *26*, 7289. [CrossRef]
29. Liu, S.; Fu, Y.; Shi, M.; Wang, H.; Guo, J. Pollution Level and Risk Assessment of Lead, Cadmium, Mercury, and Arsenic in Edible Mushrooms from Jilin Province, China. *J. Food Sci.* **2021**, *86*, 3374–3383. [CrossRef] [PubMed]
30. Nowakowski, P.; Markiewicz-Żukowska, R.; Soroczyńska, J.; Puścion-Jakubik, A.; Mielcarek, K.; Borawska, M.H.; Socha, K. Evaluation of Toxic Element Content and Health Risk Assessment of Edible Wild Mushrooms. *J. Food Compos. Anal.* **2021**, *96*, 103698. [CrossRef]
31. Sarikurkcu, C.; Popović-Djordjević, J.; Solak, M.H. Wild Edible Mushrooms from Mediterranean Region: Metal Concentrations and Health Risk Assessment. *Ecotoxicol. Environ. Saf.* **2020**, *190*, 110058. [CrossRef] [PubMed]
32. Su, J.; Zhang, J.; Li, J.; Li, T.; Liu, H.; Wang, Y. Determination of Mineral Contents of Wild *Boletus edulis* Mushroom and Its Edible Safety Assessment. *J. Environ. Sci. Health Part B Pestic. Food Contam. Agric. Wastes* **2018**, *53*, 454–463. [CrossRef]
33. Demirbas, A. Levels of Trace Elements in the Fruiting Bodies of Mushrooms Growing in the East Black Sea Region. *Energy Edu. Sci. Technol.* **2001**, *7*, 67–81.
34. Kumar, P.; Kumar, V.; Eid, E.M.; AL-Huqail, A.A.; Adelodun, B.; Abou Fayssal, S.; Goala, M.; Arya, A.K.; Bachheti, A.; Andabaka, Ž.; et al. Spatial Assessment of Potentially Toxic Elements (PTE) Concentration in *Agaricus bisporus* Mushroom Collected from Local Vegetable Markets of Uttarakhand State, India. *J. Fungi* **2022**, *8*, 452. [CrossRef] [PubMed]
35. Kumar, P.; Kumar, V.; Adelodun, B.; Bedeković, D.; Kos, I.; Širić, I.; Alamri, S.A.M.; Alrumman, S.A.; Eid, E.M.; Abou Fayssal, S.; et al. Sustainable Use of Sewage Sludge as a Casing Material for Button Mushroom (*Agaricus bisporus)* Cultivation: Experimental and Prediction Modeling Studies for Uptake of Metal Elements. *J. Fungi* **2022**, *8*, 112. [CrossRef] [PubMed]
36. Melgar, M.J.; Alonso, J.; García, M.A. Cadmium in Edible Mushrooms from NW Spain: Bioconcentration Factors and Consumer Health Implications. *Food Chem. Toxicol.* **2016**, *88*, 13–20. [CrossRef] [PubMed]
37. Meisch, H.-U.; Schmitt, J.A.; Reinle, W. Schwermetalle in Höheren Pilzen Cadmium, Zink Und Kupfer Heavy Metals in Higher Fungi Cadmium, Zinc, and Copper. *Z. Für Nat. C* **1977**, *32*, 172–181. [CrossRef]
38. Strapáč, I.; Baranová, M. Content of Chemical Elements in Wood-Destroying Fungi. *Folia Vet.* **2016**, *60*, 29–36. [CrossRef]
39. Wang, S.H.; Zhang, J.D.; Xu, H.; Li, D.H. Metal Content of *Armillaria mellea* in the Tumen River Basin. *Int. J. Food Prop.* **2017**, *20*, 2052–2059. [CrossRef]
40. Karmanska, A.; Wedzisz, A. Content of Selected Macro- and Microelements in Various Species of Large Fruiting Body Mushroom-scollected in the Province of Łódz. *Chem. Toksykol.* **2010**, *43*, 124–129.
41. Anna Adamiak, E.; Kalembasa, S.; Kuziemska, B. Contents of Heavy Metals in Selected Species of Edible Mushrooms. *Acta Agrophysica* **2013**, *20*, 7–16.
42. Chiocchetti, G.M.; Latorre, T.; Clemente, M.J.; Jadán-Piedra, C.; Devesa, V.; Vélez, D. Toxic Trace Elements in Dried Mushrooms: Effects of Cooking and Gastrointestinal Digestion on Food Safety. *Food Chem.* **2020**, *306*, 125478. [CrossRef]
43. Zavastin, D.E.; Biliută, G.; Dodi, G.; Macsim, A.M.; Lisa, G.; Gherman, S.P.; Breabăn, I.G.; Miron, A.; Coseri, S. Metal Content and Crude Polysaccharide Characterization of Selected Mushrooms Growing in Romania. *J. Food Compos. Anal.* **2018**, *67*, 149–158. [CrossRef]
44. Rasalanavho, M.; Moodley, R.; Jonnalagadda, S.B. Elemental Bioaccumulation and Nutritional Value of Five Species of Wild Growing Mushrooms from South Africa. *Food Chem.* **2020**, *319*, 126596. [CrossRef] [PubMed]

45. Kalač, P.; Svoboda, L. A Review of Trace Element Concentrations in Edible Mushrooms. *Food Chem.* **2000**, *69*, 273–281. [CrossRef]
46. Falandysz, J.; Chudzińska, M.; Barałkiewicz, D.; Drewnowska, M.; Hanć, A. Toxic Elements and Bio-Metals in *Cantharellus* Mushrooms from Poland and China. *Environ. Sci. Pollut. Res.* **2017**, *24*, 11472–11482. [CrossRef]
47. Árvay, J.; Tomáš, J.; Hauptvogl, M.; Kopernická, M.; Kováčik, A.; Bajčan, D.; Massányi, P. Contamination of Wild-Grown Edible Mushrooms by Heavy Metals in a Former Mercury-Mining Area. *J. Environ. Sci. Health Part B Pestic. Food Contam. Agric. Wastes* **2014**, *49*, 815–827. [CrossRef] [PubMed]
48. Coroian, A.; Odagiu, A.; Marchis, Z.; Miresan, V.; Raducu, C.; Oroian, C.; Longodor, A.L. Heavy Metals and the Radioactivity in Boletus (*Boletus edulis*), and Chanterelle Mushrooms (*Cantharellus cibarius*) in Transylvanian Area. *Agrolife Sci. J.* **2018**, *7*, 17–21.
49. Borovička, J.; Braeuer, S.; Sácký, J.; Kameník, J.; Goessler, W.; Trubač, J.; Strnad, L.; Rohovec, J.; Leonhardt, T.; Kotrba, P. Speciation Analysis of Elements Accumulated in *Cystoderma carcharias* from Clean and Smelter-Polluted Sites. *Sci. Total Environ.* **2019**, *648*, 1570–1581. [CrossRef]
50. Falandysz, J.; Kunito, T.; Kubota, R.; Gucia, M.; Mazur, A.; Falandysz, J.J.; Tanabe, S. Some Mineral Constituents of Parasol Mushroom (*Macrolepiota procera*). *J. Environ. Sci. Health Part B Pestic. Food Contam. Agric. Wastes* **2008**, *43*, 187–192. [CrossRef]
51. Yamaç, M.; Yildiz, D.; Sarikürkcü, C.; Çelikkollu, M.; Solak, M.H. Heavy Metals in Some Edible Mushrooms from the Central Anatolia, Turkey. *Food Chem.* **2007**, *103*, 263–267. [CrossRef]
52. Malinowska, E.; Szefer, P.; Falandysz, J. Metals Bioaccumulation by *Bay bolete, Xerocomus badius*, from Selected Sites in Poland. *Food Chem.* **2004**, *84*, 405–416. [CrossRef]
53. Yilmaz, F.; Isiloglu, M.; Merdivan, M. Heavy Metals Levels in Some Macrofungi. *Turk. J. Bot.* **2003**, *27*, 45–56.
54. Heilmann-Clausen, J.; Christensen, M.; Frøslev, T.G.; Kjøller, R. Taxonomy of Tricholoma in Northern Europe Based on ITS Sequence Data and Morphological Characters. *Pers. Mol. Phylogeny Evol. Fungi* **2017**, *38*, 38–57. [CrossRef]
55. Wheater, C.P.; Bell, J.R.; Cook, P.A. *Practical Field Ecology: A Project Guide*, 2nd ed.; John Willey & Sons Inc.: Hoboken, NJ, UAS, 2011; ISBN 978-1-119-41322-6.
56. Cui, Y.J.; Zhu, Y.G.; Zhai, R.H.; Chen, D.Y.; Huang, Y.Z.; Qiu, Y.; Liang, J.Z. Transfer of Metals from Soil to Vegetables in an Area near a Smelter in Nanning, China. *Environ. Int.* **2004**, *30*, 785–791. [CrossRef] [PubMed]
57. USEPA. *A Review of the Reference Dose and Reference Concentration Processes*; EPA/630/P-02/002F; U.S. Environmental Protection Agency, Risk Assessment Forum: Washington, DC, USA, 2002.
58. Soylak, M.; Saraçoğlu, S.; Tüzen, M.; Mendil, D. Determination of Trace Metals in Mushroom Samples from Kayseri, Turkey. *Food Chem.* **2005**, *92*, 649–652. [CrossRef]
59. Sarikurkcu, C.; Copur, M.; Yildiz, D.; Akata, I. Metal Concentration of Wild Edible Mushrooms in Soguksu National Park in Turkey. *Food Chem.* **2011**, *128*, 731–734. [CrossRef]
60. Gucia, M.; Jarzyńska, G.; Rafał, E.; Roszak, M.; Kojta, A.K.; Osiej, I.; Falandysz, J. Multivariate Analysis of Mineral Constituents of Edible Parasol Mushroom (*Macrolepiota procera*) and Soils beneath Fruiting Bodies Collected from Northern Poland. *Environ. Sci. Pollut. Res.* **2012**, *19*, 416–431. [CrossRef] [PubMed]
61. Saba, M.; Falandysz, J.; Nnorom, I.C. Accumulation and Distribution of Mercury in Fruiting Bodies by Fungus *Suillus luteus* Foraged in Poland, Belarus and Sweden. *Environ. Sci. Pollut. Res.* **2016**, *23*, 2749–2757. [CrossRef]
62. Buruleanu, L.C.; Radulescu, C.; Antonia Georgescu, A.; Dulama, I.D.; Nicolescu, C.M.; Lucian Olteanu, R.; Stanescu, S.G. Chemometric Assessment of the Interactions Between the Metal Contents, Antioxidant Activity, Total Phenolics, and Flavonoids in Mushrooms. *Anal. Lett.* **2019**, *52*, 1195–1214. [CrossRef]
63. JECFA. *Joint FAO/WHO Expert Committee on Food Additives Seventy-Second Meeting: Summary and Conclusions*; World Health Organization: Geneva, Switzerland, 2010; pp. 1–16.
64. Leung, A.O.W.; Duzgoren-Aydin, N.S.; Cheung, K.C.; Wong, M.H. Heavy Metals Concentrations of Surface Dust from E-Waste Recycling and Its Human Health Implications in Southeast China. *Environ. Sci. Technol.* **2008**, *42*, 2674–2680. [CrossRef]
65. Tchounwou, P.B.; Yedjou, C.G.; Patlolla, A.K.; Sutton, D.J. Heavy Metal Toxicity and the Environment. *Exs* **2012**, *101*, 133–164. [CrossRef]
66. de la Guardia, M.; Garrigues, S. *Handbook of Mineral Elements in Food*; de la Guardia, M., Garrigues, S., Eds.; John Wiley & Sons, Ltd.: Chichester, UK, 2015; ISBN 9781118654316.
67. Zhang, L.; Lv, J.; Liao, C. Dietary Exposure Estimates of 14 Trace Elements in Xuanwei and Fuyuan, Two High Lungcancer Incidence Areas in China. *Biol. Trace Elem. Res.* **2012**, *146*, 287–292. [CrossRef]
68. Barea-Sepúlveda, M.; Espada-Bellido, E.; Ferreiro-González, M.; Bouziane, H.; López-Castillo, J.G.; Palma, M.; Barbero, G.F. Exposure to Essential and Toxic Elements via Consumption of Agaricaceae, Amanitaceae, Boletaceae, and Russulaceae Mushrooms from Southern Spain and Northern Morocco. *J. Fungi* **2022**, *8*, 545. [CrossRef] [PubMed]

Journal of *Fungi*

Article

Fungus–Fungus Association of *Boletus griseus* and *Hypomyces chrysospermus* and Cadmium Resistance Characteristics of Symbiotic Fungus *Hypomyces chrysospermus*

Zhen Tian, Yunan Wang, Yongliang Zhuang, Chunze Mao, Yujia Shi and Liping Sun *

Faculty of Food Science and Engineering, Kunming University of Science and Technology, No. 727, Jingming South Road, Chenggong District, Kunming 650500, China; tianzhen2020@yeah.net (Z.T.); wyn9708@163.com (Y.W.); ylzhuang@kmust.edu.cn (Y.Z.); kmczmao@163.com (C.M.); kmyjshi@163.com (Y.S.)
* Correspondence: lpsun@kmust.edu.cn; Tel./Fax: +86-871-65920216

Abstract: Fungi bioaccumulation of heavy metals is a promising approach to remediate polluted soil and water. *Boletus griseus* could accumulate high amounts of Cd, even in a natural habitat with low Cd contents. This study found a symbiotic association of *B. griseus* with a fungus. The symbiotic fungus was isolated and identified as *Hypomyces chrysospermus*. The isolated strain had a strong ability to tolerate Cd. The minimum inhibitory concentration of Cd of fungal growth was 200 mg·L^{-1}. The Cd bioaccumulation capacity of the fungus reached 10.03 mg·g^{-1}. The biomass production of the fungus was promoted by 20 mg·L^{-1} Cd. However, high concentrations of Cd suppressed fungal growth and significantly altered the morphology and fine texture of fungal hyphae and chlamydospores. The immobilization effects of the cell wall and acid compounds and antioxidant enzymes were employed by the fungus to alleviate the toxic effects of Cd. The results not only demonstrate a new insight into the Cd bioconcentration mechanisms of *B. griseus* but also provide a potential bioremediation fungus for Cd contamination.

Keywords: *Boletus griseus*; *Hypomyces chrysospermus*; cadmium; symbiotic association

Citation: Tian, Z.; Wang, Y.; Zhuang, Y.; Mao, C.; Shi, Y.; Sun, L. Fungus–Fungus Association of *Boletus griseus* and *Hypomyces chrysospermus* and Cadmium Resistance Characteristics of Symbiotic Fungus *Hypomyces chrysospermus*. *J. Fungi* **2022**, *8*, 578. https://doi.org/10.3390/jof8060578

Academic Editor: Miha Humar

Received: 26 April 2022
Accepted: 22 May 2022
Published: 27 May 2022

1. Introduction

Boletus griseus is a common wild-grown edible mushroom in Yunnan Province, and it is a member of the genus *Boletus* in the family *Boletaceae* [1]. Studies have confirmed that *B. griseus* can accumulate high amounts of Cd in fruiting bodies, even from natural habitats with low Cd contents in the soil matrix [2–4]. The Cd contents of 153 samples of *B. griseus* were found to range from 1.61–42.67 mg·kg^{-1} while those in soil ranged from 0.03–0.57 mg·kg^{-1}. The bioconcentration factors of *B. griseus* for Cd were 24–386 [4]. The factors affecting Cd migration from soil to *B. griseus* were investigated. Principal component analysis elucidated that the soil physical-chemical properties, such as the Cd content in soil, electrical conductivity, total carbon, total nitrogen, pH, and dissolved organic carbon, were factors that affected the Cd accumulation of *B. griseus* [4]. Macrofungi are known to effectively accumulate higher concentrations of metals and metalloids than vascular plants [5]. Therefore, the accumulation of elements in mushrooms has attracted significant research attention [6–8]. Considering the natural values of Cd in mushrooms, *B. griseus* showed significantly higher Cd contents than most reported mushrooms, except for the family Agaricaceae [9].

B. griseus is a kind of ectomycorrhizal mushroom, and is symbiotically associated with a large number of trees and shrubs. It can be found along the forest edge of pine-oak mixed forests in Yunnan Province [10]. When sampling, two phenotypes of the sporocarps of *B. griseus* were found, namely, normal-developed sporocarps (Figure 1) and symbiotic fruiting bodies of *B. griseus*–mycoparasitic fungus (Figure 2). Owing to the geographical location and ecological, climatic, topographic, and geological factors, Yunnan Province is rich in fungal biodiversity, with more than 882 edible species and an average annual yield of 500,000 tons. Many studies have focused on the resource survey, annual yield, bioactivities and bioactive components, and food safety of wild edible mushrooms [2,11–16]. However, few reports have studied the fungicolous fungi and host–parasite relationships in wild edible mushrooms, except for the genus *Hypomyces* [17,18].

Figure 1. Normal-developed sporocarps of *B. griseus*.

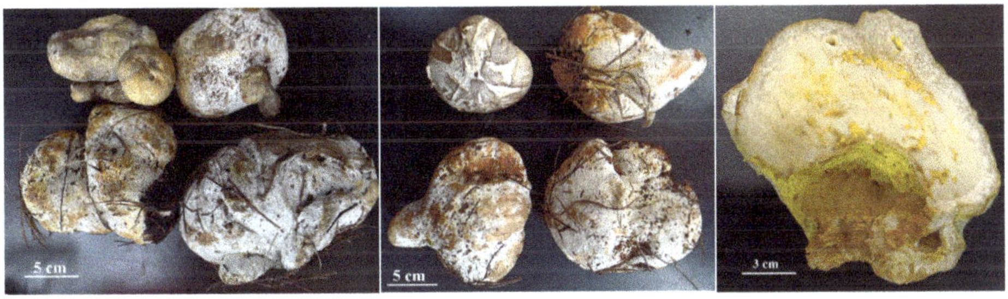

Figure 2. The symbiotic fruiting bodies of *B. griseus*–mycoparasitic fungus and their vertical section.

Hypomyces is a genus in the family *Hypocreaceae* with more than 150 species, including the most characteristic mycoparasites of diverse fungal hosts of agarics, boletes, russules, thelephores, and polypores [17]. Most species of *Hypomyces* are obligatory parasites, growing only on a specific host. The parasites of *Hypomyces* usually cause systematic infection and mummification of the host fruiting bodies [18]. Douhan and Rizzo collected 22 isolates from 21 infected bolete and one infected agaric in California oak woodlands, among which 20 isolates from individual bolete hosts were identified as *H. microspermus* and *H. chrysospermus* [17]. Kaygusuz et al. described *H. chrysospermus* isolated from infected *Suillus luteus* as a new record in Turkey [18]. Thus, the genus *Hypomyces* exclusively comprises boleticolous species.

Up to now, few studies about the *B. griseus*–mycoparasitic fungus association have been carried out. The symptoms of mycoparasitic fungus infection of *B. griseus* in this study differed from those of other bolete mushrooms, as studies found that infections of the genus *Hypomyces* led to total necrosis of bolete hosts [18]. Considering the strong Cd bioaccumulation capacity of *B. griseus*, even in natural habitats without Cd contamination, the unique biological strategy of *B. griseus* to take up Cd from the soil matrix was assumed, such as being parasitized by fungicolous fungi and forming a symbiote. Thus, this study aimed (1) to test and analyze the Cd contents in normal-developed sporocarps and symbiotic fruiting bodies of *B. griseus*–mycoparasitc fungus, (2) to isolate and identify the mycoparasitc fungus from normal-developed sporocarps (Figure 1) and symbiotic fruiting bodies of *B. griseus*–mycoparasitc fungus (Figure 2), and (3) to assess the Cd tolerance and Cd-accumulating traits of isolated mycoparasitic fungus. This study could help to exploit the possible mechanism of Cd bioaccumulation of *B. griseus*, from the perspective of the roles of the fungi-associated fungi in Cd uptake.

2. Material and Methods

2.1. Sampling

Samples of normal-developed sporocarps and symbiotic fruiting bodies of *B. griseus*–mycoparasitic fungus (marked as deformed sporocarps) were collected from four locations separated by approximately 20–40 km, being Guishan Town, Shilin Town, Banqiao Town, and Changhu Town in Shilin City, Yunnan Province, China, in June and July of 2020 and 2021. Five to six areas that were at a distance from the communication routes and other sources of environmental pollution and rich in *B. griseus* were selected from each location. During the sampling, attention was paid to make sure that the normal-developed and deformed sporocarps were collected from the same area (within approximately 10 ha^{-1}) and at the same time (a single day). In total, 22 samples of normal-developed sporocarps and 22 samples of deformed sporocarps were collected from 22 studied areas. From each of the studied areas, 10–15 fruiting bodies were collected and pooled for each sample.

The samples were brought to Kunming University of Science and Technology, Wild Edible Mushroom Research Laboratory, where the samples were photographed. Then, three normal-developed sporocarps and one deformed sporocarp were randomly selected for the isolation and identification of mycoparasitic fungus. The rest of the sporocarps were cleaned to remove forest debris and washed successively with running water and distilled water. The cleaned sporocarps were freeze dried and ground into a fine powder (40 meshes) and stored in sealed polyethylene bugs in a vacuum dryer for Cd determination.

2.2. Cd Determination

The Cd contents of normal-developed and deformed sporocarps were determined by atomic absorption spectrometry [2]. Briefly, 0.1 g samples were soaked overnight in open polytetrafluoroethylene (PTFE) vessels with 8 mL of concentrated nitric acid (65%). Then, after pre-digestion was conducted using a digestion apparatus at 120 °C for 60 min, the PTFE vessels were closed and heated in a microwave oven (MARS6, CEM Corp., Charlotte, NC, USA). The digestion conditions were set at 1.6 kW power by a 3-step heating program, being 8 min ramp, temperature of 120 °C, and 5 min hold; 5 min of ramp, temperature of 150 °C, and 5 min hold; 8 min ramp, temperature of 190 °C, and 25 min hold. The digest was diluted and filtered on 0.45 μm polyethersulfone membrane filters. Then, the solution was determined by an atomic absorption spectrophotometer (AAS400G, Analytik Jena AG, Jena, Germany) with a deuterium background corrector. The analytical wavelengths were determined at 228.8 nm. The pyrolysis and atomization temperatures were 350 and 1300 °C, respectively. Signals were measured as the peak area. The assurance quality of the analytical method was investigated through analysis of the certified reference materials of GBW10025 (CRM spirulina) supplied by Geophysiochemistry Prospecting Institute of Academy of Geological Science of China. The certified value for Cd of CRM spirulina was 0.37 $mg \cdot kg^{-1}$. CRM was analyzed for Cd with each analytical batch of samples in triplicate, and the recovery of Cd was 92.5–106.04%. Three replicates were conducted for the Cd determination of the samples.

2.3. Isolation of Mycoparasitic Fungus

All selected normal and deformed sporocarps from Section 2.1 were cleaned to remove forest debris, washed with running water to remove dirt, subjected to surface sterilization sequentially with 75% ethanol for 90 s and 5% NaCl 3 times, and then rinsed in sterile water 3 times.

For normal sporocarps, small pieces (0.5 cm × 0.5 cm × 0.5 cm) from under the epidermis were taken from the base of the stipes. The pieces were placed on potato dextrose agar (PDA) plates and incubated at 28 °C in the dark for 5 days.

For deformed sporocarps, a flame-sterilized loop was used to pick out spores (Figure 2) on PDA plates. The plates were incubated at 28 °C in the dark for 5 days.

After 5 days, the leading edges of the fungal colonies was transferred to PDA plates and incubated at 28 °C in the dark for 5 days for growth and isolation.

2.4. Morphological Identification and Taxonomic Analyses of Mycoparasitic Fungus

The morphological characteristics were recorded, including the colony shape, size, color, surface state, and edge. The microstructures of the endophyte mycelium and spores were examined using a light microscope after being stained bright blue with gossypol blue dye.

The genomic DNA of the isolated fungus was extracted using a Trelief Plant Genomic DNA Kit (Tsingke Biotechnology Co., Ltd., Beijing, China). The ITS1-ITS4 region was amplified using PCR primers ITS1 (5′-TCCGTAGGTGAACCTGCGG-3′) and ITS4 (5′-TCCTCCGCTTATTGATATGC-3′) [19]. Electrophoresis was carried out using agarose gel. The PCR products were purified using PCR purification kits (Tsingke Biotechnology Co., Ltd., Beijing, China) and sequenced using a BigDye Terminator v3.1 Cycle Sequencing Kit (Applied Biosystems, Waltham, MA, USA). Raw sequences were aligned using ContigExpress (Vector NTI Suite8.0; Invitrogen) and BLAST searched for the best match in NCBI (blast.ncbi.nlm.nih.gov accessed on 2 March 2019).

2.5. Determination of the Cd Resistance of Mycoparasitic Fungus

The Cd resistance of the isolated mycoparasitic fungus was examined using the PDA plates with gradually increasing concentrations of $CdCl_2$, with Cd^{2+} of 0 (control), 80, 120, 160, and 400 mg·L^{-1}. Six replicates were conducted for each Cd treatment. The plates were incubated at 28 °C in the dark for 5 days. The concentration of Cd^{2+} at which no visible fungal growth was observed was considered the minimum inhibitory concentration (MIC) of the mycoparasitic fungus. The Cd^{2+} level just below the MIC was considered as the highest Cd^{2+} concentration tolerated by the mycoparasitic fungus. Then, the inhibition percentage (IP, %) of Cd^{2+} of the fungus was measured by the ratio of the colony diameter with Cd^{2+} treatment to the colony diameter of the control [20]. EC_{50} refers to the Cd^{2+} concentration required to inhibit 50% of the fungal growth [21].

2.6. Effects of Cd^{2+} on the Morphology of Mycoparasitic Fungus

Scanning electron microscopy (SEM) was employed to investigate the micromorphology of the mycelia from the colonies of the 0–160 mg·L^{-1} Cd^{2+} treatments [22]. Briefly, the colonies of the mycoparasitic fungus from PDA plates with 0–160 mg·L^{-1} Cd^{2+} were collected, and the mycelia was fixed and used as the samples. Images of the mycelia under different Cd^{2+} concentrations were obtained via SEM analysis (VEGA3-SBH, Tescan) under the following analytical conditions: HW = 25.0 kW, WD = 14.77 mm, and SignalA = SE.

2.7. Effects of Cd^{2+} on the Growth of Mycoparasitic Fungus

In accordance with the highest Cd^{2+} concentration tolerated by the fungus, the effects of Cd^{2+} on the growth of the fungus were examined using shake flasks with 100 mL potato dextrose broth (PDB) added with gradually increased concentrations of $CdCl_2$, with Cd^{2+} of 0 (control), 20, 40, 60, 80, 100, 120, 140, and 160 mg·L^{-1}. Six replicates were conducted for each Cd^{2+} treatment. Cultures were incubated at 28 °C and 120 rpm for 7 days. Then, the pH values of the fermentation mixtures were determined using a pH meter. The mixture was filtered, and the mycelia were collected and washed three times with deionized water. The fresh mycelia from the three fermentation mixtures were freeze dried. The dried mycelia were accurately weighed to determine the effects of Cd^{2+} on the biomass yields. The fresh mycelia from the rest of the three fermentation mixtures was pooled for the investigation of the contents of soluble protein and the activities of antioxidant enzymes, such as superoxide dismutase (SOD), peroxidase (POD), and catalase (CAT) [22].

2.8. Cd^{2+} Absorption by Mycoparasitic Fungus

The dried mycelia from Section 2.7 were digested, and the Cd contents in the mycelia were determined, as described in Section 2.2. The values of Cd^{2+} absorption by the mycoparasitic fungus during the incubation were calculated as mg Cd per g dried mycelia.

2.9. Data Analysis

Data were analyzed by SPSS 16.0 (SPSS Inc., Chicago, IL, USA). The differences between means were analyzed using Duncan's multiple range test or Student's t-test at the 0.05 probability level. Graphical work was conducted on Excel 2007 (Microsoft, Redmond, WA, USA). The results were expressed as mean ± SD.

3. Results and Discussion

3.1. Cd Contents in Normal and Deformed Sporocarps of *B. griseus*

As shown in Table 1, the Cd contents in 22 normal-developed sporocarps of *B. griseus* ranged from 13.71–43.40 mg·kg^{-1}, with an average of 24.45 mg·kg^{-1}. According to a previous study [2], the Cd contents in 227 *Boletaceae* mushrooms from Yunnan Province ranged from not detected in *B. aereus* to 25.01 mg·kg^{-1} DW in *B. griseus*. The results of the present paper are consistent with those of the previous study [2]. Researchers have reported the contents of Cd in *Boletaceae* mushrooms, such as 1.2 mg·kg^{-1} in *B. edulis* [23]; 2.77 mg·kg^{-1} in *B. edulis* [13]; 0.58–1.31 mg·kg^{-1} in different parts of *B. aereus*, *B. aestivalis*, *B. edulis*, and *B. pinophilius* [24]; 1.44–2.01 mg·kg^{-1} in different parts of *B. badius* [25]; 0.24 mg·kg^{-1} in *B. aereus* [26]; and 0.28 mg·kg^{-1} in *Neoboletus erythropus* [27].

Table 1. Cadmium contents of 22 samples of normal and deformed sporocarps of *B. griseus* (mg·kg^{-1} DW).

Sample ID	Normal	Deformed	Sample ID	Normal	Deformed
1	26.1 ± 0.0 [a]	14.9 ± 0.0 [b]	13	16.1 ± 0.1 [b]	35.4 ± 0.0 [a]
2	21.1 ± 4.9 [a]	13.5 ± 0.8 [b]	14	21.3 ± 0.0 [a]	16.2 ± 0.1 [b]
3	18.8 ± 1.4 [b]	25.3 ± 0.5 [a]	15	43.4 ± 0.0 [b]	44.4 ± 0.1 [a]
4	27.7 ± 1.7 [a]	25.6 ± 1.6 [a]	16	27.5 ± 0.2 [b]	49.3 ± 0.9 [a]
5	15.6 ± 0.1 [b]	27.6 ± 1.0 [a]	17	19.3 ± 0.0 [b]	19.6 ± 0.0 [a]
6	33.4 ± 0.8 [b]	40.3 ± 0.0 [a]	18	27.2 ± 0.1 [b]	36.5 ± 0.6 [a]
7	24.7 ± 0.6 [a]	6.4 ± 0.2 [b]	19	21.3 ± 0.1 [b]	24.1 ± 0.1 [a]
8	15.2 ± 0.0 [a]	8.4 ± 0.6 [b]	20	17.8 ± 0.2 [b]	33.4 ± 0.1 [b]
9	37.2 ± 1.3 [a]	30.2 ± 2.8 [b]	21	13.7 ± 0.0 [a]	10.8 ± 0.1 [b]
10	31.6 ± 2.9 [b]	45.8 ± 0.3 [a]	22	21.4 ± 0.3 [a]	21.3 ± 0.2 [a]
11	35.3 ± 0.5 [a]	17.9 ± 0.1 [b]	Average	24.5 ± 7.9 [a]	26.3 ± 12.4 [a]
12	22.1 ± 0.0 [b]	31.7 ± 0.3 [a]			

Footnote of [a] and [b] in each line indicate significant differences ($p < 0.05$).

The Cd contents in 22 deformed sporocarps of *B. griseus* ranged from 6.35–49.29 mg·kg^{-1}, with an average of 26.75 mg·kg^{-1}. All the normal-developed and deformed sporocarps of *B. griseus* significantly accumulated high amounts of Cd. According to Student's t-test, among the 22 samples, 13 deformed sporocarps had significantly higher Cd contents than the normal-developed sporocarps.

In the last decade, extensive studies have shown that symbiotic association, such as mycorrhizas of fungi and plants, could enhance the tolerance and accumulation abilities of hosts regarding heavy metals (HMs) [28]. Mutualistic symbiotic fungi, such as ectomycorrhizal fungi (EMF), arbuscular mycorrhizal fungi (AMF), dark septate endophytes (DSEs), and endophytic fungi, are considered to be the remarkable factors of metal-accumulating plant species [20,29–32]. Co-culturing of fungi–fungi was recently established to produce new secondary metabolites. Wang et al. reported that mycelial pellets (*Aspergillus fumigatus*) and *Synechocystis* sp. PCC6803 comprised a fungus–microalgae symbiotic system, assisting microalgae flocculation and immobilization and the adsorption behavior for HMs [33].

The co-cultivation of microalgae with filamentous fungi is a superior method to efficiently accumulate and harvest the total biomass, and to remove pollutants from water [34]. However, the interactions of fungi associated with fungi have not yet been fully investigated, except for morphological and anatomical studies and molecular phylogenetic analyses of mycoparasitic fungi [35].

3.2. Isolation and Identification of Mycoparasitic Fungus

Fungal outgrowths from the surface-sterilized tissues of normal sporocarps and the spores of deformed sporocarps were observed after incubation on PDA plates. Only one filamentous fungal colony was isolated on the basis of unique phenotypic characteristics, as shown in Figure 3A.

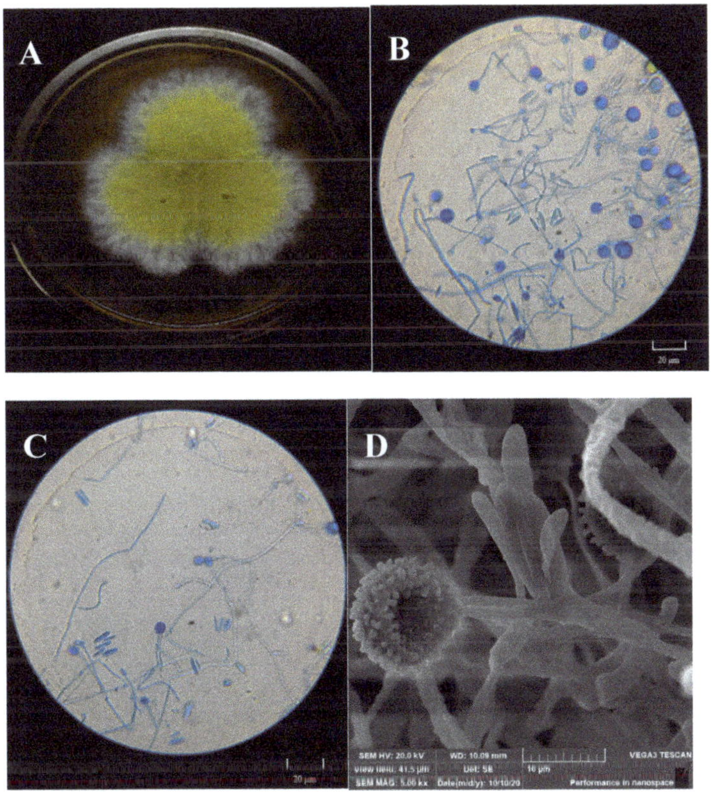

Figure 3. Macroscopic features of the isolated mycoparasitic fungus, *Hypomyces chrysospermus*. (**A**) Colonial morphology on PDA, (**B**,**C**): Optical microscopy images of the morphology of mycelia and conidiophore and chlamydospores; (**D**) scanning electron micrographs of mycelia and chlamydospores).

The colony on PDA medium was white, and it had a flat appearance with a regular edge. Then, the mycelia differentiated into yellow spores after approximately 3 days. As shown in Figure 3B,C, the mycelium was thin-walled and hyaline. Conidia were observed to be 10–20 × 3–8 μm in size, elliptical, aseptate, and single-celled, with smooth and thin walls, and hyaline. Globular tissues were observed to be approximately 10 μm in diameter under optical microscopy. SEM showed that the globular tissues were chlamydospores with thick walls and were prominently verrucose (Figure 3D).

According to the sequence of the ITS1-5.8S-ITS4 region, the isolated fungus belonged to the Ascomycota (classes: *Sordariomycetes*, *Hypocreales*, *Hypocreaceae*, and *Hypomyces*), and it was identified as *Hypomyces chrysospermus* based on a blast analysis in the NCBI GenBank database (MK560123.1).

H. chrysospermus is a cosmopolitan parasite of many boletes. According to Kaygusuz et al. [18], *Hypomyces* is easy to identify as the species in this genus usually produce bright-colored perithecia, conidiophores, and chlamydospores, and infection of *H. chrysospermus* causes decay of the host. However, *H. chrysospermus* seems to play an essential role in the development of sporocarps of *B. griseus*. *H. chrysospermus*-infected or uninfected *B. griseus* could develop normal fruiting bodies that had a distinguished gray color and were cap- and stipe-shaped. Meanwhile, under certain environmental conditions (temperature and relative humidity), deformed fruiting bodies with a significantly bigger size and weight usually formed, being a symbiont of *B. griseus*–*H. chrysospermus*. The symbiont had no shape features of bolete mushrooms, and it was usually spherical with a gray-white appearance, as shown in Figure 2.

3.3. Cd Tolerance of H. chrysospermus

The Cd tolerance of *H. chrysospermus* was determined on Cd-enriched PDA plates. As shown in Figure 4, *H. chrysospermus* grew very rapidly when cultured on PDA. When cultured for 5 days on PDA without Cd^{2+}, the diameter of the colonies was 7 cm on average (Figure 4A). Cd significantly inhibited *H. chrysospermus*. Colonies showed an inverse relationship with the Cd^{2+} concentrations, and the inhibition percentage showed a direct positive correlation with the Cd^{2+} concentrations, with R^2 being 0.7132. The highest Cd^{2+} concentration tolerated by *H. chrysospermus* was 160 $mg \cdot L^{-1}$. The MIC value of Cd^{2+} to *H. chrysospermus* was determined to be 200 $mg \cdot L^{-1}$, and the EC_{50} was calculated to be 52 $mg \cdot L^{-1}$, according to the IP of Cd to *H. chrysospermus* (Figure 5). However, when cultured for a prolonged time, growing mycelium could be observed on the PDA of 200 $mg \cdot L^{-1}$, as shown in Figure 4B. Traxler et al. described a similar phenomenon; that is, the growth of *Schizophyllum commune* was still visible even at the highest metal concentrations used in PDA medium cultivated for 14 days [36].

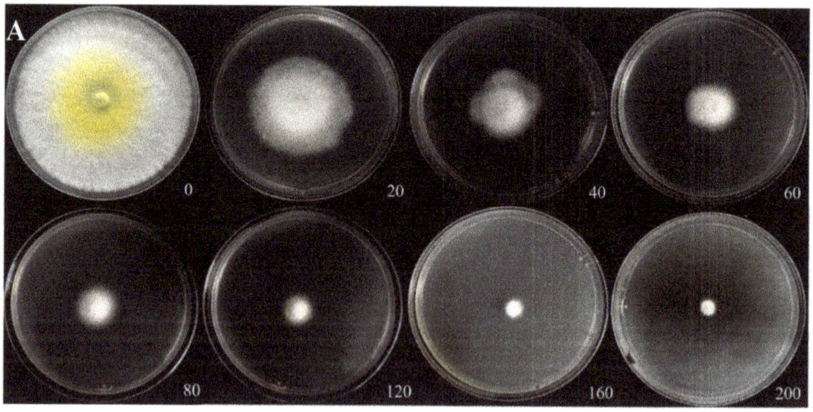

Figure 4. *Cont.*

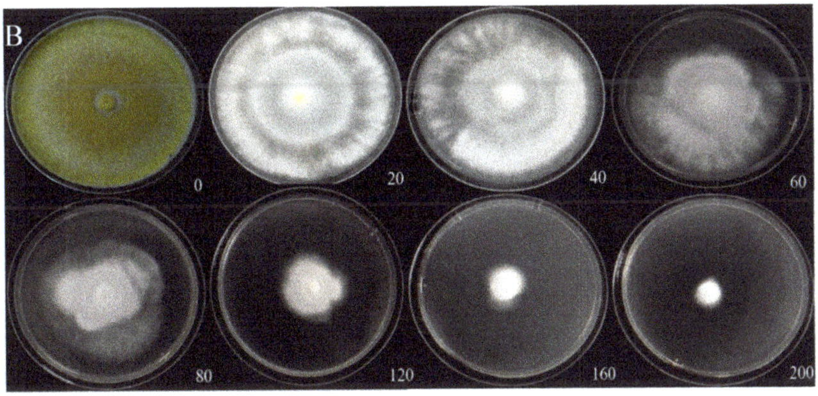

Figure 4. Cadmium tolerance of *H. chrysospermus* on PDA plates. (**A**) 5 days; (**B**) 10 days. Data in the figure indicate Cd^{2+} concentrations ($mg \cdot L^{-1}$) in PDA.

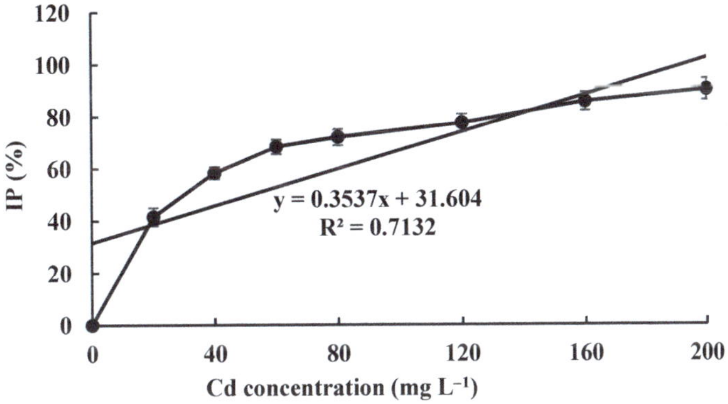

Figure 5. The inhibition percentage of Cd^{2+} at different concentrations on *H. chrysospermus*.

The effect of Cd on the micromorphology of the mycelia was investigated (Figure 6). The SEM images of control *H. chrysospermus* showed long and rope-like fungal hyphae, which were highly branched and intertwined with one another, and no physical damage was observed (Figure 6A). Meanwhile, fully developed chlamydospores with thick walls and prominent verrucose features were observed, as shown in Figures 3D and 6B. The presence of Cd resulted in locally twisted (Figure 6C) and deformed chlamydospores (Figure 6D). The increased Cd treatment distorted and shrunk the cell walls of the fungal mycelium (Figure 6E,F). The mycelia were partly swollen when treated with 160 $mg \cdot L^{-1}$ of Cd (Figure 6F). This finding may be due to the immobilization effect of the cell wall via Cd^{2+} binding by functional groups, such as carboxyl (-COOH), hydroxyl (-OH), carbonyl (-COH), and amino ($-NH_2$), to form precipitation on the cell surface [37,38]. Studies have shown that HMs have various effects on the composition of microbial cell components. The uptake and accumulation of HMs often causes significant damage at the morphological, cellular, physiological, and molecular levels. Cd is one of the most toxic metals among HMs. Sharma et al. reported that the fungal mycelium was slightly broken with some visual deformities [39]. However, the shape was distinguishable and well-regulated in the presence of Pb and Ni, whereas the presence of Cd distorted and shrunk the cell walls of *Phlebia brevispora*.

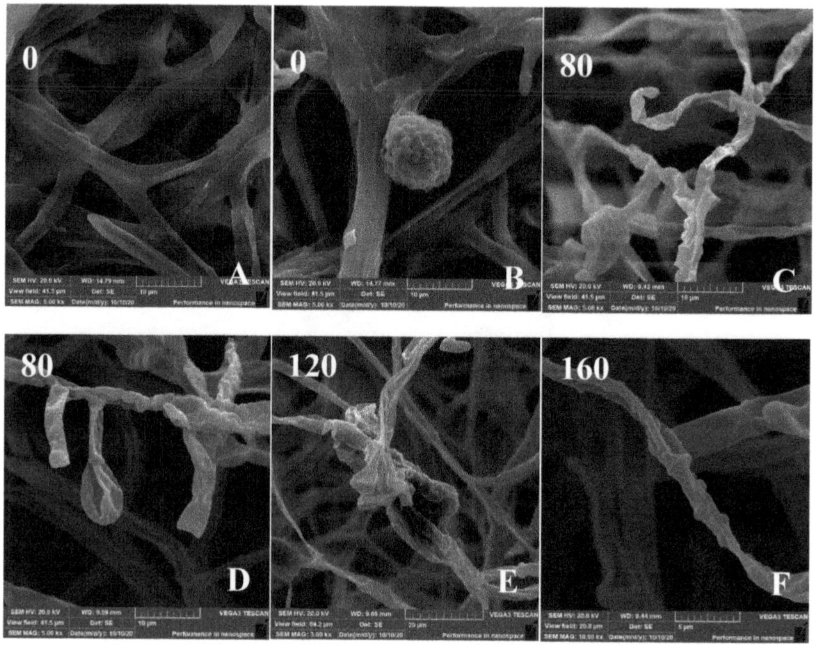

Figure 6. (**A–F**) Scanning electron micrographs of *H. chrysospermums* grown for 7 days on PDA with gradually increasing concentrations of Cd^{2+}. Data of 0–160 in the figures indicate the Cd^{2+} concentrations in PDA ($mg \cdot L^{-1}$).

Deng et al. characterized the features of a Cd-, Pb-, and Zn-resistant endophytic fungus *Lasiodiplodia* sp. MXSF31 from metal-accumulating *Portulaca oleracea*, and the strain was resistant to 5 mM Cd [40]. Mohammadian et al. studied different filamentous fungi isolated from contaminated mining soil and showed different profiles [41]. Among them, *Trichoderma harzianum* showed the maximum MIC value for Cd, being 35 $mg \cdot L^{-1}$. Albert et al. evaluated the tolerance of the soil fungus *Absidia cylindrospora* against three trace metals, namely, Cd, Cu, and Pb, before considering a possible use to treat contaminated soils. The concentration that inhibited 50% of fungal growth (IC_{50}) was 100 $mg \cdot L^{-1}$ for Cd [21].

Cd is toxic to cells, and organisms show visible toxicity symptoms under Cd stress higher than 5 $mg \cdot kg^{-1}$. However, mushrooms are characterized by a high trace element accumulation capacity, especially in the case of Cd, Pb, Cu, and Zn [24,42]. Numerous papers showed that the contents of many trace elements, especially Cd and Hg, increased in mushrooms from polluted areas compared with those from unpolluted rural sites [24]. Some plant species, such as *Solanum nigrum*, *Solanum photeinocarpum*, and *Siegesbeckia orientalis*, were studied as Cd hyperaccumulators and accumulators, which could be used for phytoremediation of Cd-contaminated soil. These plant species were usually found in different ecotypes with quite different morphology and Cd accumulation, including the mining ecotype (accumulating ecotype) and the farmland ecotype (non-accumulating ecotype) [43,44]. However, *B. griseus* showed a ubiquitous Cd-accumulating capacity, which seemed unaffected by the natural habitat. Considering the Cd tolerance of *H. chrysospermus* isolated from normal and deformed sporocarps of *B. griseus*, the Cd absorption of *H. chrysospermus* could be assumed.

3.4. Cd Absorption by H. chrysospermus

The Cd bioaccumulation of *H. chrysospermus* was assessed in relation to the initial Cd^{2+} concentrations (Figure 7). The Cd contents in dried biomass of *H. chrysospermus* increased with rising Cd^{2+} concentrations in PDB medium. The maximum uptake value was 10.03 mg·g^{-1}. Li et al. described a similar phenomenon and reported that the Cd concentrations that accumulated in the hyphae of *Pleurotus ostreatus* HAU-2 increased with increasing concentrations of Cd in the liquid culture [22].

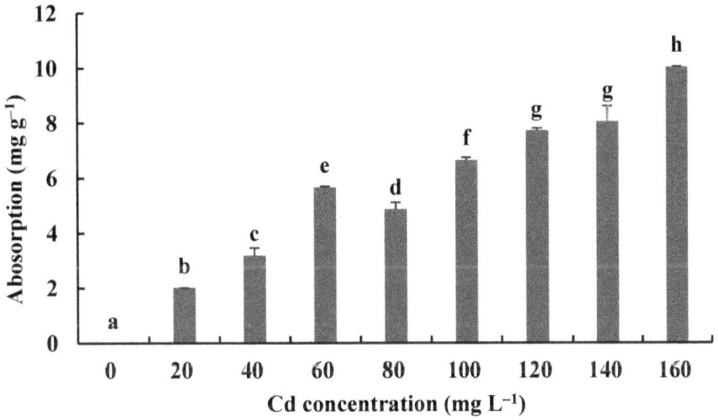

Figure 7. Cd^{2+} absorption by *H. chrysospermus*. Different letters indicated significant difference ($p < 0.05$).

In comparison, the maximum values of Cd uptake by fungi in the liquid culture have been reported to be 4.6×10^4 mg·kg^{-1} for *Lasiodiplodia* sp. MXSF31 [40], 1.89 mg·g^{-1} for *P. ostreatus* [22], 9 μg·mg^{-1} for *A. cylindrospora* [21], and 2.3–11.9 mg·g^{-1} for 5 fungal strains, with the highest Cd tolerance shown by soybean and barley [45]. During the experiment described here, *H. chrysospermus* isolated from *B. griseus* showed a significant absorption capacity for Cd^{2+}, and it may be a promising Cd bioabsorbent for bioremediation.

3.5. Effects of Cd on the Growth of H. chrysospermus

Figures 8 and 9 show the biomass production and pH measurements of *H. chrysospermus* cultured for 7 days in PDB with different Cd^{2+} concentrations. When the Cd^{2+} concentration in the PDB medium increased from 0 to 20 mg·L^{-1}, the biomass of *H. chrysospermus* increased significantly, being 0.73 mg DW. Then, the biomass decreased with increased Cd^{2+} concentrations up to 160 mg·L^{-1}. Li et al. revealed that low-concentration Cd (<20 mg·L^{-1}) and Cr (<150 mg·L^{-1}) did not notably suppress the hypha growth of *P. ostreatus* HAU 2, but higher concentrations of Cd and Cr evidently suppressed it [22]. Sharma et al. demonstrated that an enhancement in the fungal biomass of *P. brevispora* was observed with an increased metal concentration up to 1, 4, and 10 μmol·L^{-1} for Cd, Ni, and Pb, respectively, which then declined with a further increase in the metal concentrations [39].

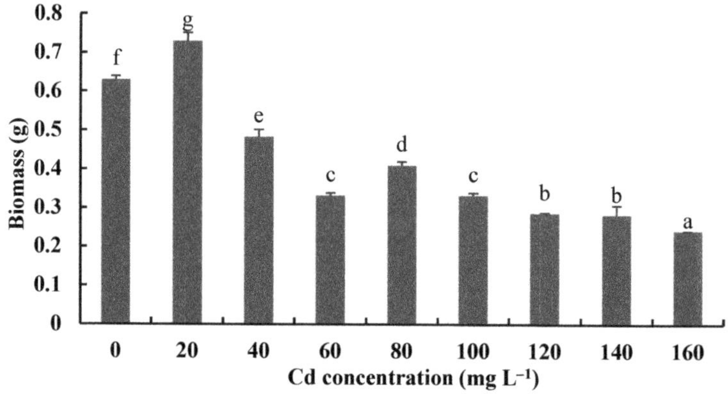

Figure 8. Effects of different concentrations of Cd^{2+} on the biomass of *H. chrysospermus*. Different letters indicated significant difference ($p < 0.05$).

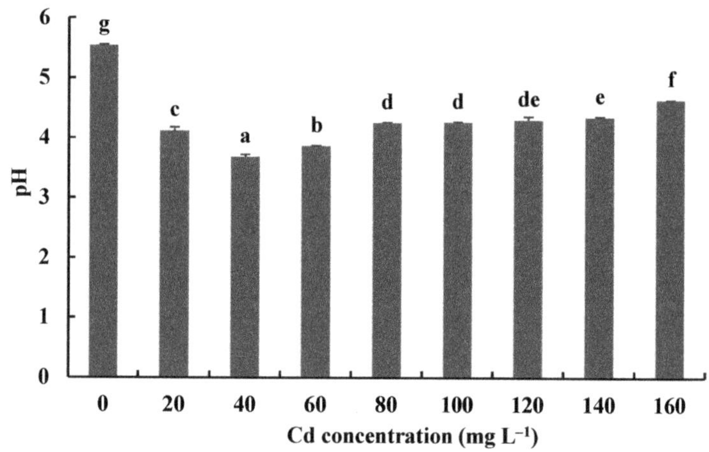

Figure 9. Effects of different concentrations of Cd^{2+} on the pH of a fermentation mixture of *H. chrysospermus*. Different letters indicated significant difference ($p < 0.05$).

The pH values of fermentation mixtures of *H. chrysospermus* with and without Cd^{2+} all decreased compared with the initial pH of 6.19. Meanwhile, the pH values of fermentation mixtures with Cd^{2+} were significantly lower than those of the control. The lowest pH value of 3.68 was shown by the fermentation mixtures with 40 mg·L^{-1} Cd^{2+}, a decrease by 2.51. Fungi can secrete inorganic acid and organic acid compounds to alleviate the toxic effects of HMs [46]. Wang et al. showed that *P. ostreatus* ISS-1 secreted oxalic acid, citric acid, and formic acid regardless of Pb exposure, but the oxalic acid content was significantly higher under Pb stress than that in the control [47]. Organic acids (citric acid and oxalic acid) chelate with toxic metal ions and form precipitation.

The responses of the soluble protein and antioxidant enzymes of the fungus with different Cd^{2+} concentrations were assayed to further investigate the tolerance factors of *H. chrysospermus* to Cd^{2+}. As shown in Figure 10, the soluble protein contents of *H. chrysospermus* significantly increased with increased Cd^{2+} concentrations. Figure 11 shows the effects of different concentrations of Cd^{2+} on the activities of the antioxidant enzymes of *H. chrysospermus*. The SOD activity was sensitive to Cd^{2+} treatment, showing a significant increasing tendency for 20 mg·L^{-1} Cd^{2+}, but it decreased with the increasing Cd^{2+} concentration. *H. chrysospermus* in this study showed a significantly lower POD activity [22], which was suppressed by Cd^{2+} treatments. The CAT activities under Cd stress were higher than those in the control group. The CAT activity reached a maximum value of 30.97 U·mg^{-1} prot with a Cd^{2+} concentration of 120 mg·L^{-1}. The exposure of fungal species to HMs induces stress conditions, resulting in the production of reaction oxygen species, such as superoxide, peroxides, and hydroxyl radicals, that damage fungal cell and organelle structures and alter metabolism [22,46]. Thus, the fungi of HM-resistant species need to adopt strategies to resist oxidative stress. The cellular immune system is the basic strategy employed to resist metal toxicity. Oxidized enzyme species, such as SOD, POD, and CAT, are important components of the cellular immune system, and they exert a significant effect on the removal of cellular active oxygen. Li et al. revealed that the concentrations of these three enzymes first increased and then decreased in the presence of Cd and Cr, indicating that oxidized enzyme species could be induced by relatively low concentrations of Cd or Cr, thus playing a role in the removal of active oxygen [22]. However, overly high concentrations of HMs may severely damage cells, suppressing enzymes.

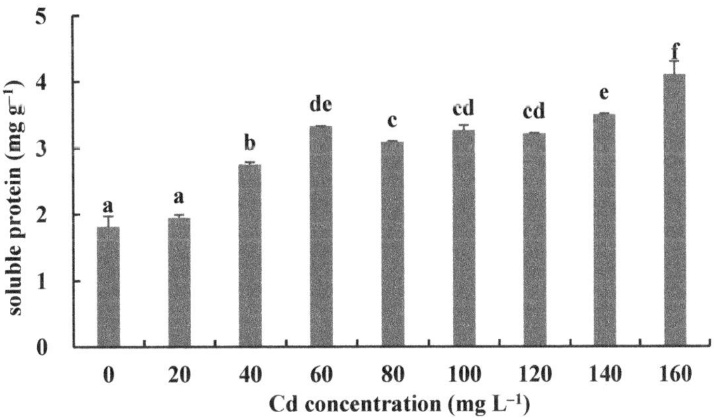

Figure 10. Effects of different concentrations of Cd^{2+} on soluble protein of *H. chrysospermus*. Different letters indicated significant difference ($p < 0.05$).

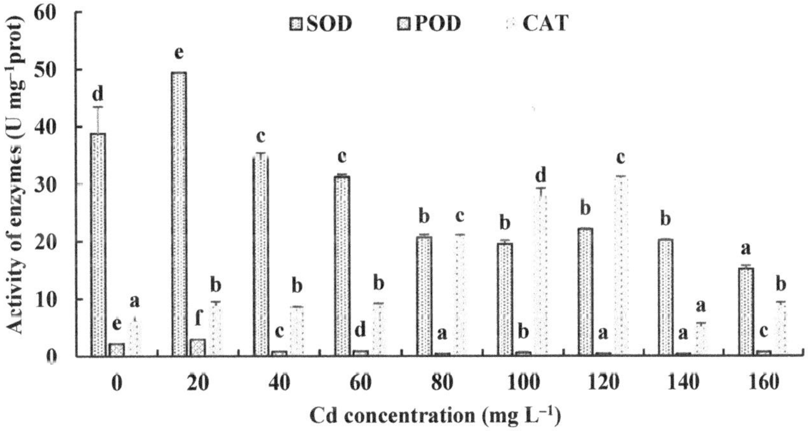

Figure 11. Effects of different concentrations of Cd^{2+} on the activities of antioxidant enzymes of *H. chrysospermus*. Different letters indicated significant difference ($p < 0.05$).

4. Conclusions

To the authors' best knowledge, this work is the first study to identify the symbiotic association of *B. griseus* and *H. chrysospermus* and elucidate the Cd-resistant characteristics of the isolated strain of *H. chrysospermus* from *B. griseus*. This study further confirms the Cd bioaccumulation capacity of *B. griseus*. The isolated strain of *H. chrysospermus* from *B. griseus* showed a strong ability to tolerate Cd, and the maximum Cd uptake reached 10.03 $mg \cdot g^{-1}$. In general, the time required for fruiting body differentiation, including induction, development, and maturation, in mushroom-forming fungi is about 4–12 days. Considering the strong Cd bioaccumulation capacity of *B. griseus*, even in natural habitats without Cd contamination, the symbiotic association of *H. chrysospermus* might represent the biological strategy adopted by *B. griseus* to promptly and efficiently take up Cd from the soil matrix. The MIC of Cd of the isolated strain of *H. chrysospermus* was 200 $mg \cdot L^{-1}$. However, the growth of the strain was still visible after prolonged culturing. The immobilization effects of the cell wall and acid compounds and antioxidant enzymes were employed by the fungi to alleviate the toxic effects of Cd. Thus, this study not only provides new insight into the Cd bioconcentration mechanisms of *B. griseus* but also provides a potential bioremediation fungus for Cd contamination.

Author Contributions: Conceptualization, Z.T. and L.S.; Methodology, Z.T., Y.Z. and L.S.; Formal analysis, Z.T. and Y.W.; Investigation, C.M. and Y.S.; Project administration, Y.Z. and L.S.; Software, Z.T.; Validation, Y.Z. and L.S.; Writing—original draft, Z.T. and L.S.; Writing—review and editing, all authors. All authors have read and agreed to the published version of the manuscript.

Funding: The research was financially supported by the National Natural Science Foundation of China (31860421, 21767014).

Conflicts of Interest: The authors declare no conflict of interest.

References

1. Mao, X.L. *The Macromycetes of China*; Science: Beijing, China, 2009; p. 816.
2. Sun, L.; Chang, W.; Bao, C.; Zhuang, Y. Metal contents, bioaccumulation, and health risk assessment in wild edible *Boletaceae* mushrooms. *J. Food Sci.* **2017**, *82*, 1500–1508. [CrossRef]
3. Bao, C.J.; Chang, W.D.; Zhuang, Y.L.; Sun, L.P. Nutritional characteristics and protein composition of fruiting bodies of *Boletus griseus*. *J. Food Sci.* **2017**, *38*, 83–89. [CrossRef]
4. Xiao, J.J.; Li, P.C.; Zhuang, Y.L.; Sun, L.P.; Sun, Y. Multivariate statistics for investigation of factors affecting migrating of cadmium from soil matrix to *Boletus griseus*. *Chin. J. Anal. Chem.* **2021**, *2*, 301–308. [CrossRef]
5. Falandysz, J.; Borovička, J. Macro and trace mineral constituents and radionuclides in mushrooms: Health benefits and risks. *Appl. Microbiol. Biotechnol.* **2013**, *97*, 477–501. [CrossRef]
6. Seeger, R. Cadmium in Pilzen. *Z. Lebensm. Unters. Forsch.* **1978**, *166*, 23–34. [CrossRef] [PubMed]
7. Drewnowska, M.; Falandysz, J. Investigation on mineral composition and accumulation by popula redible mushroom common chanterelle (*Cantharellus cibarius*). *Ecotox. Environ. Safe* **2015**, *113*, 9–17. [CrossRef] [PubMed]
8. Borovička, J.; Braeuer, S.; Walenta, M.; Hršelová, H.; Leonhardt, T.; Sácký, J.; Kaňa, A.; Goessler, W. A new mushroom hyperaccumulator: Cadmium and arsenic in the ectomycorrhizal basidiomycete. *Thelephora penicillata. Sci. Total Environ.* **2022**, *826*, 154227. [CrossRef] [PubMed]
9. Borovička, J.; Braeuer, S.; Sáckýd, J.; Kameník, J.; Goessler, W.; Trubač, J.; Strnad, L.; Rohovec, J.; Leonhardt, T.; Kotrba, P. Speciation analysis of elements accumulated in *Cystoderma carcharias* from clean and smelter-polluted sites. *Sci. Total Environ.* **2019**, *648*, 1570–1581. [CrossRef] [PubMed]
10. Yu, F.Q.; Liu, P.G. Species diversity of wild edible mushrooms from *Pinus yunnanensis* forests and conservation strategies. *Biodivers. Sci.* **2005**, *13*, 58–69. [CrossRef]
11. Liu, Y.T.; Sun, J.; Luo, Z.Y.; Rao, S.Q.; Su, Y.J.; Xu, R.R.; Yang, Y.J. Chemical composition of five wild edible mushrooms collected from Southwest China and their antihyperglycemic and antioxidant activity. *Food Chem. Toxicol.* **2012**, *50*, 1238–1244. [CrossRef]
12. Wang, X.M.; Zhang, J.; Wu, L.H.; Zhao, Y.L.; Li, T.; Li, J.Q.; Wang, Y.Z.; Liu, H.G. A mini-review of chemical composition and nutritional value of edible wild-grown mushroom from China. *Food Chem.* **2014**, *151*, 279–285. [CrossRef] [PubMed]
13. Liu, B.; Huang, Q.; Cai, H.; Guo, X.; Wang, T.; Gui, M. Study of heavy metal concentrations in wild edible mushrooms in Yunnan Province, China. *Food Chem.* **2015**, *188*, 294–300. [CrossRef] [PubMed]
14. Dowlati, M.; Sobhi, H.R.; Esrafili, A.; FarzadKia, M.; Yeganeh, M. Heavy metals content in edible mushrooms: A systematic review, meta-analysis and health risk assessment. *Trends Food Sci. Technol.* **2021**, *109*, 527–535. [CrossRef]

15. Zheng, J.; Zhang, T.; Fan, J.; Zhuang, Y.; Sun, L. Protective effects of a polysaccharide from *Boletus aereus* on S180 tumor-bearing mice and its structural characteristics. *Int. J. Biol. Macromol.* **2021**, *188*, 1–10. [CrossRef] [PubMed]
16. Wei, Y.; Li, L.; Liu, Y.; Xiang, S.; Zhang, H.; Yi, L.; Shang, Y.; Xu, W. Identification techniques and detection methods of edible fungi species. *Food Chem.* **2022**, *374*, 131803. [CrossRef]
17. Douhan, G.W.; Rizzo, D.M. Host-parasite relationships among bolete infecting *Hypomyces species*. *Mycol. Res.* **2003**, *107*, 1342–1349. [CrossRef]
18. Kaygusuz, O.; Çolak, Ö.F.; Türkekul, İ. A new genus record for the macrofungi of Turkey of a fungicolous and mycoparasitic species, *Hypomyces chrysospermus* Tul. & C. Tul. (*Hypocreaceae* De Not.). *Feddes Repert.* **2018**, *129*, 241–246. [CrossRef]
19. White, T.J.; Bruns, T.; Lee, S.; Taylor, J. Amplification and Direct Sequencing of Fungal Ribosomal RNA Genes for Phylogeentics. In *PCR Protocols: A Guide to Methods and Applications*; Innis, M.A., Gelfand, D.H., Sninsky, J.J., White, T.J., Eds.; Academic Press: San Diego, CA, USA, 1990; pp. 315–322.
20. Khan, A.R.; Waqas, M.; Ullah, I.; Khan, A.L.; Khan, M.A.; Lee, I.-J.; Shin, J.-H. Culturable endophytic fungal diversity in the cadmium hyperaccumulator *Solanum nigrum* L. and their role in enhancing phytoremediation. *Environ. Exp. Bot.* **2017**, *135*, 126–135. [CrossRef]
21. Albert, Q.; Leleyter, L.; Lemoine, M.; Heutte, N.; Rioult, J.P.; Sage, L.; Baraud, F.; Garon, D. Comparison of tolerance and biosorption of three trace metals (Cd, Cu, Pb) by the soil fungus *Absidia cylindrospora*. *Chemosphere* **2018**, *196*, 386–392. [CrossRef]
22. Li, X.; Wang, Y.; Pan, Y.; Yu, H.; Zhang, X.; Shen, Y.; Jiao, S.; Wu, K.; La, G.; Yuan, Y.; et al. Mechanisms of Cd and Cr removal and tolerance by macrofungus *Pleurotus ostreatus* HAU-2. *J. Hazard. Mater.* **2017**, *330*, 1–8. [CrossRef]
23. Vinichuk, M.M. Copper, zinc, and cadmium in various fractions of soil and fungi in a Swedish forest. *J. Environ. Sci. Health A Tox Hazard. Subst. Environ. Eng.* **2013**, *48*, 980–987. [CrossRef] [PubMed]
24. Melgar, M.J.; Alonso, J.; Garcia, M.A. Cadmium in edible mushrooms from NW Spain: Bioconcentration factors and consumer health implications. *Food Chem. Toxicol.* **2016**, *88*, 13–20. [CrossRef] [PubMed]
25. Proskura, N.; Podlasinska, J.; Skopicz-Radkiewicz, L. Chemical composition and bioaccumulation ability of *Boletus badius* (Fr.) Fr. collected in western Poland. *Chemosphere* **2017**, *168*, 106–111. [CrossRef] [PubMed]
26. Fu, Z.; Liu, G.; Wang, L. Assessment of potential human health risk of trace element in wild edible mushroom species collected from Yunnan Province, China. *Environ. Sci. Pollut. Res. Int.* **2020**, *27*, 29218–29227. [CrossRef] [PubMed]
27. Sarikurkcu, C.; Yildiz, D.; Akata, I.; Tepe, B. Evaluation of the metal concentrations of wild mushroom species with their health risk assessments. *Environ. Sci. Pollut. Res. Int.* **2021**, *28*, 21437–21454. [CrossRef]
28. Riaz, M.; Kamran, M.; Fang, Y.; Wang, Q.; Cao, H.; Yang, G.; Deng, L.; Wang, Y.; Zhou, Y.; Anastopoulos, I.; et al. Arbuscular mycorrhizal fungi-induced mitigation of heavy metal phytotoxicity in metal contaminated soils: A critical review. *J. Hazard. Mater.* **2021**, *402*, 123919. [CrossRef]
29. Tang, Y.; Shi, L.; Zhong, K.; Shen, Z.; Chen, Y. Ectomycorrhizal fungi may not act as a barrier inhibiting host plant absorption of heavy metals. *Chemosphere* **2019**, *215*, 115–123. [CrossRef]
30. Liu, B.; Wang, S.; Wang, J.; Zhang, X.; Shen, Z.; Shi, L.; Chen, Y. The great potential for phytoremediation of abandoned tailings pond using ectomycorrhizal *Pinus sylvestris*. *Sci. Total Environ.* **2020**, *719*, 137475. [CrossRef]
31. Yu, P.; Sun, Y.; Huang, Z.; Zhu, F.; Sun, Y.; Jiang, L. The effects of ectomycorrhizal fungi on heavy metals' transport in *Pinus massoniana* and bacteria community in rhizosphere soil in mine tailing area. *J. Hazard. Mater.* **2020**, *381*, 121203. [CrossRef]
32. Su, Z.Z.; Dai, M.D.; Zhu, J.N.; Liu, X.H.; Li, L.; Zhu, X.M.; Wang, J.Y.; Yuan, Z.L.; Lin, F.C. Dark septate endophyte *Falciphora oryzae*-assisted alleviation of cadmium in rice. *J. Hazard. Mater.* **2021**, *419*, 126435. [CrossRef] [PubMed]
33. Wang, J.; Chen, R.; Fan, L.; Cui, L.; Zhang, Y.; Cheng, J.; Wu, X.; Zeng, W.; Tian, Q.; Shen, L. Construction of fungi-microalgae symbiotic system and adsorption study of heavy metal ions. *Sep. Purif. Technol.* **2021**, *268*, 118689. [CrossRef]
34. Chu, R.; Li, S.; Zhu, L.; Yin, Z.; Hu, D.; Liu, C.; Mo, F. A review on co-cultivation of microalgae with filamentous fungi: Efficient harvesting, wastewater treatment and biofuel production. *Renew. Sustain. Energy Rev.* **2021**, *139*, 110689. [CrossRef]
35. Liu, J.W.; Ge, Z.W.; Horak, E.; Vizzini, A.; Halling, R.E.; Pan, C.L.; Yang, Z.L. *Squamanitaceae* and three new species of *Squamanita* parasitic on *Amanita* basidiomes. *IMA Fungus* **2021**, *12*, 4. [CrossRef] [PubMed]
36. Traxler, L.; Shrestha, J.; Richter, M.; Krause, K.; Schafer, T.; Kothe, E. Metal adaptation and transport in hyphae of the wood-rot fungus *Schizophyllum commune*. *J. Hazard. Mater.* **2022**, *425*, 127978. [CrossRef]
37. Gu, S.; Lan, C.Q. Biosorption of heavy metal ions by green alga *Neochloris oleoabundans*: Effects of metal ion properties and cell wall structure. *J. Hazard. Mater.* **2021**, *418*, 126336. [CrossRef]
38. Soto-Ramirez, R.; Lobos, M.G.; Cordova, O.; Poirrier, P.; Chamy, R. Effect of growth conditions on cell wall composition and cadmium adsorption in *Chlorella vulgaris*: A new approach to biosorption research. *J. Hazard. Mater.* **2021**, *411*, 125059. [CrossRef]
39. Sharma, K.R.; Giri, R.; Sharma, R.K. Lead, cadmium and nickel removal efficiency of white-rot fungus *Phlebia brevispora*. *Lett. Appl. Microbiol.* **2020**, *71*, 637–644. [CrossRef]
40. Deng, Z.; Zhang, R.; Shi, Y.; Hu, L.; Tan, H.; Cao, L. Characterization of Cd-, Pb-, Zn-resistant endophytic *Lasiodiplodia* sp. MXSF31 from metal accumulating *Portulaca oleracea* and its potential in promoting the growth of rape in metal contaminated soils. *Environ. Sci. Pollut. Res. Int.* **2014**, *21*, 2346–2357. [CrossRef]
41. Mohammadian, E.; Babai Ahari, A.; Arzanlou, M.; Oustan, S.; Khazaei, S.H. Tolerance to heavy metals in filamentous fungi isolated from contaminated mining soils in the Zanjan Province, Iran. *Chemosphere* **2017**, *185*, 290–296. [CrossRef]

42. Zsigmond, A.R.; Kantor, I.; May, Z.; Urak, I.; Heberger, K. Elemental composition of *Russula cyanoxantha* along an urbanization gradient in Cluj-Napoca (Romania). *Chemosphere* **2020**, *238*, 124566. [CrossRef]
43. Wang, J.; Lin, L.; Luo, L.; Liao, M.; Lv, X.; Wang, Z.; Liang, D.; Xia, H.; Wang, X.; Lai, Y.; et al. The effects of abscisic acid (ABA) addition on cadmium accumulation of two ecotypes of *Solanum photeinocarpum*. *Environ. Monit. Assess.* **2016**, *188*, 182. [CrossRef] [PubMed]
44. Tao, Q.; Liu, Y.; Li, M.; Li, J.; Luo, J.; Lux, A.; Kovac, J.; Yuan, S.; Li, B.; Li, Q.; et al. Cd-induced difference in root characteristics along root apex contributes to variation in Cd uptake and accumulation between two contrasting ecotypes of Sedum alfredii. *Chemosphere* **2020**, *243*, 125290. [CrossRef] [PubMed]
45. Ignatova, L.; Kistaubayeva, A.; Brazhnikova, Y.; Omirbekova, A.; Mukasheva, T.; Savitskaya, I.; Karpenyuk, T.; Goncharova, A.; Egamberdieva, D.; Sokolov, A. Characterization of cadmium-tolerant endophytic fungi isolated from soybean (*Glycine max*) and barley (*Hordeum vulgare*). *Heliyon* **2021**, *7*, e08240. [CrossRef] [PubMed]
46. Priyadarshini, E.; Priyadarshini, S.S.; Cousins, B.G.; Pradhan, N. Metal-Fungus interaction: Review on cellular processes underlying heavy metal detoxification and synthesis of metal nanoparticles. *Chemosphere* **2021**, *274*, 129976. [CrossRef]
47. Wang, Y.; Yi, B.; Sun, X.; Yu, L.; Wu, L.; Liu, W.; Wang, D.; Li, Y.; Jia, R.; Yu, H.; et al. Removal and tolerance mechanism of Pb by a filamentous fungus: A case study. *Chemosphere* **2019**, *225*, 200–208. [CrossRef]

Journal of *Fungi*

MDPI

Article

Exposure to Essential and Toxic Elements via Consumption of *Agaricaceae, Amanitaceae, Boletaceae,* and *Russulaceae* Mushrooms from Southern Spain and Northern Morocco

Marta Barea-Sepúlveda [1], Estrella Espada-Bellido [1,*], Marta Ferreiro-González [1], Hassan Bouziane [2], José Gerardo López-Castillo [3], Miguel Palma [1] and Gerardo F. Barbero [1]

[1] Department of Analytical Chemistry, Faculty of Sciences, University of Cadiz, Agrifood Campus of International Excellence (ceiA3), IVAGRO, 11510 Puerto Real, CA, Spain; marta.barea@gm.uca.es (M.B.-S.); marta.ferreiro@uca.es (M.F.-G.); miguel.palma@uca.es (M.P.); gerardo.fernandez@uca.es (G.F.B.)

[2] Laboratory of Applied Botany, Department of Biology, Faculty of Sciences, University Abdelmalek Essaâdi, Mhannech II, Tetouan 2121, Morocco; hasbouz@hotmail.com

[3] Unidad de Protección de la Salud, Distrito Sanitario Granada-Metropolitano, Consejería de Salud y Familias, Junta de Andalucía, 18150 Gójar, GR, Spain; joseg.lopez.sspa@juntadeandalucia.es

* Correspondence: estrella.espada@uca.es (E.E-B.); Tel.: +34-956-016355

Abstract: The demand and interest in mushrooms, both cultivated and wild, has increased among consumers in recent years due to a better understanding of the benefits of this food. However, the ability of wild edible mushrooms to accumulate essential and toxic elements is well documented. In this study, a total of eight metallic elements and metalloids (chromium (Cr), arsenic (As), cadmium (Cd), mercury (Hg), lead (Pb), copper (Cu), zinc (Zn), and selenium (Se)) were determined by ICP-MS in five wild edible mushroom species (*Agaricus silvicola, Amanita caesarea, Boletus aereus, Boletus edulis,* and *Russula cyanoxantha*) collected in southern Spain and northern Morocco. Overall, Zn was found to be the predominant element among the studied species, followed by Cu and Se. The multivariate analysis suggested that considerable differences exist in the uptake of the essential and toxic elements determined, linked to species-intrinsic factors. Furthermore, the highest Estimated Daily Intake of Metals (EDIM) values obtained were observed for Zn. The Health Risk Index (HRI) assessment for all the mushroom species studied showed a Hg-related cause of concern due to the frequent consumption of around 300 g of fresh mushrooms per day during the mushrooming season.

Keywords: wild edible mushrooms; metallic elements; metalloids; risk assessment; human health; organic food

Citation: Barea-Sepúlveda, M.; Espada-Bellido, E.; Ferreiro-González, M.; Bouziane, H.; López-Castillo, J.G.; Palma, M.; F. Barbero, G. Exposure to Essential and Toxic Elements via Consumption of *Agaricaceae, Amanitaceae, Boletaceae,* and *Russulaceae* Mushrooms from Southern Spain and Northern Morocco. *J. Fungi* **2022**, *8*, 545. https://doi.org/10.3390/jof8050545

Academic Editors: Ivan Širić and Miha Humar

Received: 22 April 2022
Accepted: 20 May 2022
Published: 23 May 2022

1. Introduction

Mushrooms are the fruiting bodies or sporocarps of a group of eukaryotic organisms classified in the division *Basidiomycota* and *Ascomycota* of the Kingdom Fungi [1–4]. Although mushrooms have traditionally been considered a delicacy in gastronomy for their palatability, advances in the scientific field have led to a new nutritional approach to their culinary use, as their properties suggest that mushroom consumption may influence the control and modulation of several functions of the organism when introduced in the diet [5,6]. Therefore, the demand and interest in mushrooms have increased among consumers in recent years due to a better understanding of their benefits and a greater awareness of the need to include more nutritious options for healthier habits. Likewise, since edible mushrooms play an important role as a source of income for many companies and local communities, the increase in its demand has also contributed to the economic growth of this market and, consequently, to that of the countries [2]. In 2020, the global mushroom market size was valued at $46.1 Bn, and it is estimated to expand at a compound annual growth rate (CAGR) of 9.5% from 2021 to 2028 [7]. In turn, in Spain, the 2020 annual household consumption panel data issued by the Ministry of Agriculture, Fisheries, and Food indicated that the quantities consumed nationwide totaled 70,221 tons [8]. Among the mushrooms marketed and consumed, edible wild mushrooms currently attract a higher level of attention, as most of them cannot be cultivated due to their complex and specific symbiotic lifestyle [3,9,10]. For this reason, freshly wild-growing mushrooms are high-valued seasonal foods, widely consumed by many people who enjoy life outside the urban areas [11,12]. Worldwide, the most consumed and appreciated wild mushroom species belong to the families *Russulaceae* (e.g., *Russula cyanoxantha* and *Lactarius deliciosus*), *Boletaceae* (e.g., *Boletus edulis* and *Boletus aereus*), *Amanitaceae* (e.g., *Amanita caesarea*), *Cantharellaceae* (e.g., *Cantharellus cibarius*), and *Agaricaceae* (e.g., *Agaricus silivicola* and *Macrolepiota procera*).

Unlike the cultivated ones, wild-growing mushrooms fructify and mature exposed to ambient conditions. Anthropogenic activities such as pesticide use, mining, and fossil fuel combustion produce large quantities of pollutants that affect environmental homeostasis when they enter the air, water, or soil [13,14]. Of all the pollutants found in the environment, metallic elements and metalloids are considered to represent one of the most dangerous for public health due to their high persistence and tendency to bioaccumulate in the trophic chain. Metallic elements and metalloids such as mercury (Hg), lead (Pb), cadmium (Cd), chromium (Cr), and arsenic (As) have no biological functions and are therefore considered non-essential and harmful to the body even at low concentrations. For example, sublethal Pb poisoning causes neurological defects and renal dysfunction [15,16]. Nonetheless, metallic elements and metalloids such as copper (Cu), zinc (Zn), and selenium (Se) are known to be essential nutrients for different physiological functions in the body, so the lack of any of them could have repercussions on biological processes, compromising human health. For example, symptoms of Zn deficiency include growth retardation, hair loss, diarrhea, eye and skin lesions, loss of appetite, and delayed wound healing [17]. The ability of wild mushrooms to accumulate metallic elements and metalloids in their tissues is well documented by other authors [18–27], showing that the concentrations of these contaminants in the wild-growing species are considerably greater than those in the soil where they grew [28]. The mechanism that enables mushrooms to accumulate metallic elements and metalloids could be explained via the network of hyphae located in the upper soil horizon [29,30]. The hyphae, consisting of elongated tubular cells enveloped by a chitin wall, are widely spread over the bioavailable area and, under specific conditions, accumulate metal ions [31]. In addition, this process seems to be influenced by environmental factors that influence the bioavailability of these pollutants in the environment (pH, organic matter, texture, and metal concentration in the soil) and intrinsic factors (taxon, morphology, size, fruiting body part, stage of development, and mycelial age) [32]. Thus, owing to the important role that mushrooms are acquiring in diets, the presence of high concentrations of metallic elements and metalloids in this food constitutes a matter of great relevance within the scientific community.

Based on this background, the present study aimed to determine the composition of five toxic elements (Cr, As, Cd, Hg, and Pb) and three essential elements (Cu, Zn, and Se) in five high-valued species of wild-growing edible mushrooms (*A. silvicola, A. caesarea, B. aereus, B. edulis,* and *R. cyanoxantha*) collected in southern Spain and northern Morocco. The latter was selected as a sampling location due to its proximity to southern Spain and the increased mycological activity during the mushrooming season in some locations in its northern region. Unsupervised multivariate techniques, such as Hierarchical Cluster Analysis (HCA) and Principal Component Analysis (PCA), were applied to study the influence of the source area and species-intrinsic factors over the metallic elements and metalloids. Furthermore, consumer exposure and health implications were assessed using two well-known safety criteria: the Estimated Daily Intake of Metals (EDIM) and the Health Risk Index (HRI).

2. Materials and Methods

2.1. Mushroom Sampling

Edible mushroom fruiting bodies were collected in 2017 and 2018 during the mushrooming season from different locations in southern Spain and northern Morocco. All sampled locations corresponded to natural forests, and therefore, the edible mushroom samples were all wild. A total of 16 samples of 5 mushroom species were collected and analyzed. Specimens studied were authenticated based on their morphological distinctive characteristics, which are unique and unmistakable from other mushroom species and families. Sample collection and preparation were carried out following the process previously described in earlier work conducted by our research group [18,33]. At least 10 complete specimens were collected *in situ* in each of the areas to form a representative set of samples from each geographic location studied. Upon arrival at the laboratory, the samples were adequately prepared for their analysis. To this end, all the fruiting bodies collected were washed with deionized water and dried in an oven at 50 °C for 48 h until a constant weight was reached. Finally, the samples were homogenized using an agate mortar and stored in clean polyethylene (PE) bottles, perfectly labeled according to species and sampling area. The habitat, year of collection, sample location, and families of the mushrooms studied are shown in Table 1.

Table 1. Sample ID, mushroom species, family, sample location, collection date, and habitat description of wild edible mushroom species studied.

Sample ID	Mushroom Species	Family	Sample Location	Date	Habitat
#1	*Russula cyanoxantha*	*Russulaceae*	Cortes de la Fra. (Malaga, Spain)	2017	Deciduous forest; *Quercus*
#2			Sendero El Palancar (Cadiz, Spain)	2018	
#3			Cortes de la Fra. (Malaga, Spain)	2018	
#4	*Amanita caesarea*	*Amanitaceae*	Puerto de Galiz (Cadiz, Spain)	2017	Deciduous forest; *Quercus suber*
#5			Puerto de Galiz (Cadiz, Spain)	2018	
#6			Sendero El Palancar (Cadiz, Spain)	2018	
#7			Cortes de la Fra. (Malaga, Spain)	2018	
#8	*Agaricus silvicola*	*Agaricaceae*	Parc Naturel Bouhachem(Chaouen, Morocco)	2017	Deciduous forest; *Quercus suber*
#9			Cortes de la Fra. (Malaga, Spain) [1]	2018	
#10			Cortes de la Fra. (Malaga, Spain) [2]	2018	
#11	*Boletus edulis*	*Boletaceae*	Puerto de Galiz (Cadiz, Spain)	2018	Deciduous forest; *Quercus suber*
#12			Valdeinfierno (Cadiz, Spain)	2018	
#13	*Boletus aereus*	*Boleataceae*	Valdeinfierno (Cadiz, Spain)	2018	Deciduous forest; *Quercus suber*
#14			Puerto de Galiz (Cadiz, Spain)	2018	
#15			Parc Naturel Bouhachem(Chaouen, Morocco) [1]	2017	
#16			Parc Naturel Bouhachem(Chaouen, Morocco) [2]	2017	

2.2. Chemicals and Solvents

The reagents used for the acid digestions of the samples were of high analytical grade and acquired from SCP Science (Montreal, Quebec, Canada): HCl PlasmaPURE (34–37%), HNO_3 PlasmaPURE (67–69%), and Sigma-Aldrich (St. Louis, MO, USA): H_2O_2 ($\geq$30%) The solutions were prepared using nanopure water obtained by passing twice-distilled water through a Milli-Q system (18 MΩ/cm, Millipore, Bedford, MA, USA).

2.3. Acid Digestion Procedure

Acid digestions were performed in a DigiPREP Jr block digestion system from SCP Science (Montreal, Quebec, Canada) equipped with a 24-position graphite heating block for 50 mL polypropylene (PE) digestion tubes (DigiTubes; SCP Science; Montreal, Quebec, Canada). The digestion procedure employed here was based on the conditions established in previous investigations by our research group [33]. The dried and powdered subsamples (0.25 g) were first placed in the digestion tubes with 5 mL of HNO_3, 2 mL of HCl, and 2 mL of nanopure water. Next, they were digested by applying a stepwise temperature increase procedure for 20 min up to 65 °C and maintaining this temperature for a total of

30 min. The second digestion was performed after a cooling step by adding 3 mL of H_2O_2, gradually increasing the temperature for 30 min up to 110 °C, and holding it for a total of 60 min. Prior to analysis, the digested samples were filtered through a 0.45-μm filter (DigiFILTER; SCP Science; Montreal, Quebec, Canada) using a –600 mbar vacuum port. They were then transferred to a clean 50-mL volumetric DigiTube, which was completed with nanopure water to 50 mL. All samples were carried out in triplicate.

2.4. Analysis

The concentrations of Cr, As, Cd, Hg, Pb, Cu, Zn, and Se in the mushroom samples were determined with an Inductively Coupled Plasma-Mass Spectrometer (Thermo X Series II ICP-MS, Waltham, MA, USA) equipped with a concentric nebulizer, cyclonic spray chamber, quadrupole mass analyzer, and collision/reaction cell. The Xt interface, kinetic energy discrimination (KED), and H_2 (7%)/He CCT were applied throughout the analyses. The instrumental conditions of the ICP-MS are shown in Table 2.

Table 2. ICP-MS instrumental conditions for the metallic elements and metalloid determination in the mushroom samples.

ICP-MS Instrumental Conditions	
CCT H2(7%)/He (mL min^{-1})	4.5
Pole Bias Voltage (V)	–17.0
Hexapole Bias Voltage (V)	–20.0
Auxiliary Ar Flow Rate (L min^{-1})	1.0
Nebulizer Ar Flow Rate (L min^{-1})	1.0
Plasma Ar Flow Rate (L min^{-1})	14.0
Sampling depth (mm)	80.0
RF Power	1380 W

An internal standard of ^{103}Rh, ^{72}Ge, ^{191}Ir, ^{209}Bi, and ^{45}Sc prepared from individual solutions (SCP Science; Montreal, Quebec, Canada) of 1000 μg mL^{-1} was used to correct temporal variations in signal intensity during the analyses. Moreover, the analytical methodology used to determine the metallic elements and metalloids concentrations were validated through triplicates, blanks, and a certified reference material (CRM), the *Boletus edulis* CS-M-3 powder control material (Institute of Nuclear Chemistry and Technology; Warsaw, Poland). No interferences were found relevant for the quantified elements in blanks. The recovery levels in the reference material (CS-M-3) were in an acceptable range of 70–130%. Limits of detection (LOD) for Cr, As, Cd, Hg, Pb, Cu, Zn, and Se were within 0.002–0.1 mg kg^{-1} dry weight (DW).

2.5. Estimated Daily Intake of Metals

For each sample analyzed, the Estimated Daily Intake of Metals (*EDIM*) was calculated using the concentration data expressed in mg kg^{-1} dry weight (DW), as shown in the following Equation (1):

$$EDIM = \frac{C_{metal} \cdot D_{food\ intake}}{BW} \quad (1)$$

where C_{metal} is the metallic element or metalloid concentration (mg kg^{-1}) in the fruiting body, $D_{food\ intake}$ refers to the mushroom daily intake, and BW is the average person's body weight in kg. Following Liu et al. (2015) and Sarikurkcu et al. (2020) [2,14], we considered a portion of 300 g of fresh mushrooms (30 g of dried mushrooms per day) and a regular consumer of 70 kg body weight.

2.6. Health Risk Index

The Health Risk Index (HRI) was calculated by following Equation (2) to assess the potential risk to human health due to exposure to the metallic elements and metalloids in the different species of edible wild-growing mushrooms studied:

$$\mathrm{HRI} = \frac{EDIM}{R_f D} \tag{2}$$

where *EDIM* is the daily intake of metals via consumption of the studied mushrooms, and R_fD is the maximum acceptable daily oral dose of a toxic substance. According to the data provided by Sarikurkcu et al. (2020) and the U.S. EPA Integrated Risk Information System (IRIS) data, the Reference Doses (R_fD) for Cr, As, Cd, Hg, Pb, Cu, Zn, and Se are 3, 0.3, 1, 0.3, 3.5, 40, 300, and 0.5 μg kg^{-1} body weight per day, respectively.

2.7. Software and Multivariate Analysis

An unsupervised multivariate analysis was applied to identify clustering trends among the species studied. To this end, the average concentrations in mg kg^{-1} of the elements determined were used as variables, obtaining a data matrix, D_{nxm}, where n is the number of samples (n = 16), and m is the number of metallic and metalloid elements determined (m = 8). The data matrix was normalized using Min–Max normalization. A hierarchical cluster analysis (HCA) and Principal Components Analysis (PCA) were performed first. HCA allows the hierarchical classification of the data according to their similarity. The Euclidean distance was selected for the inter-individual similarity matrix calculation and Ward's method as the inter-cluster measure. In this study, the choice of Ward's method was settled by comparing the agglomerative coefficient obtained with different linkage methods (Average, Complete, Unique, and Ward). This coefficient enables us to find the linkage method that identifies the strongest clustering structure. An agglomerative coefficient equal to 1 is the highest value, indicating a strong clustering structure. In this case, Ward's method was the linkage method that presented an agglomerative coefficient closest to 1, being 0.76. PCA is an unsupervised multivariate technique that allows for reducing the dimensionality of the data, so it is used together with HCA as an exploratory technique.

The multivariate analysis was performed with Rstudio (R version 4.0.5, Boston, MA, USA). The hierarchical Cluster Analysis (HCA) was carried out using the *hclust* function from the *stast* package. The selection of the Linkage method for the HCA was established by using the *agnes* function of the *clust* package. The HCA results were plotted in a graphical combination of the resulting sample and variable dendrograms with a heatmap using the *heatmap.2* function of the *gplots* package. The Principal Components Analysis (PCA) was performed using the *prcomp* function from the *stats* package. The *fviz_eig* function from the *factoextra* package was used to extract and visualize the output of this multivariate data analysis. The scores and loadings obtained from the PCA were graphically displayed using the *ggplot* function from the *ggplot2* package. Furthermore, the *ggplot* function from the *ggplot2* was also applied for the bar chart representation of the Health Risk Index (HRI) results.

3. Results and Discussion

3.1. Metallic Elements and Metalloids Content in Mushrooms

The metallic and metalloid concentrations of the wild edible mushrooms studied are given in Table 3, expressed as the mean of triplicates in mg kg^{-1} dry weight (DW). All the elements included in our study were detected in all the mushroom samples, except for As, which was below the Limit of Detection (LOD) in samples #15 and #16, which were both *Boletus aereus* from Parc Naturel of Bouhachem (Chaouen Morocco). In general, it was observed that Zn was the predominant element in most of the cases, followed by Cu and Se. The other metallic elements and metalloids were detected in relatively lower concentrations.

Table 3. Metallic elements and metalloid concentrations (mg kg^{-1} DW) in the analyzed wild edible mushroom species from southern Spain and northern Morocco.

Sample ID	Cu	Zn	Se	Cr	As	Cd	Hg	Pb
#1	23.7 ± 0.035	51.2 ± 0.873	0.613 ± 0.876	7.39 ± 0.951	0.147 ± 0.008	0.408 ± 0.027	2.14 ± 0.024	0.397 ± 0.032
#2	43.1 ± 2.81	61.4 ± 0.972	2.55 ± 0.102	5.01 ± 0.119	0.302 ± 0.004	0.223 ± 0.030	4.12 ± 0.069	0.325 ± 0.007
#3	35.3 ± 0.254	63.4 ± 1.15	1.49 ± 0.059	10.2 ± 0.079	0.111 ± 0.003	1.16 ± 0.045	3.21 ± 0.091	0.305 ± 0.017
#4	18.9 ± 0.231	117 ± 2.03	2.28 ± 0.002	1.84 ± 0.093	0.113 ± 0.009	4.03 ± 0.006	2.49 ± 0.002	0.199 ± 0.009
#5	123 ± 1.47	85.6 ± 0.624	2.13 ± 0.004	0.828 ± 0.011	0.113 ± 0.009	2.79 ± 0.023	2.27 ± 0.038	0.115 ± 0.013
#6	22.0 ± 0.500	74.1 ± 0.155	41.1 ± 0.761	3.57 ± 0.181	0.436 ± 0.010	2.09 ± 0.039	1.37 ± 0.018	0.342 ± 0.013
#7	25.5 ± 0.512	67.6 ± 1.24	0.958 ± 0.037	10.0 ± 0.026	0.529 ± 0.040	1.10 ± 0.013	3.09 ± 0.064	0.591 ± 0.020
#8	101 ± 1.63	213 ± 3.62	0.750 ± 0.029	1.38 ± 0.010	0.453 ± 0.048	13.5 ± 0.100	1.19 ± 0.023	0.354 ± 0.004
#9	147 ± 2.07	105 ± 1.81	1.57 ± 0.002	2.77 ± 0.991	3.78 ± 0.005	25.9 ± 0.418	5.16 ± 0.153	1.21 ± 0.019
#10	102 ± 5.90	104 ± 5.95	1.38 ± 0.049	0.531 ± 0.051	1.05 ± 0.046	20.8 ± 1.39	2.96 ± 0.175	0.405 ± 0.030
#11	18.6 ± 0.257	144 ± 2.21	30.3 ± 0.686	1.13 ± 0.308	0.570 ± 0.002	1.25 ± 0.034	4.70 ± 0.126	0.293 ± 0.013
#12	33.2 ± 0.034	199 ± 0.245	54.7 ± 0.201	0.675 ± 0.032	0.439 ± 0.002	1.50 ± 0.035	6.85 ± 0.147	0.094 ± 0.008
#13	32.1 ± 0.181	155 ± 1.34	0.278 ± 0.003	0.750 ± 0.023	0.278 ± 0.003	1.13 ± 0.001	6.66 ± 0.073	0.078 ± 0.001
#14	50.3 ± 0.737	160 ± 1.78	76.8 ± 1.46	0.660 ± 0.007	0.328 ± 0.007	1.30 ± 0.021	6.49 ± 0.050	0.047 ± 0.003
#15	41.4 ± 0.509	112 ± 5.69	1.13 ± 0.154	4.03 ± 0.010	<0.200	0.248 ± 0.007	11.1 ± 0.489	0.090 ± 0.003
#16	33.4 ± 0.065	133 ± 0.729	29.7 ± 0.774	1.00 ± 0.020	<0.200	0.272 ± 0.037	4.32 ± 0.007	0.144 ± 0.002

3.1.1. Essential Metallic Elements and Metalloids

In the present study, the highest Cu concentrations (Table 3) were observed in sample #9 *A. silvicola* (147 ± 2.07 mg kg^{-1} DW) collected in Cortes de la Frontera (Malaga, Spain), while the lowest levels determined for this metallic element were found in sample #11 *B. edulis* (18.6 ± 0.257 mg kg^{-1} DW) collected in Puerto de Galiz (Cadiz, Spain). The Cu results obtained in this study were compared with those reported by other authors. For *R. cyanoxantha*, A. R. Zsigmond et al. (2020) indicated concentrations of 37.1– 55 mg kg^{-1} DW for samples collected in Romania [34]. Sarikurkcu et al. (2010) established Cu levels of 16.4 mg kg^{-1} DW in A. *caeasera* mushrooms from Turkey [35]. On the other hand, A. Demirbaş (2001) reported concentrations of 6.24 mg kg^{-1} DW for A. *silvicola* samples collected in Turkey [29]. M. G. Alaimo et al. (2018) indicated Cu concentrations of 31 mg kg^{-1} DW (cap) and 16 mg kg^{-1} DW (steam) in *B. aereus* mushrooms from Italy [36]. Meanwhile, J. Falandysz et al. (2008) reported concentrations in a range of 26–57 mg kg^{-1} DW (caps) for *B. edulis* samples collected in Poland [37]. The Cu results obtained in the present research were in general above the data observed in the literature.

The highest/lowest Zn concentrations (Table 3) determined were observed in samples #8 *A. silvicola* (213 ± 3.62 mg kg^{-1} DW) from Parc Naturel Bouhachem (Chaouen, Morocco) and #2 *R. cyanoxantha* (51.2 ± 0.873 mg kg^{-1} DW) from Cortes de la Frontera (Cadiz, Spain), respectively. The Zn results obtained in this research were compared with those reported by previous studies. A. R. Zsigmond et al. (2020) indicated concentrations in the range of 74.4–108 mg kg^{-1} DW for *R. cyanoxantha* samples collected from geographical locations in Romania [34]. For *A. caesarea*, Sarikurkcu et al. (2010) reported 123.8 mg kg^{-1} DW concentrations for samples collected in Turkey [35]. Meanwhile, A. Demirbaş (2001) established Zn concentrations of 25.6 mg kg^{-1} DW for *A. silvicola* mushrooms from Turkey [29]. On the other hand, M. G. Alaimo et al. (2018) indicated concentrations of 125 mg kg^{-1} DW (cap) and 31 mg kg^{-1} DW (steam) in *B. aereus* mushrooms from Italy [36]. J. Falandysz et al. (2008) reported Zn concentrations in a range of 150–210 mg kg^{-1} DW (caps) for *B. edulis* samples collected in Poland [37]. The results obtained for Zn through our analysis were in most of the cases above those reported by other researchers.

Regarding Se (Table 3), the highest concentrations were obtained in sample #14 *B. aereus* (76.8 ± 1.46 mg kg^{-1} DW) from Puerto de Galiz (Cadiz, Spain) and the lowest in sample #13 *B. aereus* (0.278 ± 0.003 mg kg^{-1} DW) from Valdeinfierno (Cadiz, Spain). The Se concentrations observed in this study were compared with those indicated by other authors in previous research. For *R. cyanoxantha*, L. Cocchi et al. (2006) reported concentrations of 2.18 mg kg^{-1} DW in samples of this species collected in Italy [38]. Likewise, L. Cocchi et al. (2006) indicated concentrations of 3.30 mg kg^{-1} DW in *A. caesarea* samples from Italy [38]. On the other hand, M. Tuzen et al. (2007) established Se concentrations of 1.23 mg kg^{-1} DW for *A. silvicola* mushrooms collected in Turkey [39]. Moreover, L. Cocchi et al. (2006) reported concentrations of 24.6 and 30.8 mg kg^{-1} DW for *B. aereus* and *B. edulis* samples, collected in Italy [38]. Comparing the results obtained with those previously reported by these authors, it has been observed that the Se concentrations here were consistent with the literature.

3.1.2. Toxic Metallic Elements and Metalloids

Among the species studied, the highest Cr concentrations (Table 3) were determined in sample #3 *R. cyanoxantha* (10.2 ± 0.079 mg kg^{-1} DW) collected in Cortes de la Frontera (Malaga, Spain), whereas the lowest were in sample #10 *A. silvicola* (0.531 ± 0.051 mg kg^{-1} DW) from Cortes de la Frontera (Malaga, Spain)[2]. The Cr results obtained in this study were compared with those reported by other authors. For *R. cyanoxantha*, A. R. Zsigmond et al. (2020) indicated concentrations of 0.40 mg kg^{-1} DW for samples collected in Romania [34]. M. Yamaç et al. (2007) reported Cr levels of 0.54 mg kg^{-1} DW in the *A. caesarea* mushroom from Turkey [40]. Meanwhile, M. A. García et al. (2013) established concentrations of 3.3 mg kg^{-1} DW in *A. silvicola* samples from northern Spain [41]. Moreover, M. A. García et al. (2013) indicated concentrations of 3.8 mg kg^{-1} DW for both *B. aereus* and *B. edulis* mushrooms collected in northern Spain regions [41]. The Cr levels found in the samples studied were by the concentrations reported in the literature.

Regarding As (Table 3), the highest concentration was registered in sample #9 *A. silvicola* (3.78 ± 0.005 mg kg^{-1} DW) from Cortes de la Frontera (Malaga, Spain)[1] and the lowest levels in sample #3 *R. cyanoxantha* (0.111 ± 0.003 mg kg^{-1} DW) collected in Cortes de la Frontera (Malaga, Spain). The As results obtained in this study were compared with those reported by other authors. L. Cocchi et al. (2006) indicated concentrations of 0.10 mg kg^{-1} DW for *R. cyanoxantha* samples collected from geographical locations in Italy [38]. For *A. caesarea*, G. M. Chiocchetti et al. (2020) established As concentrations in a range of 0.275 — 0.706 mg kg^{-1} DW for samples from Spain [42]. A. Demirbaş (2001) reported concentrations of 0.76 mg kg^{-1} DW in *A. silvicola* mushrooms collected in Turkey [29]. Moreover, L. Cocchi et al. (2006) indicated concentrations of 0.39 and 0.10 mg kg^{-1} DW for *B.s aereus* and *B. edulis* samples, collected in Italy [38]. Thus, the As results obtained through our analysis agree with those reported in previous studies.

Concerning Cd (Table 3), the highest/lowest concentrations were determined in samples #9 A. silvicola (25.9 ± 0.418 mg kg^{-1} DW) collected from Cortes de la Frontera (Malaga, Spain)[1] and #2 *R. cyanoxantha* (0.223 ± 0.030 mg kg^{-1} DW) from Sendero El Palancar (Cadiz, Spain), respectively. The Cd results obtained in this research were compared with those reported by previous studies. M. J. Melgar et al. (2016) indicated concentrations of 0.28 mg kg^{-1} DW for *R. cyanoxantha* samples collected in northern Spain regions [27]. On the other hand, Sarikurkcu et al. (2010) reported Cd concentrations of 0.54 mg kg^{-1} DW in A. caesaera samples collected in Turkey [35]. For A. silvicola, A. Demirbaş (2001) indicated Cd levels of 1.04 mg kg^{-1} DW in samples from Turkey [29]. Furthermore, M. J. Melgar et al. (2016) established concentrations of 0.93 and 0.64 mg kg^{-1} DW for *B. aereus* and *B. edulis* samples, respectively, collected in northern Spain regions [27]. The Cd results obtained in the present research were generally above the data observed in the literature.

The maximum Hg concentrations (Table 3) in this study were observed in sample #15 *B. aereus* (11.1 ± 0.489 mg kg^{-1} DW) collected in Parc Naturel Bouhachem (Chaouen, Morocco) and the minimum in sample #8 A. silvicola (1.19 ± 0.023 mg kg^{-1} DW) from Parc Naturel Bohuachem (Chaouen, Morocco). The Hg concentrations observed in this study were compared with those indicated by other authors in previous research. For *R. cyanoxantha*, L. Cocchi et al. (2006) reported concentrations of 1.31 mg kg^{-1} DW in samples of this species from Italy [38]. Ostos et al. (2016) reported Hg concentrations of 0.81 mg kg^{-1} DW (stem) and 2.03 mg kg^{-1} DW (caps) in *A. caesarea* samples collected in southern Spain [24]. A. Demirbaş (2001) reported concentrations of 0.15 mg kg^{-1} A. silvicola mushrooms from

Turkey [29]. On the other hand, M. J. Melgar et al. (2009) established concentrations of 3.0 and 2.0 mg kg^{-1} DW for *B. aereus* and *B. edulis* samples, collected in northern Spain [43]. Therefore, the Hg content levels determined in our study for these species were found to be above the data presented by the mentioned studies.

The highest Pb concentrations (Table 3) were found in sample #9 A. silvicola (1.21 $\pm$ 0.019 mg kg^{-1} DW) collected from Cortes de la Frontera (Malaga, Spain) and the lowest in sample #15 *B. aereus* (0.047 $\pm$ 0.003 mg kg^{-1} DW) from Parc Naturel Bouhachem (Chaouen, Morocco). The Pb results obtained in this research were compared with those reported by previous studies. A. Demirbaş (2001) indicated concentrations in the range of 2.05 mg kg^{-1} DW for *R. cyanoxantha* samples collected from geographical locations in Turkey [29]. For *A. caesaera*, Sarikurkcu et al. (2010) reported concentrations of 5.0 mg kg^{-1} DW in samples from Turkey [35]. A. Demirbaş (2001) reported concentrations of 0.92 mg kg^{-1} DW in A. silvicola samples collected in Turkey [29]. On the other hand, M. A. García et al. (2009) established concentrations of 0.70 and 0.67 mg kg^{-1} DW for *B. aereus* and *B. edulis* samples collected in northern Spain [44]. We observed that our Pb concentrations for these species were mostly lower than those reported in the literature.

3.2. Multivariate Analysis

The results obtained through HCA using the Wards method and the Euclidean distance are shown in Figure 1 using a graphic representation based on the combination of the resulting dendrograms of variables and samples, together with a heatmap. As can be seen in Figure 1, a color scale ranging from purple to yellow was established to visualize more intuitively the levels of Cr, As, Cd, Hg, Pb, Cu, Zn, and Se in the mushroom samples. Thus, darker shades of purple would indicate a low concentration of the given element, and gradual changes to yellow would indicate an increase in concentration. As can be seen, three main clusters (A, B, and C) were obtained. Based on the information provided by the heatmap, these three clusters are differentiated from each other according to the Cr, As, Cd, Hg, Pb, Cu, Zn, and Se content. Specifically, clusters A and B are separated from cluster C due to a lower content of Se, Hg, and Zn and a higher content of Cd and Cu in the mushroom samples grouped therein. On the other hand, cluster A differs from cluster B mainly because the samples included in it have a lower Cr content and a higher Cd and Cu content. From the results obtained in the HCA, it was observed that there was a strong tendency of metal and metalloid accumulation related to the mushroom species and family, regardless of the geographical location where they were collected. In this sense, cluster A is formed by all the *A. silvicola* samples studied, while cluster B is constituted by all the *R. cyanoxantha* samples analyzed. Cluster C is formed by both *B. aereus* and *B. edulis* samples, which are both species belonging to the *Boletaceae* family. Thus, it is possible to observe that although environmental factors affect the bioavailability of metallic elements and metalloids in the medium, intrinsic factors such as species and family have a greater influence on the phenomenon of metal accumulation in mushrooms. However, this clustering is not completely consistent, since the *A. caesarea* samples were grouped among the three main clusters and not in an exclusive cluster like the rest of the species studied.

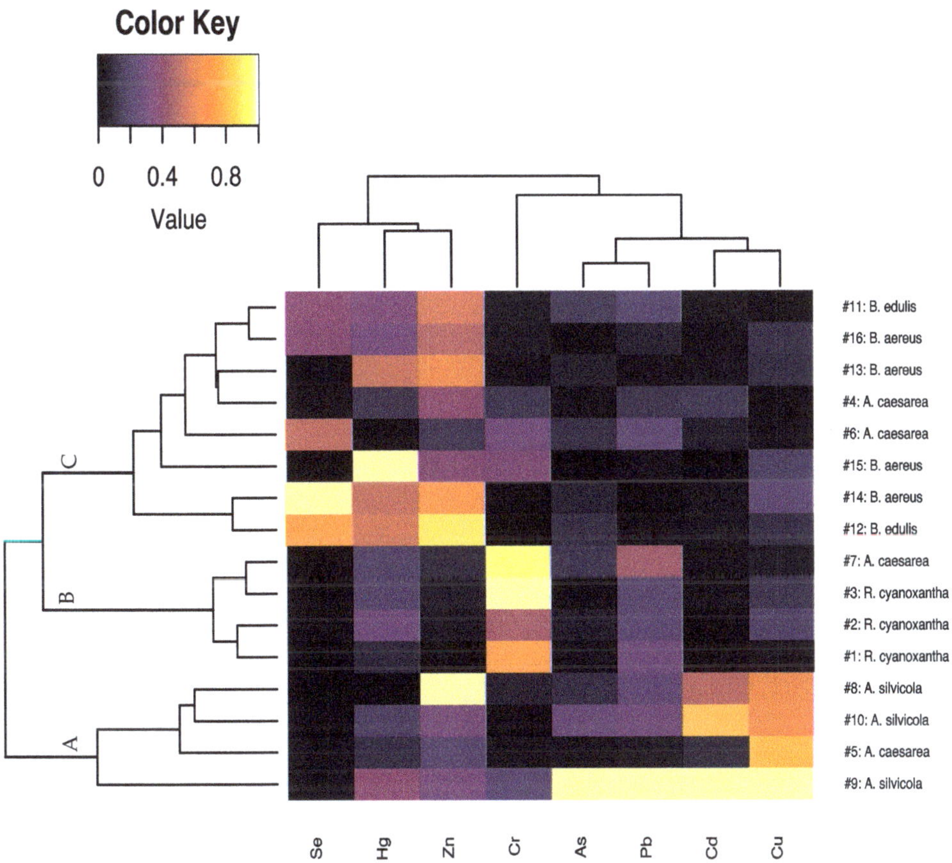

Figure 1. Graphical combination of the resulting HCA dendrograms with a heatmap to identify clustering trend patterns among the studied mushroom species based on the content of the eight elements determined.

A Principal Component Analysis (PCA) was also carried out to corroborate this clustering trend. Figure 2A shows the plot of the scores obtained for the first two principal components (PC1 and PC2) for all samples (n = 16), and Figure 2B shows the plot of the loadings obtained for each principal component (PCs). Principal components 1 and 2 explained 39.0% and 32.6% of the variance of the data, respectively, which implies a total accumulated variance of 71.6% between both. It can be observed that PC1 (Figure 2A) allowed the separation of the *A. silivicola* samples from the rest of the species studied. For their part, Cd and Cu were the metallic elements that had a greater weight on this PC (Figure 2B). Meanwhile, PC2 (Figure 2A) enabled the separation of the samples of *R. cyanoxantha* and those of the genus *Boletus* amongst them. In this case, Cr and Zn were the metallic elements with the major weight on this PC. Based on the PCA results, the *A. caesarea* samples (Figure 2A) were not grouped and separated from the rest of the species. The results obtained from the PCA agreed with those obtained by HCA, indicating a greater tendency for clustering linked to the mushroom species in terms of accumulation of metallic elements and metalloids determined in this study.

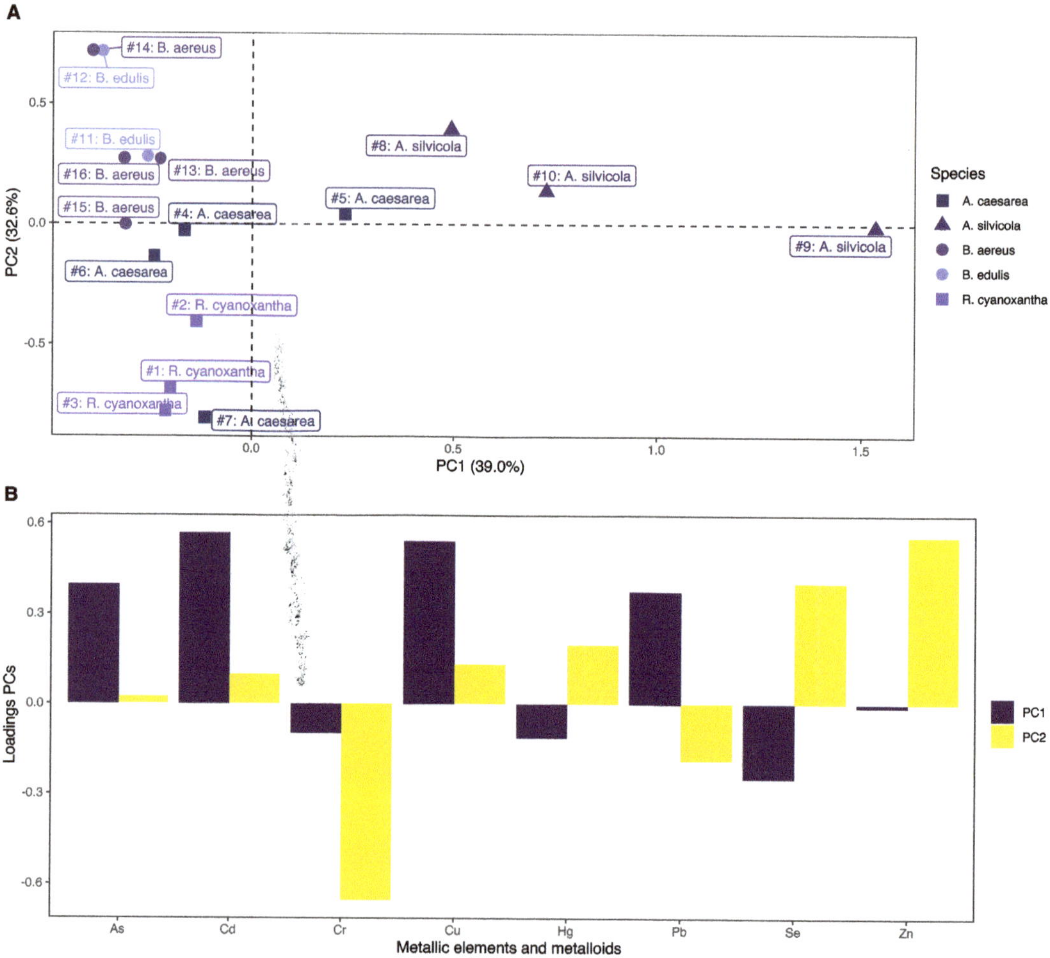

Figure 2. (**A**) Score obtained for PC1 and PC2 for all the samples (n = 16); (**B**) Loadings obtained in PC1 and PC2.

3.3. Estimated Daily Intake of Metals

To evaluate the potential human health risks associated with the consumption of the wild edible mushroom species studied (*R. cyanoxantha*, *A. caesarea*, *A. silvicola*, *B. aereus*, and *B. edulis*), the estimated daily intake of metals (EDIM) was calculated for the five toxic elements (Cr, As, Cd, Hg, and Pb), as well as for the three essential elements (Cu, Zn, and Se). For this purpose, a 300-g portion of fresh mushrooms (30 g of dried mushrooms) and an average consumer of 70 kg were assumed. The EDIM results obtained (Table 4) were compared with the values established by the Joint FAO/WHO Expert Committee on Food Additives (JECFA) for the Provisional Tolerable Maximum Daily Intake (PTMDI) and the Provisional Tolerable Daily Intake (PTDI) for Cu, Zn, As, Cd, and Hg, and with the R_fD values established for Se, Cr, and Pb.

Table 4. Estimated Daily Intake of Metals (EDIM) expressed as μg kg body weight^{-1} per day for the analyzed wild edible mushrooms species from southern Spain and northern Morocco.

Sample ID	Cu	Zn	Se	Cr	As	Cd	Hg	Pb
#1	10.2	21.9	0.263	3.17	0.0629	0.175	0.918	0.170
#2	18.5	26.3	1.09	2.15	0.129	0.0957	1.76	0.139
#3	15.1	27.2	0.640	4.40	0.0475	0.496	1.38	0.131
#4	8.12	50.1	0.976	0.787	0.0483	1.73	1.07	0.085
#5	52.5	36.7	0.912	0.355	0.0485	1.20	0.971	0.049
#6	9.45	31.7	17.62	1.53	0.187	0.895	0.586	0.146
#7	10.9	29.0	0.410	4.29	0.227	0.472	1.32	0.253
#8	43.2	91.4	0.322	0.59	0.194	5.79	0.512	0.152
#9	62.9	45.0	0.671	1.19	1.62	11.1	2.21	0.520
#10	43.5	44.6	0.590	0.228	0.451	8.89	1.27	0.173
#11	7.96	61.6	13.0	0.483	0.244	0.535	2.01	0.126
#12	14.2	85.3	23.5	0.289	0.188	0.644	2.94	0.0401
#13	13.8	66.3	0.119	0.322	0.119	0.486	2.85	0.0332
#14	21.6	68.8	32.9	0.283	0.141	0.556	2.78	0.0203
#15	17.8	48.0	0.485	1.73	n.d.	0.106	4.74	0.0385
#16	14.3	56.9	12.7	0.429	n.d.	0.117	1.85	0.0618
RfD [a] (μg kg body weight-1 per day)	40 [d]	300 [e]	0.5 [e]	3 [d]	0.3 [d]	1 [d]	0.3 [d]	3.5 [e]
PTDI [b] (μg kg body weight-1 per day)	-	-	-	-	2.14 [f]	0.82 [f]	0.57 [f]	-
PTMDI [c] (μg kg body weight^{-1} per day)	5000 [f]	300–1000 [f]	-	-	-	-	-	-

[a] R_fD: Reference dose. [b] PTDI: Provisional tolerable daily intake. [c] PTMDI: provisional maximum tolerable daily intake. [d] Sarikurkcu, C. et al. (2020) [2]. [e] USEPA: U.S. Environmental Protection Agency [45]. [f] JECEFA:The Joint FAO/WHO Expert Committee on Food Additives [46].

On a global basis, the highest EDIM values observed were for Zn, and specifically, for sample #8 *A. silvicola* collected in Parc Naturel Bouhachem (Chaouen, Morocco; 91.4 μg kg body weight^{-1} per day). Nevertheless, when comparing the EDIMs obtained for this metallic element with the PTMDI (300–1000 μg kg body weight^{-1} per day), it was noted that all were found to be below the established value. Similarly, it was observed that the EDIMs for Cu were generally below the established PTMDI (5000 μg kg body weight^{-1} per day). On the other hand, the EDIMs values obtained for Se were found to be generally above the R_fD established for this metalloid (0.5 μg kg body weight^{-1} per day), except for samples #1 *R. cyanoxantha* from Cortes de la Frontera (Malaga, Spain; 0.263 μg kg body weight^{-1} per day), #7 *A. caesarea* from Cortes de la Frontera (Malaga, Spain; 0. 410 μg kg body weight^{-1} per day), #8 *A. silvicola* from Parc Naturel Bouhachem (Chaouen, Morocco; 0.322 μg kg body weight^{-1} per day), #13 *B. aereus* from Valdeinfierno (Cadiz, Spain; 0.119 μg kg body weight^{-1} per day), and #15 *B. aereus* from Parc Naturel Bouhachem (Chaouen, Morocco; 0.485 μg kg body weight^{-1} per day). Regarding Cr, the EDIMs obtained were mainly under R_fD (3 μg kg body weight^{-1} per day), except for samples #1 *R. cyanoxantha* from Cortes de la Frontera (Malaga, Spain; 3.17 μg kg body weight^{-1} per day), #3 *R. cyanoxantha* from Cortes de la Frontera (Malaga, Spain; 4.40 μg kg body weight^{-1} per day), and #7 *A. caesarea* from Cortes de la Frontera (Malaga, Spain; 4.29 μg kg body weight ^{1} per day). From the results obtained through this study, it is worth noting that the highest EDIMs for Cr have been detected in all cases in samples collected in Cortes de la Frontera (Malaga, Spain). The EDIMs obtained for As were found to be below the established PTDI (2.14 μg kg body weight^{-1} per day) for all the samples studied. Likewise, EDIMs for Pb were also under the established R_fD value (3.5 μg kg body weight^{-1} per day) for all mushroom species analyzed in this study. Additionally, the EDIMs obtained for Cd were found to be below the established PTDI (0.82 μg kg body weight ^{1} per day), except for samples #4 *A. caesarea* from Puerto de Galiz (Cadiz, Spain; 1.73 μg kg body weight^{-1} per day), #5 *A. cesarea* from Puerto de Galiz (Cadiz, Spain; 1.20 μg kg body weight^{-1} per day), #6 *A. caesarea* from Sendero El Palancar (Cadiz, Spain; 0. 895 μg kg body weight^{-1} per day), #8 *A. silvicola* from Parc Naturel Bouhachem (Chaouen, Morocco; 5.79 μg kg body weight^{-1} per day), #9 *A. silvicola* (11.1 μg kg body weight^{-1} per day), and #10 *A. silvicola* (8.89 μg kg body weight^{-1} per day) both from Cortes de la Frontera (Malaga, Spain). In contrast, the EDIMS obtained for Hg were found to be above the PTDI (0.57 μg kg body weight^{-1} per day), apart from sample #8 *A. silvicola* from Parc Naturel Bouhachem (Chaouen, Morocco; 0.512 μg kg body weight^{-1} per day).

3.4. Health Risk Assessment

The assessment of the potential health risk was performed using the Health Risk Index (HRI) calculation, which is the ratio between the EDIM for each of the samples studied and the R_fD for each element. The results have been graphically displayed in bar charts as shown in Figure 3. HRI values equal to or less than 1 for a given element would indicate that the consumption of a particular mushroom species collected from a given geographic location is considered safe for the consumer [2,14]. This limit has been represented with the help of a vertical black line in the bar charts. According to the results shown in Figure 3, it should be highlighted that only Zn and Pb presented HRIs ≤ 1 for all the samples analyzed, indicating that the consumption of the studied mushrooms from these sample sites would be exempt from health risks, and especially for Pb, which is a probable human carcinogen, Group 2A, according to the International Agency for Research on Cancer (IARC) [47]. Regarding the remaining metallic elements and metalloids, the HRIs were found to be above the criteria for safe consumption (HRI ≤ 1) established for all or part of the mushrooms studied. Specifically, the highest HRI values in this study were observed for Se and, secondly, for Hg. For both cases, the calculated HRIs were considerably above the established safe consumption criterion. In general, the HRIs calculated for Se (Figure 3H) were found to be above 1, except for samples #1 *R. cyanoxantha* from Cortes de la Frontera (Malaga, Spain), #7 *A. caesarea* from Cortes de la Frontera (Malaga, Spain), #8 *A. silvicola* from Parc Naturel Bouhachem (Chaouen, Morocco), #13 *B. aereus* from Valdeinfierno (Cadiz, Spain), and #15 *B. aereus* from Parc Naturel Bouhachem (Chaouen, Morocco), of which the HRI values were found to be below 1. Se is a metalloid classified as an essential trace element, necessary for the organism's normal functioning. Selenium intoxication due to an overdose is generally rare, especially if it comes from food sources. However, selenium intake above the recommended dose may contribute to the prevention of prostate cancer [48,49]. Therefore, there is no evidence that the consumption of the mushroom species from the geographical areas studied with HRIs greater than 1 may pose a health threat related to this metalloid.

Regarding Hg, all the samples studied showed HRI values higher than 1 (Figure 3D). Hg is a highly toxic metallic element that is present in mercury-based organic compounds, including methylmercury, and in many inorganic forms, such as metallic mercury (Hg^0) and mercurous (Hg^{2++}) or mercuric (Hg^{++}) salts. The main organs that are potentially affected by mercury and mercurial salts are the intestinal lining and the kidneys, whereas methylmercury is widely distributed throughout the organism [50]. Nowadays, metallic Hg and inorganic Hg compounds are classified by IARC in Group 3 as not classifiable as to their carcinogenicity to humans and methylmercury compounds as possibly carcinogenic to humans [51]. Since no chemical speciation was carried out in this work and due to the toxicity of the different forms in which Hg occurs, the consumption of the mushrooms studied from the geographic locations sampled may represent a risk to human health in terms of exposure to this metallic element during the mushrooming season. On the other hand, the HRIs calculated for Cu (Figure 3F) were mostly below 1, except for samples #5 *A. caesarea* from Puerto de Galiz (Cadiz, Spain); #8 *A. silvicola* from Parc Naturel Bouhachem (Chaouen, Morocco); and #9 and #10, which were both *A. silvicola* from Cortes de la Frontera (Malaga, Spain), which all had HRIs was above 1. Cu is a metallic element classified as an essential trace element for humans occurring in many enzymes which are important in various systems, such as the immune and nervous systems. Nonetheless, it may still pose somewhat of a risk to human health at elevated levels of exposure, mainly in the gastrointestinal tract [52]. According to the results obtained in this research, it cannot be excluded that prolonged exposure to this metallic element through consumption of the mushrooms with an HRI > 1 may have health repercussions for the consumers. Nevertheless, moderate consumption throughout the mushrooming season should not necessarily have adverse health implications associated with this essential metallic element, since it is necessary for the proper functioning of the organism at safe concentrations.

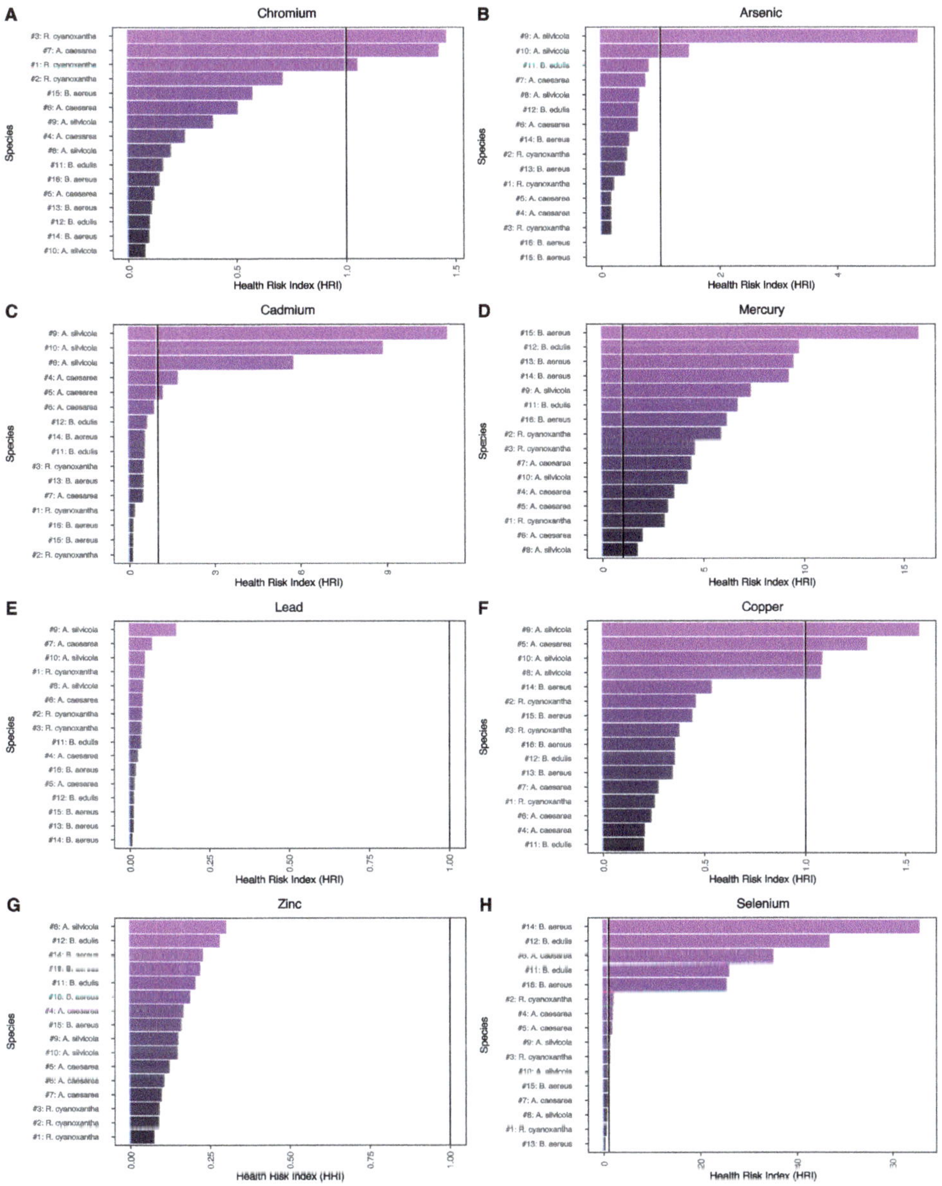

Figure 3. Bar charts of the Health Risk Index (HRI) result according to metallic elements and metalloids determined for all the wild edible mushroom species studied: (**A**) Cr HRIs; (**B**) As HRIs; (**C**) Cd HRIs; (**D**) Hg HRIs; (**E**) Pb HRIs; (**F**) Cu HRIs; (**G**) Zn HRIs; (**H**) Se HRIs. The HRI limit has been represented with the help of a vertical black line in the bar charts.

The HRI values for Cr (Figure 3A) were observed below 1, except in samples #1 and #3 of *R. cyanoxantha* and # 7 of *A. caesarea*, with all of them collected in Cortes de la Frontera (Malaga, Spain). Cr is a metallic element that can occur in two forms: trivalent chromium (Cr^{3+}) and hexavalent chromium (Cr^{6+}). The trivalent form is considered an essential element for the proper functioning of the organism, since it participates in the metabolism of glucose, cholesterol, and fatty acids. However, its hexavalent form is highly reactive, constituting a form of toxicological concern due to its adverse effects on the body, such as disorders in the respiratory and digestive systems [53,54]. Consequently, Cr^{3+} has been classified by the IARC in Group 3 as unclassifiable regarding its carcinogenicity to humans and Cr^{6+} in Group 1 as potentially carcinogenic to humans [53]. In the present study, Cr was determined as the total Cr. Notwithstanding, due to toxicological concerns associated with one of the forms of this metallic element and considering that the mutual transitions from one form to the other are quite dynamic, it is not possible to exclude health risks associated with Cr exposure through the consumption of mushrooms with HRIs > 1. For As, the HRIs values (Figure 3B) were found to be below 1, except for samples #9 and #10 *A. silvicola*, which were both collected in Cortes de la Frontera (Malaga, Spain). Arsenic is a metalloid element causing adverse health effects due to its high toxicity in its inorganic form, specifically in its trivalent (As^{3+}) and pentavalent (As^{5+}) states. Inorganic arsenic has been classified by the Agency for Toxic Substances and Disease Registry (ASTDR) as a Group 1 carcinogen for humans [55]. Therefore, the consumption of such mushrooms that present an HRI < 1 may pose a risk to human health in terms of exposure to this metalloid. Regarding Cd (Figure 3C), the HRIs calculated were below 1, except for samples #4 and #5 *A. caesarea* from Puerto de Galiz (Cadiz, Spain), #8 *A. silvicola* from Parc Naturel Bouhachem (Chaouen, Morocco), and #9 and #10 *A. silvicola* from Cortes de la Frontera (Malaga, Spain). Cd is a metallic element that is mainly found in its bivalent state (Cd^{2+}) and accumulates in tissues and organs and can cause serious diseases, such as cancer [56]. Thus, Cd and its compounds have been classified as Group 1 carcinogenic to humans by the IARC [57]. Due to the toxicological attention concerning Cd, the consumption of those samples with HRI > 1 may pose a health risk in terms of this metallic element.

4. Conclusions

The total content of five toxic elements (Cr, As, Cd, Hg, and Pb) and three essential elements (Cu, Zn, and Se) was determined in wild-growing edible mushrooms (*A. silvicola, A. caesarea, B. aereus, B. edulis, and R. cyanoxantha*) collected in southern Spain and northern Morocco. Among the metallic elements and metalloids determined, Zn was the most abundant, followed by Cu and Se. The rest of the elements were found in relatively lower concentrations. The application of multivariate techniques, such as Hierarchical Cluster Analysis (HCA) and Principal Component Analysis (PCA), indicated that there are considerable differences in the uptake of metallic elements and metalloids related to species-intrinsic factors. On the other hand, consumer exposure and health implications were assessed using two well-known safety criteria: the Estimated Daily Intake of Metals (EDIM) and the Health Risk Index (HRI). The highest EDIM values obtained were observed for Zn. The HRI assessment for all mushroom species studied showed a concern related to Hg due to the frequent consumption of between 300 g of fresh mushrooms per day. Meanwhile, mushroom species that presented HRIs > 1 for Cr, As, and Cd also pose a cause for concern in terms of exposure to these metallic elements and metalloids.

As is well recognized, mushroom consumers normally assume that the intake of edible wild mushrooms is risk-free. However, the findings reported in the present study indicate that excessive consumption of the mushroom species studied during the mushrooming season could have health implications and that mushrooms should therefore be consumed moderately.

Author Contributions: Conceptualization, M.B.-S., E.E.-B. and G.F.B.; Data curation, M.B.-S., E.E.-B. and G.F.B.; Formal analysis, M.B.-S. and M.F.-G.; Funding acquisition, E.E.-B., M.P. and G.F.B.; Investigation, M.B.-S., E.E.-B. and G.F.B.; Methodology, M.B.-S.; Project administration, E.E.-B. and G.F.B.; Resources, H.B., J.G.L.-C. and M.P.; Software, M.B.-S. and M.F.-G.; Supervision, E.E.-B. and G.F.B.; Validation, M.B.-S.; Visualization, M.B.-S.; Writing—original draft, M.B.-S.; Writing—review & editing, E.E.-B. and G.F.B. All authors have read and agreed to the published version of the manuscript.

Funding: This work was financially supported by the University of Cadiz, partially through the "Plan Propio UCA 2022-2023 de Apoyo y Estímulo a la Investigación y la Transferencia" and through the Aula Universitaria del Estrecho within the framework of grants for international collaboration projects (Ref. UCA/R62REC/2021): "Mapa del contenido metálico en setas silvestres comercializables del sur de Andalucía y norte de Marruecos" and "Profundización en el estudio y comparativa de compuestos bioactivos en setas", and (Ref. UCA/R96REC/2018): "001ENE2019" and "002ENE2019".

Institutional Review Board Statement: Not applicable.

Informed Consent Statement: Not applicable.

Data Availability Statement: Please refer to suggested Data Availability Statements in section "MDPI Research Data Policies" at https://www.mdpi.com/ethics (accessed on 22 April 2022).

Acknowledgments: The authors are grateful to the Plan Propio de Apoyo y Estímulo a la Investigación y la Transferencia and Aula Universitaria del Estrecho (University of Cadiz) for funding support. A special acknowledgment is extended to Asociación Micológica del Estrecho (Mairei) for their invaluable help during the sample stage in Cadiz and Morocco, and Spectroscopy/ICP Division from the Central Research Services for Science and Technology (SC-ICYT) of the University of Cadiz, especially to Antonio Benítez Rodríguez for his collaboration throughout the analysis of the samples.

Conflicts of Interest: The authors declare no conflict of interest.

References

1. Tang, X.; Mi, F.; Zhang, Y.; He, X.; Cao, Y.; Wang, P.; Liu, C.; Yang, D.; Dong, J.; Zhang, K.; et al. Diversity, population genetics, and evolution of macrofungi associated with animals. *Mycology* **2015**, *6*, 94–109. [CrossRef] [PubMed]
2. Sarikurkcu, C.; Popović-Djordjević, J.; Solak, M.H. Wild edible mushrooms from Mediterranean region: Metal concentrations and health risk assessment. *Ecotoxicol. Environ. Saf.* **2020**, *190*, 110058. [CrossRef] [PubMed]
3. Ainsworth, G.C. *Ainsworth and Bisby's Dictionary of the Fungi*, 10th ed.; Kirk, P.M., Cannon, P.F., David, J.C., Stalpers, J.A., Eds.; CABI Publishing: Wallingford, UK, 2008.
4. Huili, L.; Yang, T.; Nelson, M.; Lei, Y.; Karunarathna, S.C.; Perez-Moreno, J.; Mahmudur Rahman, M.; Md Harunur, R.; Pheng, P.; Rizal, L.; et al. Reviewing the world's edible mushroom species: A new evidence-based classification system. *Compr. Rev. Food Sci. Food Saf.* **2021**, *20*, 1982–2014. [CrossRef]
5. Valverde, M.E.; Hernández-Pérez, T.; Paredes-López, O. Edible mushrooms: Improving human health and promoting quality life. *Int. J. Microbiol.* **2015**, *2015*, 376387. [CrossRef] [PubMed]
6. Fogarasi, M.; Diaconeasa, Z.M.; Pop, C.R.; Fogarasi, S.; Semeniuc, C.A.; Fărcaş, A.C.; Ţibulcă, D.; Sălăgean, C.-D.; Tofană, M.; Socaci, S.A. Elemental Composition, Antioxidant and Antibacterial Properties of Some Wild Edible Mushrooms from Romania. *Agronomy* **2020**, *10*, 1972. [CrossRef]
7. Global Mushroom Market Size, Market Share, Application Analysis, Regional Outlook, Growth Trends, Key Players, Competitive Strategies and Forecasts, 2021 to 2028. Available online: https://www.grandviewresearch.com/industry-analysis/mushroom-market (accessed on 23 January 2022).
8. Ministerio De Agricultura, Pesca Y Alimentacion, Statistical Services From the Spanish Government. Available online: https://www.mapa.gob.es/es/alimentacion/temas/consumo-tendencias/panel-de-consumo-alimentario/series-anuales/default.aspx (accessed on 23 January 2022).
9. Milenge Kamalebo, H.; Nshimba Seya Wa Malale, H.; Masumbuko Ndabaga, C.; Degreef, J.; De Kesel, A. Uses and importance of wild fungi: Traditional knowledge from the Tshopo province in the Democratic Republic of the Congo. *J. Ethnobiol. Ethnomed.* **2018**, *14*, 17464269. [CrossRef]
10. Chelela, B.L.; Chacha, M.; Matemu, A. Wild Mushrooms from Tanzania: Characterization and their Importance to the Rural Communities. *Curr. Res. Environ. Appl. Mycol.* **2015**, *5*, 307–321. [CrossRef]
11. Haro, A.; Trescastro, A.; Lara, L.; Fernández-Fígares, I.; Nieto, R.; Seiquer, I. Mineral elements content of wild growing edible mushrooms from the southeast of Spain. *J. Food Compos. Anal.* **2020**, *91*, 103504. [CrossRef]
12. Falandysz, J.; Drewnowska, M.; Jarzyńska, G.; Zhang, D.; Zhang, Y.; Wang, J. Mineral constituents in common chanterelles and soils collected from a high mountain and lowland sites in Poland. *J. Mt. Sci.* **2012**, *9*, 697–705. [CrossRef]
13. Krupa, P.; Kozdrój, J. Accumulation of Heavy Metals by Ectomycorrhizal Fungi Colonizing Birch Trees Growing in an Industrial Desert Soil. *World J. Microbiol. Biotechnol.* **2004**, *20*, 427–430. [CrossRef]

14. Liu, B.; Huang, Q.; Cai, H.; Guo, X.; Wang, T.; Gui, M. Study of heavy metal concentrations in wild edible mushrooms in yunnan province, China. *Food Chem.* **2015**, *188*, 294–300. [CrossRef] [PubMed]
15. Schaffer, S.J.; Campbell, J.R. Lead Poisoning. *Pediatr. Clin. Advis.* **2007**, 332. [CrossRef]
16. Canfield, R.L.; Jusko, T.A. Lead Poisoning. *Encycl. Infant Early Child. Dev.* **2008**, 200–213. [CrossRef]
17. Cai, L.; Li, X.-K.; Song, Y.; Cherian, M. Essentiality, Toxicology and Chelation Therapy of Zinc and Copper. *Curr. Med. Chem.* **2005**, *12*, 2753–2763. [CrossRef]
18. Barea-Sepúlveda, M.; Espada-Bellido, E.; Ferreiro-González, M.; Benítez-Rodríguez, A.; López-Castillo, J.G.; Palma, M.; Barbero, G.F. Metal concentrations in Lactarius mushroom species collected from Southern Spain and Northern Morocco: Evaluation of health risks and benefits. *J. Food Compos. Anal.* **2021**, *99*, 103859. [CrossRef]
19. Zsigmond, A.R.; Varga, K.; Kántor, I.; Urák, I.; May, Z.; Héberger, K. Elemental composition of wild growing Agaricus campestris mushroom in urban and peri-urban regions of Transylvania (Romania). *J. Food Compos. Anal.* **2018**, *72*, 15–21. [CrossRef]
20. Türkmen, M.; Budur, D. Heavy metal contaminations in edible wild mushroom species from Turkey's Black Sea region. *Food Chem.* **2018**, *254*, 256–259. [CrossRef]
21. Mleczek, M.; Rzymski, P.; Budka, A.; Siwulski, M.; Jasińska, A.; Kalač, P.; Poniedziałek, B.; Gąsecka, M.; Niedzielski, P. Elemental characteristics of mushroom species cultivated in China and Poland. *J. Food Compos. Anal.* **2018**, *66*, 168–178. [CrossRef]
22. Igbiri, S.; Udowelle, N.A.; Ekhator, O.C.; Asomugha, R.N.; Igweze, Z.N.; Orisakwe, O.E. Edible Mushrooms from Niger Delta, Nigeria with Heavy Metal Levels of Public Health Concern: A Human Health Risk Assessment. *Recent Pat. Food Nutr. Agric.* **2018**, *9*, 31–41. [CrossRef]
23. Falandysz, J.; Sapkota, A.; Mędyk, M.; Feng, X. Rare earth elements in parasol mushroom Macrolepiota procera. *Food Chem.* **2017**, *221*, 24–28. [CrossRef]
24. Ostos, C.; Pérez-Rodríguez, F.; Arroyo, B.M.; Moreno-Rojas, R. Study of mercury content in wild edible mushrooms and its contribution to the Provisional Tolerable Weekly Intake in Spain. *J. Food Compos. Anal.* **2015**, *37*, 136–142. [CrossRef]
25. Melgar, M.J.; Alonso, J.; García, M.A. Total contents of arsenic and associated health risks in edible mushrooms, mushroom supplements and growth substrates from Galicia (NW Spain). *Food Chem. Toxicol.* **2014**, *73*, 44–50. [CrossRef] [PubMed]
26. Wang, X.; Liu, H.; Zhang, J.; Li, T.; Wang, Y. Evaluation of heavy metal concentrations of edible wild-grown mushrooms from China. *J. Environ. Sci. Health Part B Pestic. Food Contam. Agric. Wastes* **2017**, *52*, 178–183. [CrossRef] [PubMed]
27. Melgar, M.J.; Alonso, J.; García, M.A. Cadmium in edible mushrooms from NW Spain: Bioconcentration factors and consumer health implications. *Food Chem. Toxicol.* **2016**, *88*, 13–20. [CrossRef] [PubMed]
28. Zhu, F.; Qu, L.; Fan, W.; Qiao, M.; Hao, H.; Wang, X. Assessment of heavy metals in some wild edible mushrooms collected from Yunnan Province, China. *Environ. Monit. Assess.* **2011**, *179*, 191–199. [CrossRef]
29. Demirbaş, A. Heavy metal bioaccumulation by mushrooms from artificially fortified soils. *Food Chem.* **2001**, *74*, 293–301. [CrossRef]
30. Lalotra, P.; Gupta, D.; Yangdol, R.; Sharma, Y.; Gupta, S. Bioaccumulation of heavy metals in the sporocarps of some wild mushrooms. *Curr. Res. Environ. Appl. Mycol.* **2016**, *6*, 159–165. [CrossRef]
31. Damodaran, D.; Vidya Shetty, K.; Raj Mohan, B. Uptake of certain heavy metals from contaminated soil by mushroom—Galerina vittiformis. *Ecotoxicol. Environ. Saf.* **2014**, *104*, 414–422. [CrossRef]
32. Kokkoris, V.; Massas, I.; Polemis, E.; Koutrotsios, G.; Zervakis, G.I. Accumulation of heavy metals by wild edible mushrooms with respect to soil substrates in the Athens metropolitan area (Greece). *Sci. Total Environ.* **2019**, *685*, 280–296. [CrossRef]
33. Barea-Sepúlveda, M.; Espada-Bellido, E.; Ferreiro-González, M.; Bouziane, H.; López-Castillo, J.G.; Palma, M.; Barbero, G.F. Toxic elements and trace elements in Macrolepiota procera mushrooms from southern Spain and northern Morocco. *J. Food Compos. Anal.* **2022**, *108*, 104419. [CrossRef]
34. Zsigmond, A.R.; Kántor, I.; May, Z.; Urák, I.; Héberger, K. Elemental composition of Russula cyanoxantha along an urbanization gradient in Cluj-Napoca (Romania). *Chemosphere* **2020**, *238*, 124566. [CrossRef] [PubMed]
35. Sarikurkcu, C.; Tepe, B.; Semiz, D.K.; Solak, M.H. Evaluation of metal concentration and antioxidant activity of three edible mushrooms from Mugla, Turkey. *Food Chem. Toxicol.* **2010**, *48*, 1230–1233. [CrossRef] [PubMed]
36. Alaimo, M.G.; Dongarrà, G.; La Rosa, A.; Tamburo, E.; Vasquez, G.; Varrica, D. Major and trace elements in Boletus aereus and Clitopilus prunulus growing on volcanic and sedimentary soils of Sicily (Italy). *Ecotoxicol. Environ. Saf.* **2018**, *157*, 182–190. [CrossRef] [PubMed]
37. Falandysz, J.; Kunito, T.; Kubota, R.; Bielawski, L.; Frankowska, A.; Falandysz, J.J.; Tanabe, S. Multivariate characterization of elements accumulated in King Bolete Boletus edulis mushroom at lowland and high mountain regions. *J. Environ. Sci. Health Part A Toxic/Hazard. Subst. Environ. Eng.* **2008**, *43*, 1692–1699. [CrossRef]
38. Cocchi, L.; Vescovi, L.; Petrini, L.E.; Petrini, O. Heavy metals in edible mushrooms in Italy. *Food Chem.* **2006**, *98*, 277–284. [CrossRef]
39. Tuzen, M.; Sesli, E.; Soylak, M. Trace element levels of mushroom species from East Black Sea region of Turkey. *Food Control* **2007**, *18*, 806–810. [CrossRef]
40. Yamaç, M.; Yildiz, D.; Sarikürkcü, C.; Çelikkollu, M.; Solak, M.H. Heavy metals in some edible mushrooms from the Central Anatolia, Turkey. *Food Chem.* **2007**, *103*, 263–267. [CrossRef]
41. García, M.A.; Alonso, J.; Melgar, M.J. Bioconcentration of chromium in edible mushrooms: Influence of environmental and genetic factors. *Food Chem. Toxicol.* **2013**, *58*, 249–254. [CrossRef]
42. Chiocchetti, G.M.; Latorre, T.; Clemente, M.J.; Jadán-Piedra, C.; Devesa, V.; Vélez, D. Toxic trace elements in dried mushrooms: Effects of cooking and gastrointestinal digestion on food safety. *Food Chem.* **2020**, *306*, 125478. [CrossRef]

43. Melgar, M.J.; Alonso, J.; García, M.A. Mercury in edible mushrooms and underlying soil: Bioconcentration factors and toxicological risk. *Sci. Total Environ.* **2009**, *407*, 5328–5334. [CrossRef]
44. García, M.Á.; Alonso, J.; Melgar, M.J. Lead in edible mushrooms: Levels and bioaccumulation factors. *J. Hazard. Mater.* **2009**, *167*, 777–783. [CrossRef] [PubMed]
45. Integrated Risk Information System | US EPA. Available online: https://www.epa.gov/iris (accessed on 23 January 2022).
46. Joint Fao WHO Expert Committee on Food Additives (JECFA). 2021. Available online: https://apps.who.int/food-additives-contaminants-jecfa-database/search.aspx?fcc=2/ (accessed on 23 January 2022).
47. Rousseau, M.-C.; Straif, K.; Siemiatycki, J. IARC Carcinogen Update. *Environ. Health Perspect.* **2005**, *113*, A580–A581. [CrossRef] [PubMed]
48. Falandysz, J. Selenium in edible mushrooms. *J. Environ. Sci. Health Part C Environ. Carcinog. Ecotoxicol. Rev.* **2008**, *26*, 256–299. [CrossRef] [PubMed]
49. Mirończuk-Chodakowska, I.; Socha, K.; Zujko, M.E.; Terlikowska, K.M.; Borawska, M.H.; Witkowska, A.M. Copper, manganese, selenium and zinc in wild-growing edible mushrooms from the eastern territory of "green lungs of Poland": Nutritional and toxicological implications. *Int. J. Environ. Res. Public Health* **2019**, *16*, 3614. [CrossRef] [PubMed]
50. Bernhoft, R.A. Mercury Toxicity and Treatment: A Review of the Literature. *J. Environ. Public Health.* **2012**, *2012*, 460508. [CrossRef]
51. Mercury and Mercury Compounds (IARC Summary & Evaluation, Volume 58, 1993). Available online: http://www.inchem.org/documents/iarc/vol58/mono58-3.html (accessed on 23 January 2022).
52. Taylor, A.A.; Tsuji, J.S.; Garry, M.R.; McArdle, M.E.; Goodfellow, W.L.; Adams, W.J.; Menzie, C.A. Critical Review of Exposure and Effects: Implications for Setting Regulatory Health Criteria for Ingested Copper. *Environ. Manag.* **2019**, *65*, 131–159. [CrossRef]
53. Naz, A.; Mishra, B.K.; Gupta, S.K. Human Health Risk Assessment of Chromium in Drinking Water: A Case Study of Sukinda Chromite Mine, Odisha, India. *Expo. Health* **2016**, *8*, 253–264. [CrossRef]
54. Tseng, C.H.; Lee, I.H.; Chen, Y.C. Evaluation of hexavalent chromium concentration in water and its health risk with a system dynamics model. *Sci. Total Environ.* **2019**, *669*, 103–111. [CrossRef]
55. ASTDR. *Toxicological Profile for Arsenic*; U.S. Department of Health and Human Services, Public Health Service Agency for Toxic Substances and Disease Registry: Chamblee, GA, USA, 2007.
56. Waalkes, M.P. Cadmium carcinogenesis. *Mutat. Res. Mol. Mech. Mutagen.* **2003**, *533*, 107–120. [CrossRef]
57. Kim, H.S.; Kim, Y.J.; Seo, Y.R. An Overview of Carcinogenic Heavy Metal: Molecular Toxicity Mechanism and Prevention. *J. Cancer Prev.* **2015**, *20*, 232–240. [CrossRef]

Journal of
Fungi

MDPI

Article

Spatial Assessment of Potentially Toxic Elements (PTE) Concentration in *Agaricus bisporus* Mushroom Collected from Local Vegetable Markets of Uttarakhand State, India

Pankaj Kumar [1,†], Vinod Kumar [1,†], Ebrahem M. Eid [2,3,†], Arwa A. AL-Huqail [4,†], Bashir Adelodun [5,6], Sami Abou Fayssal [7,8], Madhumita Goala [9], Ashish Kumar Arya [10], Archana Bachheti [10], Željko Andabaka [11], Kyung Sook Choi [7] and Ivan Širić [11,*]

[1] Agro-Ecology and Pollution Research Laboratory, Department of Zoology and Environmental Science, Gurukula Kangri (Deemed to be University), Haridwar 249404, Uttarakhand, India; rs.pankajkumar@gkv.ac.in (P.K.); drvksorwal@gkv.ac.in (V.K.)
[2] Biology Department, College of Science, King Khalid University, Abha 61321, Saudi Arabia; ebrahem.eid@sci.kfs.edu.eg
[3] Botany Department, Faculty of Science, Kafrelsheikh University, Kafr El-Sheikh 33516, Egypt
[4] Department of Biology, College of Science, Princess Nourah bint Abdulrahman University, P.O. Box 84428, Riyadh 11671, Saudi Arabia; aaalhuqail@pnu.edu.sa
[5] Department of Agricultural and Biosystems Engineering, University of Ilorin, PMB 1515, Ilorin 240103, Nigeria; adelodun.b@unilorin.edu.ng
[6] Department of Agricultural Civil Engineering, Kyungpook National University, Daegu 41566, Korea
[7] Department of Agronomy, Faculty of Agronomy, University of Forestry, 10 Kliment Ohridski Blvd, 1797 Sofia, Bulgaria; sami.aboufaycal@st.ul.edu.lb (S.A.F.); ks.choi@knu.ac.kr (K.S.C.)
[8] Department of Plant Production, Faculty of Agriculture, Lebanese University, Beirut 1302, Lebanon
[9] Nehru College, Pailapool, Affiliated Assam University, Cachar, Silchar 788098, Assam, India; madhumitagoalap@gmail.com
[10] Department of Environmental Science, Graphic Era (Deemed to be University), Dehradun 248002, Uttarakhand, India; ashishkumararya_20941076.evs@geu.ac.in (A.K.A.); bachheti.archana@gmail.com (A.B.)
[11] University of Zagreb, Faculty of Agriculture, Svetosimunska 25, 10000 Zagreb, Croatia; zandabaka@agr.hr
* Correspondence: isiric@agr.hr
† These authors contributed equally to this work.

Citation: Kumar, P.; Kumar, V.; Eid, E.M.; AL-Huqail, A.A.; Adelodun, B.; Abou Fayssal, S.; Goala, M.; Arya, A.K.; Bachheti, A.; Andabaka, Ž.; et al. Spatial Assessment of Potentially Toxic Elements (PTE) Concentration in *Agaricus bisporus* Mushroom Collected from Local Vegetable Markets of Uttarakhand State, India. *J. Fungi* **2022**, *8*, 452. https://doi.org/10.3390/jof8050452

Academic Editor: Nuria Ferrol

Received: 5 April 2022
Accepted: 26 April 2022
Published: 27 April 2022

Abstract: This study presents a spatial assessment of eight potentially toxic elements (PTE: Cd, Cr, Cu, Fe, Pb, Ni, Mn, and Zn) in white button (*Agaricus bisporus* J.E. Lange) mushroom samples collected from the local vegetable markets of Uttarakhand State, India. Fresh *A. bisporus* samples were collected from thirteen districts and fifteen sampling locations (M1-M15) and analyzed for the concentration of these PTE using atomic absorption spectroscopy (AAS). The results revealed that *A. bisporus* contained all eight selected PTE in all sampling locations. Based on the inverse distance weighted (IDW) interpolation, principal component (PC), and hierarchical cluster (HC) analyses, the areas with a plane geographical distribution showed the highest PTE concentrations in the *A. bisporus* samples as compared to those in hilly areas. Overall, the decreasing order of PTE concentration in *A. bisporus* was recognized as Fe > Zn > Mn > Cr > Cu > Ni > Cd > Pb. The Kruskal–Wallis ANOVA tests displayed a highly significant ($p < 0.05$) difference among the sampling locations. However, the concentration of PTE was below permissible limits, indicating no potential hazard in consuming the *A. bisporus*. Similarly, the health risk assessment studies using the target hazard quotient (THQ) also showed no significant health risk associated with the consumption of *A. bisporus* being sold in the local mushroom markets of Uttarakhand, India. This study is the first report on state-level monitoring of PTE in *A. bisporus* mushrooms, which provides crucial information regarding the monitoring and occurrence of potentially toxic metallic elements.

Keywords: cluster analysis; health risk; potentially toxic elements; mushroom; spatial analysis; principal component

1. Introduction

Around the globe, the consumption of mushrooms is gaining attention due to their unique flavor and high nutritional value compared to meat and vegetables [1]. Their nutritional value is mainly related to the composition of substrates on which they are grown, including the type of added supplements. Therefore, substrates (e.g., composts), including raw and casing materials, play a crucial role in the production, preservation, and marketing of quality mushrooms [2]. The close interrelationship between substrates and mushrooms prompts researchers to investigate possible dangers to human health as a result of mushroom consumption. *Agaricus bisporus* (button mushroom) is considered a strong bio-accumulator of potentially toxic elements (PTE), especially mercury (Hg), lead (Pb), and cadmium (Cd) [3,4]. The mushroom usually grows on composted lignocellulosic materials, which can contain considerable levels of PTE [5]. Accordingly, *A. bisporus* could be a natural "treatment plant" for different agro-industrial, dairy, and domestic wastewaters [6], resulting in reduced hazardous dispersion of PTE in the environment. However, not all button mushrooms that are consumed are a product of human cultivation, especially in developing countries where the local population tends to forage and sell the mushroom in rural markets as their main source of income [7,8]. The PTE concentrations of these wild mushrooms may be very high if they are collected near mines or industrial factories [9].

Potentially toxic elements are non-degradable elements that can accumulate and pose high risks to human health when found in high concentrations in the food chain [10–14]. Although several PTE (e.g., Cu, Fe, Zn, Ni) are essential elements for the human body, when consumed at high concentrations, they can cause fatal neurological disorders, allergic reactions, abnormal hemoglobin content, growth retardation in children, hemolysis, and nephrotic effects [3,15,16]. On this basis, several countries and commissions, such as the USEPA, WHO, and FAO/WHO Codex, have specified maximum limits of PTE in mushrooms intended for human consumption [17,18].

Uttarakhand, one of the 28 states of India, contains 45.44% forest area, making it a plentiful location for several saprophytic and mycorhizal fungal species [19–21]. Uttarakhand State has been contributing to increasing mushroom production in India since its establishment in the year 2000. Recent figures found the annual mushroom production in Uttarakhand to be 10236 metric tons [22]. Out of this total, *A. bisporus* is the most commonly produced, and is obtained from three sources: (i) collected from wild forest areas, (ii) commercially grown on farms, and (iii) locally grown by small farmers [23]. In this state, the majority of cultivated mushrooms are sold in the local vegetable markets. For the cultivation of *A. bisporus*, the farmers use agricultural residues of wheat crop (straw). Since it is observed that PTE contents can be transferred to the edible parts of mushrooms through the process of bioaccumulation, it is therefore necessary to monitor their levels in mushrooms sold in the local vegetable markets in Uttarakhand State.

Sinha and co-workers previously investigated the PTE contents and their availability in button mushrooms collected from two markets in other states of India, including Maharashtra [3]. They observed copper (Cu), iron (Fe), zinc (Zn), manganese (Mn), cadmium (Cd), and lead (Pb) contents in *A. bisporus* lower than the safe limits reported by USEPA, Indian Standards, WHO, and FAO [18,24]. However, a large spatial assessment of PTE concentrations in mushrooms collected from Uttarakhand State markets within the districts has not yet been conducted. Therefore, the current study investigated the spatial PTE contents in *A. bisporus* mushrooms collected from fifteen local markets located in different areas of Uttarakhand State, India.

2. Materials and Methods

2.1. Description of the Study Area and Sample Collection

The current study was conducted in the Uttarakhand State of India. Uttarakhand State is one of the 28 states of India with a total area of 53,483 km^2 and a population of 11.5 million. Over 86% of the total area is covered by mountains while 45.44% is covered by forests. This makes Uttarakhand a rich habitat for many edible and non-edible macrofungi [20,21]. The mushroom samples were collected from fifteen sampling sites (M1 to M15) located in each of the thirteen districts in Uttarakhand State. Figure 1 and Table 1 show the geographical distribution of *A. bisporus* sample collection sites. The sampling locations were divided into two groups, namely hilly and plain areas, based on their elevation. Specifically, the fresh fruiting bodies of *A. bisporus* mushroom were collected from local vegetable markets of the selected locations. The collected mushroom samples ($n = 3$ for each market) were washed thoroughly to remove any adhering dirt or soil particles and dried using blotting paper.

The mushroom samples were collected from November 2021 to January 2022, which is the most suitable period for seasonal *A. bisporus* cultivation by local growers. The samples were immediately transported to the laboratory in a polyurethane foam insulated ice cooler box (11 L, PinnacleThermo, Ahemdabad, Gujrat, India) and oven-dried (KI-181, Khera Instruments Pvt. Ltd., Delhi, India) at 60 °C until the constant weight of biomass was achieved. The samples were converted into a fine powder using a mechanical grinder (HL7576/00 600 W, Philips Amaze Ltd., Solan, India).

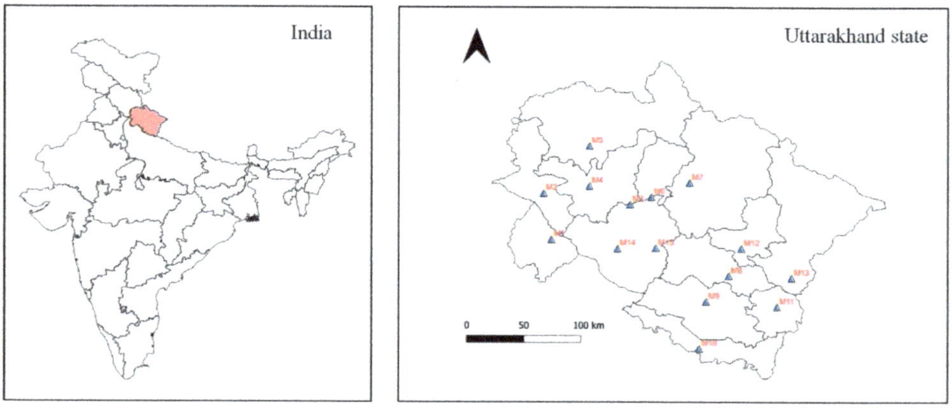

Figure 1. Map view of the study area (▲ sampling sites).

Table 1. Description of different locations (M1–M15) selected for *A. bisporus* collection in Uttarakhand state, India.

Site Code	Site Name (Vegetable Market)	District	Longitude	Latitude	Elevation (m)
M1	Jwalapur Sabji Mandi	Haridwar	78.10285	29.91545	281
M2	Dehradun Sabji Mandi	Dehradun	78.03421	30.31900	659
M3	Pauri Garhwal Sabji Mandi	Pauri Garhwal	78.79454	30.22143	626
M4	Tehri Garhwal Sabji Mandi	Tehri Garhwal	78.43813	30.38010	1050
M5	Uttarkashi Bus Stand	Uttarkashi	78.43798	30.72878	1141
M6	Rudraprayag Bridge Market	Rudraprayag	78.98489	30.28644	723
M7	Chamoli Gopeshwar Market	Chamoli	79.31739	30.41037	1474
M8	Almora Chandni Chowk	Almora	79.66361	29.60164	1602
M9	Nainital Bus Stand	Nainital	79.46428	29.37957	1936
M10	Udham Singh Nagar Rudrapur	Udham Singh Nagar	79.40320	28.97121	215
M11	Champawat Naad Bora	Champawat	80.08405	29.33380	1691
M12	Bageshwar Saryu Bridge	Bageshwar	79.77404	29.83861	877
M13	Pithoragarh Bus Stand	Pithoragarh	80.21207	29.58334	1503
M14	Lansdown Trishakti Chowk	Pauri Garhwal	78.68650	29.83720	1687
M15	Bironkhal	Pauri Garhwal	79.02541	29.84294	1545

2.2. Analytical Methods

The concentration of PTE in mushroom samples was analyzed by using atomic absorption spectroscopy (AAS: A-Analyst 800, PerkinElmer, Waltham, USA). For this, 1 g dehydrated mushroom powder was mixed in a di-acid mixture (5 mL HNO_3 + 2.5 mL $HClO_4$) and left overnight for self-digestion (12 h). Further, the sample was adjusted to 50 mL using a 3% HNO_3 solution followed by heating digestion on a hot plate (150 °C for 1 h) until a 10 mL sample was left. Finally, the digested sample was filtered through Whatman filter paper no. 41 and supernatants were used for PTE quantification (i.e., Cd, Cr, Cu, Fe, Pb, Mn, Ni, and Zn) using AAS. The detection limits of the instrument for Cd, Cr, Cu, Fe, Pb, Mn, Ni, and Zn were 3, 4, 4, 5, 20, 3, 10, and 3 μg/L, respectively [25,26]. Selective hollow cathode lamps of PTE were cast off at optimum current and operated by following the standard operating procedures (SOPs) recommended by the manufacturer. The slit width of the instrument was adjusted to 0.5 nm and a mixture of air/acetylene gas was used to run the AAS. Calibration curves were prepared using standard solutions (0 as control, 0.5, 1, 5, 10, 50, and 100 mg/L) of PTE. Qualitative assurance of the PTE analysis results was performed based on the maximum recovery percentage (>98%). All analyses were conducted in triplicate.

2.3. Data Analysis

The data of PTE concentration were analyzed using principal component analysis (PCA) and hierarchical cluster analysis (HCA) tools. The principal component analysis is a statistical technique widely used to study the relative contribution of participating data groups based on their correlation or covariance coefficients. By using PCA, eigenvectors and variance values are extracted from the matrix, which allows for a comparison of the dominance of selected variables [27]. These values are extracted as two or more groups based on data coverage known as principal components (PCs). A biplot is drawn from the computed covariance matrix, which reflects the coordinates of the original variable and their relation to participating data groups. Due to its high efficiency in estimating spatial and temporal patterns of environmental and agriculture data, PCA has been explored by numerous researchers [4,28]. Similarly, cluster analysis is a useful technique for identifying data groups having the highest or lowest pairwise similarities [29]. Agglomerative nesting (AGNES) is one of the best hierarchical algorithms for developing similarity that utilizes single element clusters for each group. The result is developed as a form of a combined tree and heatmap that represents similarities between all participating groups [30]. Therefore, PCA and cluster analysis were performed in the current study to draw a covariance matrix, biplots, and clustered heatmaps to understand the relationship between sampling locations and their influence on PTE availability in *A. bisporus* mushroom samples collected from the Uttarakhand State. Moreover, the map-based graphical visualization of the PTE data was conducted using the inverse distance weighted (IDW) interpolation method of the geographical information system (GIS) approach [31].

The health risk associated with the consumption of PTE contaminated *A. bisporus* by the consumers of Uttarakhand State, India was computed using the Target Hazard Quotient (THQ) approach. By using THQ, a health risk index (HRI) is developed, which is used to assess the occurrence of health hazards [31–33]. An HRI value above 1 indicates a potential health risk in consuming the contaminated mushroom [34]. For this, THQ and HRI values of *A. bisporus* samples collected from selected sites were calculated by using the following equations (Equations (1) and (2)):

$$THQ = 10^{-3} \times (PTE_{EE} \times I_{EA} \times PTE_{CF} \times PTE_{C})/(I_{BW} \times PTE_{ACP} \times PTE_{RD}) \quad (1)$$

where, PTE_{EE} is the exposure efficiency of PTE (365 days/year), I_{EA} is the exposure age of an individual (70 years), PTE_{CF} is the consumption frequency of PTE (2.2 g/day), PTE_{C} is the PTE concentration in *A. bisporus* sample (fresh weight basis), I_{ABW} is the average body weight of the vegetable consumer (70 kg and 16 kg for adult and child groups), and PTE_{ACP} is the average consumption period of PTE (25,550 days). PTE_{RD} represents the PTE reference doses of Cd, Cu, Cr, Fe, Pb, Mn, Ni, and Zn, namely, 5.0×10^{-4}, 4.2×10^{-2},

3.0×10^{-3}, 7.0×10^{-1}, 2.0×10^{-2}, 1.4×10^{-2}, 3.5×10^{-3}, and 3.0×10^{-1} mg/kg/day, respectively [34]. Afterward, the HRI of PTE intake was computed [18,35]:

$$\text{HRI} = \sum \text{THQs} \quad (2)$$

2.4. Software and Tools

The data were analyzed using Microsoft Excel 2019 (Microsoft Corporation, Redmond, WA, USA), OriginPro 2021b (OriginLab Corporation, Northampton, MA, USA), and QGIS Desktop (3.22.3-Białowieża, Open Source, Gispo Ltd., Helsinki, Finland) software packages. All values presented in the current study were calculated as mean followed by standard deviation (SD).

3. Results and Discussion

3.1. Concentration of PTE in A. bisporus Mushroom

The results of the PTE concentration in *A. bisporus* mushroom samples collected from different locations (M1–M15) of Uttarakhand State, India are summarized in Table 2. The statistical results of the Kruskal–Wallis (K-W) ANOVA revealed that the mean PTE concentration in *A. bisporus* samples varied significantly ($p < 0.05$) between the selected sampling locations. The concentration of Cd ranged from 0.06 ± 0.01 to 0.09 ± 0.1 mg/kg at all sites, which is very close to the safe limit (0.10 mg/kg) of Indian Standards [3]. The mean concentration of Cr was recorded as 16.21 ± 2.87 mg/kg. Similarly, the mean concentrations of Cu, Fe, Pb, Mn, Ni, and Zn were recorded as 15.07 ± 2.48, 37.37 ± 8.59, 0.05 ± 0.02, 20.00 ± 3.44, 1.05 ± 0.43, and 36.37 ± 3.39 mg/kg, respectively. Based on the location variation, the highest concentration of PTE was found at multiple sites such as Cd (M3, M9, M10, M11), Cr (M1), Cu (M12), Fe (M1, M3), Pb (M1, M2), Mn (M1, M13), Ni (M1, M12), and Zn (M1). In general, the M1 site showed a relatively higher concentration for Cr, Fe, Pb, Mn, Ni, and Zn PTE. The coefficient of variance (CV < 37.37%) also showed a relatively low error rate in state-wise monitoring of PTE in *A. bisporus* mushroom samples. Overall, no sample was found to have any of the selected PTE below the detection limits of the AAS. The presence of PTE in mushrooms is due to their vital role in fungal growth, metabolism, and reproduction. Elements such as Cu, Cr, Fe, Mn, Ni, and Zn are taken up by fungal mycelia as micro or trace nutrients, which further support their effective growth [36]. However, Cd and Pb are not essential nutrients and may harm mushroom growth if present in the substrate at high concentrations. However, these toxic elements may be taken up by mushroom cell walls as a substitute for other elements and to bring chemical equilibrium in the mycelial growth zone [4].

Based on the contour maps generated using the IDW interpolation tool of QGIS (Figures 2 and 3), the eight districts of Uttarakhand State (M1: Haridwar; M3: Pauri Garhwal; M8: Almora; M9: Nainital; M10: Udham Singh Nagar; M11: Chapawat; M12: Bageshwar, and M13: Pithoragarh) were more affected with Cd contamination, whereas Fe and Cr were highest in the case of the M1 and M2 locations. Similarly, these two locations showed the occurrence of other PTE in the highest concentrations, which might be due to their location in plane areas where the majority of agricultural and industrial activities occur. The mushroom growers use locally available substrates such as wheat straw, wheat bran, animal manures, chemical fertilizers (urea, di-ammonium phosphate, super-phosphate, etc.), and soil-compost mixtures for casing material, which might be the source of the PTE absorbed by the *A. bisporus*.

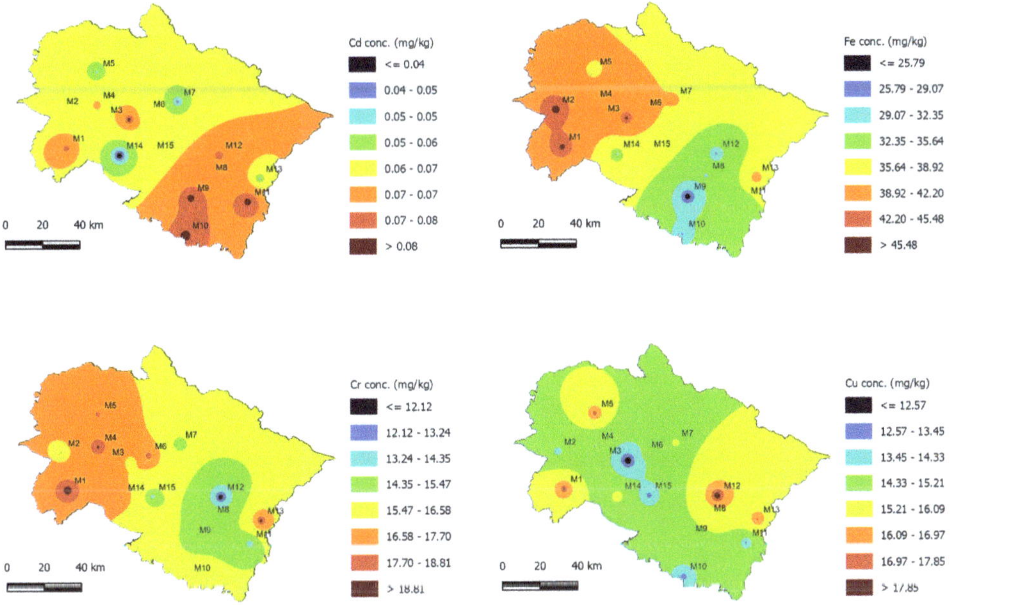

Figure 2. Spatial concentration of Cd, Cr, Cu, and Fe in *A. bisporus* samples collected from different locations (M1–M15) of Uttarakhand state, India.

Table 2. Concentration (Mean ± SD) of PTE in *A. bisporus* samples collected from different locations (M1–M15) of Uttarakhand state, India.

Site Code	Potentially Toxic Elements (PTE) Concentration (mg/kg)							
	Cd	Cr	Cu	Fe	Pb	Mn	Ni	Zn
M1	0.08 ± 0.01	20.35 ± 0.16	18.26 ± 0.08	48.28 ± 0.62	0.08 ± 0.01	26.35 ± 0.15	1.65 ± 0.10	42.34 ± 1.52
M2	0.06 ± 0.01	15.19 ± 0.28	13.69 ± 0.56	49.35 ± 0.45	0.08 ± 0.01	20.84 ± 0.37	1.51 ± 0.23	40.07 ± 0.80
M3	0.09 ± 0.01	16.02 ± 0.42	10.36 ± 1.30	48.16 ± 0.28	0.06 ± 0.02	17.18 ± 0.40	1.07 ± 0.09	37.20 ± 2.43
M4	0.07 ± 0.01	19.46 ± 0.80	14.21 ± 0.96	42.03 ± 1.34	0.01 ± 0.03	20.07 ± 0.24	0.68 ± 0.06	41.93 ± 1.30
M5	0.05 ± 0.01	18.08 ± 1.07	17.82 ± 0.25	37.28 ± 2.70	0.03 ± 0.02	16.66 ± 0.16	0.47 ± 0.07	34.45 ± 0.65
M6	0.06 ± 0.02	19.06 ± 0.71	14.67 ± 0.30	41.73 ± 1.08	0.03 ± 0.01	24.50 ± 0.70	0.39 ± 0.05	36.85 ± 1.05
M7	0.04 ± 0.03	14.38 ± 1.02	15.44 ± 0.46	40.26 ± 0.55	0.04 ± 0.02	18.15 ± 1.30	0.66 ± 0.12	40.10 ± 0.30
M8	0.07 ± 0.01	15.10 ± 0.45	15.35 ± 0.27	31.82 ± 0.38	0.07 ± 0.01	15.94 ± 0.83	0.90 ± 0.16	34.42 ± 2.73
M9	0.09 ± 0.01	14.72 ± 0.30	14.84 ± 0.02	20.10 ± 3.91	0.01 ± 0.01	16.49 ± 0.32	0.65 ± 0.10	36.00 ± 0.50
M10	0.09 ± 0.01	16.05 ± 0.21	12.73 ± 0.15	28.13 ± 0.56	0.06 ± 0.01	18.76 ± 0.60	1.30 ± 0.24	31.08 ± 0.78
M11	0.09 ± 0.01	13.29 ± 0.59	13.08 ± 0.80	34.19 ± 0.10	0.07 ± 0.01	21.30 ± 1.30	1.10 ± 0.08	35.19 ± 1.35
M12	0.08 ± 0.02	10.23 ± 0.28	19.70 ± 0.11	27.32 ± 0.43	0.02 ± 0.01	23.42 ± 0.45	1.66 ± 0.19	33.87 ± 0.82
M13	0.05 ± 0.01	19.75 ± 0.14	17.36 ± 1.09	43.12 ± 0.22	0.03 ± 0.02	25.10 ± 0.20	1.32 ± 0.14	34.03 ± 0.12
M14	0.02 ± 0.01	18.03 ± 0.53	16.02 ± 0.26	30.54 ± 0.70	0.05 ± 0.01	17.05 ± 0.36	1.54 ± 0.08	32.30 ± 0.36
M15	0.06 ± 0.01	13.45 ± 0.42	12.55 ± 0.08	38.27 ± 0.86	0.06 ± 0.02	18.16 ± 0.30	0.87 ± 0.12	35.71 ± 0.10
Mean ± SD	0.07 ± 0.02	16.21 ± 2.87	15.07 ± 2.48	37.37 ± 8.59	0.05 ± 0.02	20.00 ± 3.44	1.05 ± 0.43	36.37 ± 3.39
CV (%)	31.40	17.68	16.48	23.00	51.05	17.20	40.97	9.32
K-W	0.047	0.001	0.002	0.003	0.042	0.009	0.025	0.019
SL	0.10	20.00	40.00	425.00	0.20	30.00	1.50	50.00
Reference	[3]	[3]	[3]	[18]	[18]	[18]	[3]	[3]

CV: coefficient of variance; SL: safe limits; K-W: Kruskal–Wallis p-value.

Based on the altitudinal variation in sampling sites, it is evident that the sites having lower elevation (plane areas) showed higher concentrations of the selected PTE. For instance, M1 is a plane region and its samples exhibited high levels of Cr, Cu, Fe, Pb, Mn, Ni, and Zn contamination. Wheat is the major crop in the M1 region, which can be attributed to the plane terrain and optimal seasonal conditions. The straw obtained after wheat harvesting is mainly used as fodder and for mushroom cultivation. As a result of the extensive utilization of chemical-based fertilizers and pesticides, and irrigation using polluted water sources, the wheat straw may also become contaminated, later affecting the elemental composition and quality of *A. bisporus*. The local mushroom growers of this region utilize wheat straw waste which might be contaminated with this PTE, which later accumulate within the fruiting bodies of *A. bisporus*. Higher elevation sites, by contrast, have limited availability of wheat straw substrate because of the unsuitable climatic conditions (i.e., they are usually cold). However, the sites with higher elevations exhibit good climatic conditions (usually cold) that support the natural occurrence of certain mushroom species including *A. bisporus*. The local markets at higher elevation sites sell *A. bisporus* sourced from both natural and commercial harvesting, which might be a reason behind the lower concentration of some PTE observed for these sites.

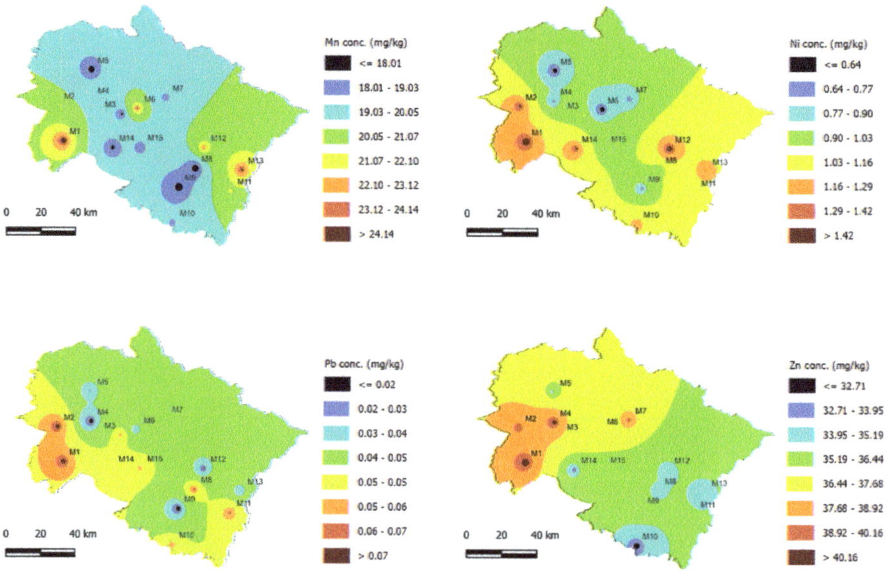

Figure 3. Spatial concentration of Pb, Mn, Ni, and Zn in *A. bisporus* samples collected from different locations (M1–M15) of Uttarakhand state, India.

Previously, no study has reported a state-level analysis of PTE concentration in *A. bisporus* in the Uttarakhand State, India. However, a study by Singh et al. [37] analyzed the elemental composition of four Ganoderma mushroom species collected from wild forest areas of Uttarakhand State, India. They reported a total of 27 elements that did not exceed the recommended dose reference (RDF) values and were found to be safe for human consumption. Another study by Gaur et al. [38] also investigated the nutritional and elemental composition of seven mushroom species, including *A. bisporus*. Their results revealed that *A. bisporus* had significant PTE contents, including Cu, Cr, Fe, Mn, and Zn as analyzed by AAS. However, they did not report any concentration of Cd, Pb, As, and Hg PTE. Thus, the findings of this study suggest occurrences of several PTE, which did not reach the safe limit but still could pose risks to the consumer's health.

3.2. PCA and Hierarchical Cluster Analysis

In this study, the interactive effect of sampling location on PTE availability in *A. bisporus* samples was analyzed using the PCA tool. The dimension reduction method of PCA helps in identifying the positive or negative interaction between the input variables [27]. Based on the PCA, the data were transformed into two different principal components (PC1 and PC2). The extracted components, namely, PC1 and PC2, had eigenvalues of 83.47 and 12.61

with the variance of 74.70% and 11.29%, respectively. These PCs were helpful in deriving the interactive effects of input locations and PTE availability in *A. bisporus* samples through the vector lengths given in Figure 4a. Moreover, the data given in Table 3 also shows the dominating PCs of PTE based on their actual vector lengths. The results indicate that Fe concentration was highest at M2, whilst Zn was highest at M1. The concentration of Cr and Mn was also highest at M1 with a positive interaction. On the other hand, Cu, Ni, and Pb exhibited negative interaction with sampling locations, where the highest Cu was observed at M12, and the highest Pb and Ni at the M1 site. Besides this, Cd showed the highest concentration at multiple sites (M3, M9, M10, and M11), indicating a potential health risk posed by this PTE at these locations.

Similarly, the hierarchical cluster analysis (HCA) is a descriptive classification method widely used in identifying the data objects or groups having the highest similarities or dissimilarities [39]. In this study, the similarities among locations were determined based on the closeness of available heavy metal concentration in *A. bisporus* mushroom samples. By this method, a heat map-based clustered diagram (Figure 4b) was produced to understand identifiable data groups. Based on the nearest neighboring method of Euclidean clustering, the minimum and maximum distance identified were 1.77 and 3.96, respectively. Further, the highest similarities were shown by Pb-Ni, Cu-Mn, and Fe-Zn, while two other PTE (Cd and Cr) appeared in quite different individual clusters.

In previous studies, PCA has been a widely used and accepted tool for big data analysis, particularly for PTE in mushroom species. In a study by Buruleanu et al. [40], heavy metal concentrations were determined in the mushroom samples (regional, wild, and cultivated) collected from different locations in Romania. They implemented the PCA tool to assess the interaction between dominating PTE through the Varimax rotation method and found that K, Mg, Cd, and Cr showed the highest positive interaction. Moreover, the findings of this study were in line with those reported by Širić et al. [41], in which they analyzed PTE in 10 saprophytic mushroom species collected from Croatia. The cluster analysis showed the highest similarities between edible mushroom species belonging to the same phenotypic groups. Similarly, Bosiacki et al. [42] investigated PTE levels in wild *A. bisporus* mushroom samples collected from Poland. The HCA analysis helped in identifying similar regions having the highest heavy metal levels in *A. bisporus* samples. Therefore, PCA and HCA were helpful to understand the interactive effects of the availability of PTE in *A. bisporus* mushroom samples collected from the Uttarakhand state of India.

Table 3. PCA matrix results for dominance of Potentially Toxic Elements (PTE) in *A. bisporus* samples collected from different locations (M1-M15) of Uttarakhand state, India.

Potentially Toxic Elements (PTE)	Principal Components	
	PC1	PC2
Variance (%)	74.70	11.29
Eigenvalue	83.47	12.61
Cd	**−0.01**	−0.02
Cr	0.15	**0.19**
Cu	−0.03	**0.55**
Fe	**0.93**	−0.16
Pb	**0.01**	−0.01
Mn	0.17	**0.79**
Ni	0.03	**0.04**
Zn	**0.26**	0.01

Bold values indicate dominating axis for specific PTE.

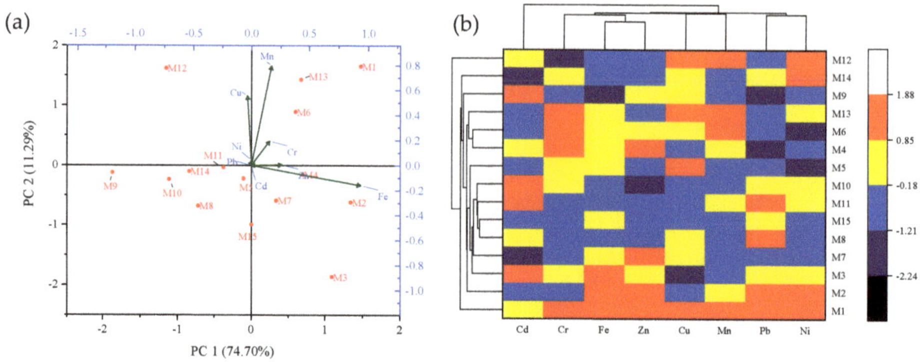

Figure 4. (**a**) PCA biplot and (**b**) clustered heatmap of Potentially Toxic Elements (PTE) concentration in *A. bisporus* samples collected from different locations (M1–M15) of Uttarakhand state, India.

3.3. Health Risk Assessment of *A. bisporus* Mushroom

With an increasing rate of edible mushroom consumption, there has been a need for adequate monitoring of associated heavy metal levels. However, the toxicity of PTE varies largely based on the type and amount consumed. For this reason, toxicity studies using the target hazard quotient (THQ) health risk index (HRI) provide a better insight into the possible health hazard posed to an individual as well as the combined intake. In the present study, THQ and HRI indices were used to obtain critical index values for adult and child human groups. The results showed that the child human group was more susceptible to all eight selected PTE compared to the adult group. As given in Table 4, THQ values vary largely with changes in the sampling location. Notably, the combined HRI values were below 1, indicating no possible health hazard associated with the consumption of *A. bisporus*. More specifically, the highest HRI values were observed at the M1 (Haridwar) site, which might be due to the occurrence of a large number of industrial units (pharmaceutical, electroplating, agro-industrial, papermaking, textile, distillery, etc.) releasing toxic wastes into the environment. Apart from that, the M1 is a plane region with extensive agricultural activities, meaning high use of chemical fertilizers and pesticides [43], so these higher levels could also be a result of bioaccumulation of PTE from agricultural wastes to edible parts of mushrooms. Overall, the decreasing order of HRI in the study area was identified as M1 > M13 > M6 > M4 > M14 > M5 > M10 > M2 > M3 > M8 > M9 > M11 > M7 > M15 > M12. The uptake of PTE by *A. bisporus* depends on several factors such as their bioavailable concentrations in the composted substrate, casing soil, and irrigation water [5]. Moreover, substrate pH and organic matter also play an important role in constructing the substrate–fungal network, which facilitates the migration of PTE to the upper edible parts of the mushroom. The most prevalent PTE, including Cd, Cr, Cu, Pb, Ni, and Zn may pose health risks to the consumer if their intake amount and exposure duration are high. Lately harvested mushrooms may also show a high PTE concentration, as time plays an important role in creating equilibrium between mushrooms and the substrate used to grow them [8].

A study by Igbiri et al. [44] investigated the THQ and HR of PTE in edible mushroom species in the Niger Delta, Nigeria. They observed that the maximum HR values were identified for Ni heavy metal, indicating serious health risks for the consumers. Similarly, Kumar et al. [5] also performed HRI studies for the uptake of PTE by *A. bisporus* cultivated on compost loaded with industrial wastewater. The values of THQ and HRI were below the specified health hazard level. Moreover, Karataş [45] also analyzed the contents of PTE in cultivated oyster mushrooms and performed THQ studies. The results revealed that the contents of six elements (Ca, Mg, Na, Zn, Cd, and Cr) were within permissible limits, indicating the edibility of the cultivated mushroom. Thus, the results of the above-mentioned studies are in agreement with the findings of the current study, thereby suggesting the importance of THQ and HRI indices in determining the health risks associated with the consumption of PTE-containing mushrooms.

Table 4. Target Hazard Quotient (TQH) and Health Risk Index (HRI) results of PTE in the *A. bisporus* samples collected from different locations (M1–M15) of Uttarakhand state, India.

Site	Age Group	Target Hazard Quotient (THQ)								Health Risk Index (HRI) ^
		Cd	Cr	Cu	Fe	Pb	Mn	Ni	Zn	
M1	Child	0.0033	0.1413	0.0101	0.0014	0.0001	0.0392	0.0098	0.0029	0.2082
	Adult	0.0008	0.0323	0.0021	0.0003	0.0001	0.0090	0.0022	0.0007	0.0474
M2	Child	0.0025	0.1055	0.0075	0.0015	0.0001	0.0310	0.0090	0.0028	0.1599
	Adult	0.0006	0.0241	0.0016	0.0003	0.0001	0.0071	0.0021	0.0006	0.0364
M3	Child	0.0038	0.1113	0.0079	0.0014	0.0001	0.0256	0.0064	0.0026	0.1590
	Adult	0.0009	0.0254	0.0012	0.0003	0.0001	0.0058	0.0015	0.0006	0.0357
M4	Child	0.0029	0.1351	0.0097	0.0013	0.0001	0.0299	0.0040	0.0029	0.1858
	Adult	0.0007	0.0309	0.0016	0.0003	0.0001	0.0068	0.0009	0.0007	0.0419
M5	Child	0.0021	0.1256	0.0090	0.0011	0.0001	0.0248	0.0028	0.0024	0.1677
	Adult	0.0005	0.0287	0.0020	0.0003	0.0001	0.0057	0.0006	0.0005	0.0383
M6	Child	0.0025	0.1324	0.0095	0.0012	0.0001	0.0365	0.0023	0.0026	0.1869
	Adult	0.0006	0.0303	0.0017	0.0003	0.0001	0.0083	0.0005	0.0006	0.0422
M7	Child	0.0017	0.0999	0.0071	0.0012	0.0001	0.0270	0.0039	0.0028	0.1436
	Adult	0.0004	0.0228	0.0018	0.0003	0.0001	0.0062	0.0009	0.0006	0.0329
M8	Child	0.0029	0.1049	0.0075	0.0009	0.0001	0.0237	0.0054	0.0024	0.1478
	Adult	0.0007	0.0240	0.0017	0.0002	0.0001	0.0054	0.0012	0.0005	0.0338
M9	Child	0.0038	0.1022	0.0073	0.0006	0.0001	0.0245	0.0039	0.0025	0.1448
	Adult	0.0009	0.0234	0.0017	0.0001	0.0001	0.0056	0.0009	0.0006	0.0331
M10	Child	0.0038	0.1115	0.0080	0.0008	0.0001	0.0279	0.0077	0.0022	0.1619
	Adult	0.0009	0.0255	0.0014	0.0002	0.0001	0.0064	0.0018	0.0005	0.0366
M11	Child	0.0038	0.0923	0.0066	0.0010	0.0001	0.0317	0.0065	0.0024	0.1444
	Adult	0.0009	0.0211	0.0015	0.0002	0.0001	0.0072	0.0015	0.0006	0.0330
M12	Child	0.0033	0.0710	0.0051	0.0008	0.0001	0.0349	0.0099	0.0024	0.1274
	Adult	0.0008	0.0162	0.0022	0.0002	0.0001	0.0080	0.0023	0.0005	0.0302
M13	Child	0.0021	0.1372	0.0098	0.0013	0.0001	0.0374	0.0079	0.0024	0.1979
	Adult	0.0005	0.0313	0.0020	0.0003	0.0001	0.0085	0.0018	0.0005	0.0450
M14	Child	0.0008	0.1252	0.0089	0.0009	0.0001	0.0254	0.0092	0.0022	0.1727
	Adult	0.0002	0.0286	0.0018	0.0002	0.0001	0.0058	0.0021	0.0005	0.0393
M15	Child	0.0025	0.0934	0.0067	0.0011	0.0001	0.0270	0.0052	0.0025	0.1385
	Adult	0.0006	0.0213	0.0014	0.0003	0.0001	0.0062	0.0012	0.0006	0.0315

^: Indicates potential health risk if the value exceeds > 1.00.

4. Conclusions

This study analyzed the spatial variations in the concentrations of potentially toxic elements (PTE) in *A. bisporus* mushroom samples collected from different locations in Uttarakhand State, India. The findings of this study reveal that *A. bisporus* samples showed varying contents and concentration levels of eight PTE (Cd, Cr, Cu, Fe, Pb, Mn, Ni, and Zn). The mushroom samples collected from plane regions showed high concentrations of PTE compared to hilly regions. The PCA and HCA tools were useful in identifying the dominance and similarity characteristics of heavy metal availability. The contents of PTE in *A. bisporus* mushroom did not exceed the safe limits, while the health risk index exhibited no potential health hazard associated with their consumption. Thus, this study suggests the *A. bisporus* mushroom being sold in the local vegetable markets were safe for human consumption. Also, this study points out the healthiness and suitability of button mushrooms sold in Uttarakhand State vegetable markets. Further studies on monitoring of other PTE (Hg, As, Co, etc.) in *A. bisporus*, as well as other commercially sold mushroom species, are highly recommended.

Author Contributions: Conceptualization, P.K., V.K. and M.G.; Formal analysis, P.K.; Funding acquisition, E.M.E., A.A.A.-H., Ž.A. and I.Š.; Investigation, P.K. and A.K.A.; Methodology, P.K.; Project administration, E.M.E. and A.A.A.-H.; Resources, V.K.; Software, B.A. and M.G.; Supervision, V.K.; Validation, V.K., E.M.E., A.A.A.-H., B.A., S.A.F., A.K.A., A.B., Ž.A., K.S.C. and I.Š.; Visualization, P.K.; Writing—original draft, P.K., M.G. and I.Š.; Writing—review & editing, V.K., E.M.E., A.A.A.-H., B.A., S.A.F., A.K.A., A.B., Ž.A. and K.S.C. All authors have read and agreed to the published version of the manuscript.

Funding: This research was funded by King Khalid University (grant number RGP.1/7/43), Abha, Saudi Arabia; and Princess Nourah bint Abdulrahman University Researchers Supporting Project (PNURSP 2022R93), Princess Nourah bint Abdulrahman University, Riyadh, Saudi Arabia.

Institutional Review Board Statement: Not applicable.

Informed Consent Statement: Not applicable.

Data Availability Statement: Not applicable.

Acknowledgments: This work is a person-to-person collaboration between P.K., E.M.E., and I.Š. The authors are grateful to their host institutes for providing the necessary facilities to conduct this study. E.M.E. extends his appreciation to King Khalid University for funding this work through the Research Group Project under grant number RGP. 1/7/43, King Khalid University, Abha, Saudi Arabia; A.A.A. express her gratitude to Princess Nourah bint Abdulrahman University Researchers Supporting Project (PNURSP 2022R93), Princess Nourah bint Abdulrahman University, Riyadh, Saudi Arabia. All individuals included in this section have consented to the acknowledgement.

Conflicts of Interest: The authors declare no conflict of interest.

References

1. Abou Fayssal, S.; Alsanad, M.A.; el Sebaaly, Z.; Ismail, A.I.H.; Sassine, Y.N. Valorization of Olive Pruning Residues through Bioconversion into Edible Mushroom Pleurotus Ostreatus (Jacq. Ex Fr.) P. Kumm. (1871) of Improved Nutritional Value. *Scientifica* **2020**, *2020*, 39503. [CrossRef] [PubMed]
2. Alsanad, M.A.; Sassine, Y.N.; el Sebaaly, Z.; Abou Fayssal, S. Spent Coffee Grounds Influence on Pleurotus Ostreatus Production, Composition, Fatty Acid Profile, and Lignocellulose Biodegradation Capacity. *CYTA-J. Food* **2021**, *19*, 11–20. [CrossRef]
3. Sinha, S.K.; Upadhyay, T.K.; Sharma, S.K. Heavy Metals Detection in White Button Mushroom (*Agaricus bisporus*) Cultivated in State of Maharashtra, India. *Biochem. Cell. Arch.* **2019**, *19*, 3501–3506. [CrossRef]
4. Kumar, P.; Kumar, V.; Adelodun, B.; Bedeković, D.; Kos, I.; Širić, I.; Alamri, S.A.M.; Alrumman, S.A.; Eid, E.M.; Abou Fayssal, S.; et al. Sustainable Use of Sewage Sludge as a Casing Material for Button Mushroom (*Agaricus bisporus*) Cultivation: Experimental and Prediction Modeling Studies for Uptake of Metal Elements. *J. Fungi* **2022**, *8*, 112. [CrossRef]
5. Kumar, V.; Kumar, P.; Singh, J.; Kumar, P. Use of Sugar Mill Wastewater for *Agaricus bisporus* Cultivation: Prediction Models for Trace Metal Uptake and Health Risk Assessment. *Environ. Sci. Pollut. Res.* **2021**, *28*, 26923–26934. [CrossRef]
6. Mohammadhasani, F.; Ahmadimoghadam, A.; Asrar, Z.; Mohammadi, S.Z. Growth Responses and Accumulation of Heavy Metals by Fungus *Agaricus bisporus*. *Acta Bot. Hung.* **2016**, *59*, 401–409. [CrossRef]
7. Kumar, V.; Goala, M.; Kumar, P.; Singh, J.; Kumar, P. Integration of Treated Agro-Based Wastewaters (TAWs) Management with Mushroom Cultivation. In *Environmental Degradation: Causes and Remediation Strategies*; Agro Environ Media-Agriculture and Ennvironmental Science Academy: Haridwar, India, 2020; pp. 63–75.
8. Kumar, P.; Kumar, V.; Goala, M.; Singh, J.; Kumar, P. Integrated Use of Treated Dairy Wastewater and Agro-Residue for *Agaricus bisporus* Mushroom Cultivation: Experimental and Kinetics Studies. *Biocatal. Agric. Biotechnol.* **2021**, *32*, 101940. [CrossRef]
9. Sithole, S.C.; Mugivhisa, L.L.; Amoo, S.O.; Olowoyo, J.O. Pattern and Concentrations of Trace Metals in Mushrooms Harvested from Trace Metal-Polluted Soils in Pretoria. *S. Afr. J. Bot.* **2017**, *108*, 315–320. [CrossRef]
10. Birla Singh, K.; Taneja, S.K. Concentration of Zn, Cu and Mn in Vegetables and Meat Foodstuffs Commonly Available in Manipur: A North Eastern State of India. *Electron. J. Environ. Agric. Food Chem.* **2010**, *9*, 610–616.
11. Guerra, F.; Trevizam, A.R.; Muraoka, T.; Marcante, N.C.; Canniatti-Brazaca, S.G. Heavy Metals in Vegetables and Potential Risk for Human Health. *Sci. Agric.* **2012**, *69*, 54–60. [CrossRef]
12. Širić, I.; Humar, M.; Kasap, A.; Kos, I.; Mioč, B.; Pohleven, F. Heavy Metal Bioaccumulation by Wild Edible Saprophytic and Ectomycorrhizal Mushrooms. *Environ. Sci. Pollut. Res.* **2016**, *23*, 18239–18252. [CrossRef]

13. Širić, I.; Poljak, M.; Tomić, D.; Markota, T.; Kos, I.; Kasap, A. Sadržaj Teških Metala i Bioakumulacijski Potencijal Nekih Samoniklih Jestivih Gljiva. *Šumarski List* **2016**, *140*, 29–37. [CrossRef]
14. Falandysz, J.; Mędyk, M.; Saba, M.; Zhang, J.; Wang, Y.; Li, T. Mercury in traditionally foraged species of fungi (macromycetes) from the karst area across Yunnan province in China. *Appl. Microbiol. Biotechnol.* **2020**, *104*, 9421–9432. [CrossRef]
15. Kapahi, M.; Sachdeva, S. Mycoremediation Potential of Pleurotus Species for Heavy Metals: A Review. *Bioresour. Bioprocess.* **2017**, *4*, 32–41. [CrossRef] [PubMed]
16. Árvay, J.; Tomáš, J.; Hauptvogl, M.; Kopernická, M.; Kováčik, A.; Bajčan, D.; Massányi, P. Contamination of Wild-Grown Edible Mushrooms by Heavy Metals in a Former Mercury-Mining Area. *J. Environ. Sci. Health Part B* **2014**, *49*, 815–827. [CrossRef]
17. Širić, I.; Falandysz, J. Contamination, Bioconcentration and Distribution of Mercury in Tricholoma Spp. Mushrooms from Southern and Northern Regions of Europe. *Chemosphere* **2020**, *251*, 126614. [CrossRef] [PubMed]
18. USEPA. Integrated Risk Information System, USEPA. Available online: https://www.epa.gov/iris (accessed on 3 April 2022).
19. Pokhriyal, P.; Rehman, S.; Areendran, G.; Raj, K.; Pandey, R.; Kumar, M.; Sahana, M.; Sajjad, H. Assessing Forest Cover Vulnerability in Uttarakhand, India Using Analytical Hierarchy Process. *Model. Earth Syst. Environ.* **2020**, *6*, 821–831. [CrossRef]
20. Kothiyal, G.; Singh, K.; Kumar, A.; Juyal, P.; Guleri, S. Wild Macrofungi (Mushrooms) Diversity Occurrence in the Forest of Uttarakhand, India. *Biodiv. Res.* **2019**, *53*, 7–32.
21. FSI. *Forest Survery of India: Uttarakhand, Technical Report*; Land Use Statistics, Ministry of Agriculture: Dehradun, India, 2017.
22. Sharma, V.P.; Annepu, S.K.; Gautam, Y.; Singh, M.; Kamal, S. Status of Mushroom Production in India. *Mushroom Res.* **2017**, *26*, 111–120.
23. Bhatt, R.P.; Singh, U.; Uniyal, P. Healing Mushrooms of Uttarakhand Himalaya, India. *Curr. Res. Environ. Appl. Mycol.* **2018**, *8*, 1–23. [CrossRef]
24. Gezahegn, W.W. Study of Heavy Metals Accumulation in Leafy Vegetables of Ethiopia. *IOSR J. Environ. Sci. Toxicol. Food Technol.* **2017**, *11*, 57–68. [CrossRef]
25. Elbagermi, M.A.; Edwards, H.G.M.; Alajtal, A.I. Monitoring of Heavy Metal Content in Fruits and Vegetables Collected from Production and Market Sites in the Misurata Area of Libya. *ISRN Anal. Chem.* **2012**, *2012*, 827645. [CrossRef]
26. Abrham, F. Analysis of Heavy Metal Concentration in Some Vegetables Using Atomic Absorption Spectroscopy. *Pollution* **2021**, *7*, 205–216.
27. Abdi, H.; Williams, L.J. Principal Component Analysis. *Wiley Interdiscip. Rev. Comput. Stat.* **2010**, *2*, 433–459. [CrossRef]
28. Gowen, A.A.; O'Donnell, C.P.; Taghizadeh, M.; Cullen, P.J.; Frias, J.M.; Downey, G. Hyperspectral Imaging Combined with Principal Component Analysis for Bruise Damage Detection on White Mushrooms (*Agaricus bisporus*). *J. Chemom.* **2008**, *22*, 259–267. [CrossRef]
29. Granato, D.; Santos, J.S.; Escher, G.B.; Ferreira, B.L.; Maggio, R.M. Use of Principal Component Analysis (PCA) and Hierarchical Cluster Analysis (HCA) for Multivariate Association between Bioactive Compounds and Functional Properties in Foods: A Critical Perspective. *Trends Food Sci. Technol.* **2018**, *72*, 83–90. [CrossRef]
30. McCarthy, R.V.; McCarthy, M.M.; Ceccucci, W. Finding Associations in Data Through Cluster Analysis. In *Applying Predictive Analytics*; Springer: Cham, Switzerland, 2022; pp. 199–232.
31. Xu, H.; Zhang, C. Development and Applications of GIS-Based Spatial Analysis in Environmental Geochemistry in the Big Data Era. *Environ. Geochem. Health* **2022**, 1–12. [CrossRef] [PubMed]
32. Lion, G.N.; Olowoyo, J.O. Population Health Risk Due to Dietary Intake of Toxic Heavy Metals from Spinacia Oleracea Harvested from Soils Collected in and around Tshwane. *S. Afr. J. Bot.* **2013**, *88*, 178–182. [CrossRef]
33. Tucaković, I.; Barišić, D.; Grahek, Ž.; Kasap, A.; Širić, I. 137 Cs in Mushrooms from Croatia Sampled 15–30 Years after Chernobyl. *J. Environ.* **2018**, *181*, 147–151. [CrossRef]
34. Mahabadi, M. Assessment of Heavy Metals Contamination and the Risk of Target Hazard Quotient in Some Vegetables in Isfahan. *Pollution* **2020**, *6*, 69–78. [CrossRef]
35. Zhong, T.; Xue, D.; Zhao, L.; Zhang, X. Concentration of Heavy Metals in Vegetables and Potential Health Risk Assessment in China. *Environ. Geochem. Health* **2018**, *40*, 313–322. [CrossRef] [PubMed]
36. Gerwien, F.; Skrahina, V.; Kasper, L.; Hube, B.; Brunke, S. Metals in Fungal Virulence. *FEMS Microbiol. Rev.* **2018**, *42*, fux050. [CrossRef]
37. Singh, R.; Kaur, N.; Shri, R.; Singh, A.P.; Dhingra, G.S. Proximate Composition and Element Contents of Selected Species of Ganoderma with Reference to Dietary Intakes. *Environ. Monit. Assess.* **2020**, *192*, 270. [CrossRef]
38. Gaur, T.; Rao, P.B.; Kushwaha, K.P.S. Nutritional and Anti-Nutritional Components of Some Selected Edible Mushroom Species. *Indian J. Nat. Prod. Resour.* **2016**, *7*, 155–161.
39. Cai, L.M.; Wang, Q.S.; Wen, H.H.; Luo, J.; Wang, S. Heavy Metals in Agricultural Soils from a Typical Township in Guangdong Province, China: Occurrences and Spatial Distribution. *Ecotoxicol. Environ. Saf.* **2019**, *168*, 184–191. [CrossRef] [PubMed]
40. Buruleanu, L.C.; Radulescu, C.; Antonia Georgescu, A.; Dulama, I.D.; Nicolescu, C.M.; Lucian Olteanu, R.; Stanescu, S.G. Chemometric Assessment of the Interactions Between the Metal Contents, Antioxidant Activity, Total Phenolics, and Flavonoids in Mushrooms. *Anal. Lett.* **2019**, *52*, 1195–1214. [CrossRef]
41. Širić, I.; Kasap, A.; Bedeković, D.; Falandysz, J. Lead, Cadmium and Mercury Contents and Bioaccumulation Potential of Wild Edible Saprophytic and Ectomycorrhizal Mushrooms, Croatia. *J. Environ. Sci. Health Part B* **2017**, *52*, 156–165. [CrossRef]

42. Bosiacki, M.; Siwulski, M.; Sobieralski, K.; Krzebietke, S. The Content of Selected Heavy Metals in Fruiting Bodies of *Agaricus bisporus* (Lange) Imbach. Wild Growing in Poland. *J. Elem.* **2018**, *23*, 875–886. [CrossRef]
43. Tandon, S.; Joshi, R.K.; Sand, N.K. Monitoring of Organochlorine Pesticides in Water Collected from District Haridwar and Dehradun of Uttarakhand. *Pestic. Res. J.* **2010**, *22*, 19–22.
44. Igbiri, S.; Udowelle, N.A.; Ekhator, O.C.; Asomugha, R.N.; Igweze, Z.N.; Orisakwe, O.E. Edible Mushrooms from Niger Delta, Nigeria with Heavy Metal Levels of Public Health Concern: A Human Health Risk Assessment. *Recent Pat. Food Nutr. Agric.* **2018**, *9*, 31–41. [CrossRef]
45. Karataş, A. Effects of Different Agro-Industrial Waste as Substrates on Proximate Composition, Metals, and Mineral Contents of Oyster Mushroom (Pleurotus Ostreatus). *Int. J. Food Sci. Technol.* **2022**, *57*, 1429–1439. [CrossRef]

Journal of *Fungi*

MDPI

Article

Sustainable Use of Sewage Sludge as a Casing Material for Button Mushroom (*Agaricus bisporus*) Cultivation: Experimental and Prediction Modeling Studies for Uptake of Metal Elements

Pankaj Kumar [1], Vinod Kumar [1,*], Bashir Adelodun [2,3], Dalibor Bedeković [4], Ivica Kos [4], Ivan Širić [4], Saad A. M. Alamri [5], Sulaiman A. Alrumman [5], Ebrahem M. Eid [5,6], Sami Abou Fayssal [7,8], Madhumita Goala [9], Ashish Kumar Arya [10], Archana Bachheti [10], Kyung Sook Choi [3], Fidelis Odedishemi Ajibade [11,12] and Luis F. O. Silva [13]

1 Agro-Ecology and Pollution Research Laboratory, Department of Zoology and Environmental Science, Gurukula Kangri (Deemed to Be University), Haridwar 249404, India; rs.pankajkumar@gkv.ac.in
2 Department of Agricultural and Biosystems Engineering, University of Ilorin, Ilorin PMB 1515, Nigeria; adelodun.b@unilorin.edu.ng
3 Department of Agricultural Civil Engineering, Kyungpook National University, Daegu 41566, Korea; ks.choi@knu.ac.kr
4 Faculty of Agriculture, University of Zagreb, Svetosimunska 25, 10000 Zagreb, Croatia; dbedekovic@agr.hr (D.B.); ikos@agr.hr (I.K.); isiric@agr.hr (I.Š.)
5 Biology Department, College of Science, King Khalid University, Abha 61321, Saudi Arabia; saralomari@kku.edu.sa (S.A.M.A.); salrumman@kku.edu.sa (S.A.A.); ebrahem.eid@sci.kfs.edu.eg (E.M.E.)
6 Botany Department, Faculty of Science, Kafrelsheikh University, Kafr El-Sheikh 33516, Egypt
7 Department of Agronomy, Faculty of Agronomy, University of Forestry, 10 Kliment Ohridski Blvd, 1797 Sofia, Bulgaria; sami.aboufaycal@st.ul.edu.lb
8 Department of Plant Production, Faculty of Agriculture, Lebanese University, Beirut 1302, Lebanon
9 Nehru College, Pailapool, Affiliated Assam University, Silchar 788098, India; madhumitagoalap@gmail.com
10 Department of Environment Science, Graphic Era (Deemed to be University), Dehradun 248002, India; ashishtyagi.gkv@gmail.com (A.K.A.); bachheti.archana@gmail.com (A.B.)
11 Department of Civil and Environmental Engineering, Federal University of Technology, Akure PMB 704, Nigeria; foajibade@futa.edu.ng
12 Key Laboratory of Urban Pollutant Conversion, Institute of Urban Environment, Chinese Academy of Sciences, Xiamen 361021, China
13 Department of Civil and Environmental Engineering, Universidad de la Costa, Calle 58 #55-66, Barranquilla 080002, Colombia; lsilva8@cuc.edu.co
* Correspondence: drvksorwal@gkv.ac.in

Citation: Kumar, P.; Kumar, V.; Adelodun, B.; Bedeković, D.; Kos, I.; Širić, I.; Alamri, S.A.M.; Alrumman, S.A.; Eid, E.M.; Abou Fayssal, S.; et al. Sustainable Use of Sewage Sludge as a Casing Material for Button Mushroom (*Agaricus bisporus*) Cultivation: Experimental and Prediction Modeling Studies for Uptake of Metal Elements. *J. Fungi* **2022**, *8*, 112. https://doi.org/10.3390/jof8020112

Academic Editor: Samantha C. Karunarathna

Received: 30 December 2021
Accepted: 24 January 2022
Published: 25 January 2022

Abstract: The present study focused on the use of sewage sludge (SS) as a casing material amendment and the potential uptake of metal elements by the cultivated white button (*Agaricus bisporus*: MS-39) mushroom. Laboratory experiments were performed under controlled environmental conditions to grow *A. bisporus* on the composted wheat straw substrate for 50 days. Different treatments (0, 50, 100, 150, and 200 g/kg) of casing material were prepared by mixing garden and dried SS and applied on the mushroom substrate after proper sterilization. The results revealed that SS application was significant ($p < 0.05$) in accelerating mushroom yield with a biological efficiency of 65.02% for the mixing rate of 200 g/kg. Moreover, the maximum bioaccumulation of selected metal elements (Cu, Cr, Cd, Fe, Mn, and Zn) was observed using the same treatment. Additionally, the multiple regression models constructed for the uptake prediction of metal elements showed an acceptable coefficient of determination ($R^2 > 0.9900$), high model efficiency (ME > 0.98), and low root mean square error (RMSE < 0.410) values, respectively. The findings of this study represent sustainable use of SS for the formulation of mushroom casing material contributing toward synergistic agro-economy generation and waste management.

Keywords: bioaccumulation; mushroom cultivation; prediction models; regression analysis; waste management

1. Introduction

Over the last decades, the generation of sewage sludge (SS) has increased dramatically in India because of the increasing population and unplanned urbanization. Historically, SS disposal was considered a secondary issue that was facing humanity [1], but it has been causing environmental pollution with dangerous impacts on all forms of life [2,3]. Sewage sludge disposal should be correctly managed especially in countries with high population density [4,5]. In India, numerous sewage wastewater (SW) treatment plants (around 234 facilities) were built along the banks of major rivers [6]. However, it still could not cope with the daily treatment capacity of SW, estimated at 62 thousand of million liters per day in urban areas. Recent reports state that the majority of the SS is improperly disposed of, with only a small volume being recycled in a sustainable manner [7].

The nutritional quality of commercially produced mushrooms is strongly determined by the chemical composition of the growing substrate [8–15] and the type of supplements added to them [12,14]. *Agaricus bisporus*, commonly known as the white button mushroom, is cultivated on a fermented substrate composed of a mixture of lignocellulosic materials, mainly horse and chicken manure [8,16,17]. During the production process of this mushroom, the casing is essential to stimulate fruit body formation. Casing act as an important layer for mushroom as it assists the carpophores in several ways, such as: providing a physical support system, acting as a water reservoir, assisting in nutrient exchange, osmotic pressure regulation, creating an aerated environment for efficient metabolism, stimulating fructification, facilitating gas interchanges, and providing zones for ion exchange [7]. The casing material used for this aim should essentially provide suitable physicochemical and biological properties that stimulate the process of pinhead formation [18,19]. A suitable casing soil is characterized by a water holding capacity above 45%, porosity above 50%, pH around 7.8, and electrical conductivity below 1.6 mhos [20]. Loam soil (surface layer organic soil) is the most widely used casing material on a commercial scale in India [21]. Nevertheless, various agro-industrial residues have been used as alternatives or as additives to loam soil, such as composted vine shoots [22], spent mushroom substrate and coconut fiber pith [23], waste paper [24,25], and many others [26–28].

The use of SS in *Agaricus bisporus* production was first reported by Block [29], who recognized this waste as a good source of nitrogen, vitamins, and trace metals for microorganisms involved in the compost, without affecting (when found in low concentrations) the composition of mushroom carpophores in terms of trace elements [30]. Later, Kāposts et al. [31] noted that mushroom composts amended with sludge had a 20–50% richer but admissible metal elements profile. More recently, Yamauchi et al. [32] found that the use of SS in a compost mixture with cow manure increased the yield and free amino acids content of mushrooms. Commonly, SS is known as valuable organic manure rich in recycled nitrogen and phosphorus [33,34] and is characterized by a relatively low C:N ratio, ranging between 6 and 9 [35,36], high bulk density, and low air-filled porosity [37]. However, reports on the use of SS at the level of casing during *A. bisporus* production processes are almost lacking. Similarly, the impact of this waste type on the metal element profile of mushrooms was mostly reported on those collected from the wild. The carpophores had high concentrations of Cd, Pb, and Hg adjacent to landfill sites of SS [38] and normal As, Cu, Se, and Zn contents in forests subjected to sludge application [30].

Therefore, the current study investigates the use of SS in mixtures with loam soil as a casing material for the production of *A. bisporus* and reports its effect on the yield. Furthermore, the potential risk associated with metal element uptake by the harvested mushrooms was investigated using bioaccumulation and multiple linear regression approaches.

2. Materials and Methods

2.1. Experimental Materials

A commercial strain of the white button (*Agaricus bisporus* MS39) mushroom was procured from Welkin Overseas Pvt. Ltd. located in Haridwar, Uttarakhand, India (29°46′43.7″ N 77°47′26.8″ E). This strain was developed by the Directorate of Mushroom

Research (DMR), Solan, Himachal Pradesh, India (30°55′20.0″ N 77°06′08.9″ E). DMR is a government organization dedicated to the development of mushroom research technologies in India. The healthy spawn of *A. bisporus* grown on sterile wheat grains was stored at 4 °C until its application. The mushroom substrate was prepared using wheat straw (WS) (chopped into 3–5 cm pieces) following a short-term method of composting [39], as recommended by DMR [40,41]. Composting was performed using a traditional pile system facilitated with aeration through percolated plastic pipes. A nitrogen, phosphorus, and potassium (N:P:K) fertilizer treatment of 33:10:25 was used to enrich the substrate with essential nutrients after moistening in tap water. The compost was pasteurized under controlled environmental conditions. For the casing material, the soil was collected from the departmental garden of Gurukula Kangri (Deemed to be University), Haridwar, India (29°55′10.3″ N 78°07′08.3″ E). Sewage sludge was collected in sterile polyethylene bags (5 kg capacity) from Sewage Treatment Plant (27 MLD capacity STP) located at Jagjeetpur, Haridwar, India (29°54′01.2″ N 78°08′14.9″ E).

2.2. Experimental Design

For the cultivation of *A. bisporus*, a total of 5 kg compost was filled in a sterile polyethylene bag (20 cm diameter). Thirty grams of healthy spawn was divided between four layers as the bag was being filled with compost to a depth of 15 cm (Figure 1). Afterward, the bags were placed in an environmentally controlled cultivation room maintained at 25 °C, relative humidity of 80%, a light intensity of 650 lux, and a CO_2 concentration of 10,000 ppm that lasted for 20 days. Sewage sludge was sieved through a mesh size of 1300 μm (No. 16, Elysian IN) and sterilized in an autoclave at 121 °C for 20 min (KI- 174, Khera Instrument, IN). There were five different treatment mixtures of loam soil (LS) and SS, i.e., 0 mg/kg (control as absolute LS), 50 g/kg (950 g LS + 50 g SS), 100 g/kg (900 g LS + 100 g SS), 150 g/kg (850 g LS + 150 g SS), and 200 g/kg (800 g LS + 200 g SS). The formulated casing material was treated using a 10% formalin solution followed by air drying for 3 days. The casing material was applied 3 cm thick to the top of the compost in each bag and watered accordingly to maintain 60% moisture content. The pinhead formation was initiated by adjusting the air temperature (12 °C), humidity (75%), light intensity (650 lux), and CO_2 (900 ppm). Finally, the fruiting bodies of *A. bisporus* were harvested in three subsequent flushes and expressed as total yield. Fruiting bodies were carefully twisted and harvested (after 5–7 days) once they reached a maximum diameter of 3–4 cm.

2.3. Analytical Methods

The mushroom substrate and formulated casing materials were analyzed for various physicochemical and metal element parameters following standard methods [8,42]. The substrate samples were analyzed for the selected physiochemical and metal element parameters immediately before the spawn inoculation, while the casing material samples were analyzed before applying it on the substrate layer after achieving a complete spawn run (20th day). Whole fruiting bodies of mushrooms were obtained from each respective treatment and air-dried for further elemental analysis. A 1:10 suspension of dried mushroom substrate or casing material in distilled H_2O was prepared to analyze pH and electrical conductivity using a calibrated ESICO 1615 m (ESICO, IN). The organic carbon was estimated by following Walkley and Black method [43]. Total nitrogen was estimated by following the digestion and distillation method using Agilent Cary 60 UV-visible spectroscopy. The contents of metal elements viz., Cu, Cr, Cd, Fe, Mn, and Zn were estimated using an Inductively coupled plasma atomic emission spectroscopy (ICP-OES: 7300 DV, Perkin Elmer, Massachusetts, USA) instrument. For the estimation of metal elements, the samples were digested using a di-acid mixture (1:6) of nitric–perchloric acids and then analyzed using an ICP-OES instrument using standardized metal solutions (0, 2.5, 5, 10, 25, 50, and 100 mg/L). Similarly, the harvested mushroom bodies were oven-dried at 105 °C and then converted into fine powder. Mushroom biomass was also analyzed for selected metal elements following the same method of the substrate and expressed as mg/kg dry weight basis (dwt.).

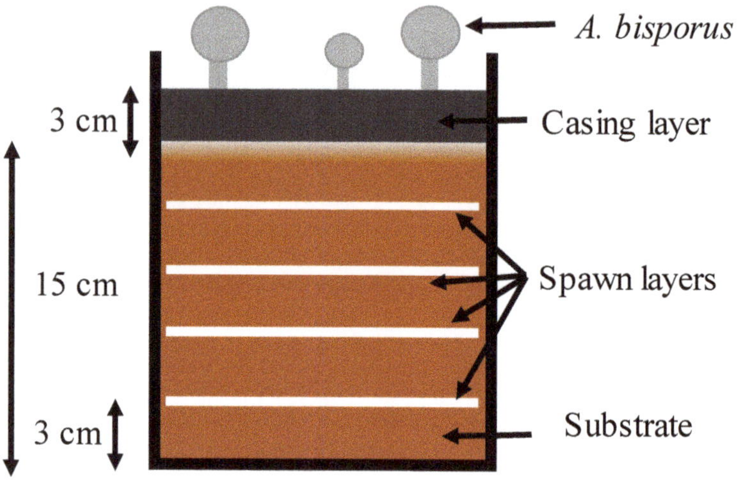

Figure 1. Illustration of each experimental unit for *A. bisporus* cultivation (the bag was 20 cm in diameter).

2.4. Bioaccumulation Factor and Prediction Modeling of Metal Elements Uptake

The uptake and localization of metal elements by living organisms can be enumerated by analyzing their concentration in the organism's tissues and substrate/media in which they grow. The bioaccumulation factor (BAF) is a widely used tool that determines the uptake efficiency of organisms. In this study, the BAF of *A. bisporus* for metal elements uptake from its SS amended substrate was calculated using the following formula (Equation (1)):

$$BAF = C_M / C_S \tag{1}$$

where C_M and C_S are the concentrations of the metal element in mushroom fruiting bodies and substrate (wheat straw + SS), respectively.

Multiple linear regression (MLR) modeling is one of the most implemented methods of determining the uptake and accumulation of these elements in different parts of crops. The generalized linear regression method was used to assess the impact of selected physico-chemical characteristics of mushroom casing material/substrate and their potential uptake by the fruiting bodies of *A. bisporus*. For this, the following MLR model was constructed to predict the uptake of Cd, Cu, Cr, Fe, Mn, and Zn elements by *A. bisporus* (Equation (2)):

$$y = \beta + (\beta \times pH_C) + (\beta \times OC_C) + (\beta \times ME_{S+C}) \tag{2}$$

where y indicates the concentration (mg/kg dwt.) of net metal element absorbed by *A. bisporus* during its vegetative growth, β indicates the regression coefficients, pH_C and OC_C are the pH and organic carbon (mg/kg) of the casing material, while ME_{S+C} indicates the sum of metal elements present in the mushroom substrate and casing material, respectively. Besides this, the constructed models were validated using two different validation tools, viz., model efficiency (ME) and root mean square error (RMSE), as suggested by Eid et al. [44].

2.5. Software and Statistics

All experiments were performed in triplicate (n = 3 replicates × 5 treatment groups = 15 experimental units) under controlled environmental conditions. The cultivation bags were placed on a vertical plastic rack in a completely randomized design. One sample from each bag was collected, i.e., three for each treatment, and the mean parameter value was used. For the statistical and modeling analyses, Microsoft Excel (2019, Microsoft Corp., Redmond, WA, USA) and OriginPro (2020b; OriginLab Corp., Northampton, MA, USA) software packages were used. The level of statistical significance for analysis of variance (ANOVA) and correlation studies was Prob. (p) < F values of 0.05.

3. Results and Discussion

3.1. Characteristics of the Mushroom Substrate and Casing Materials

The results showed that the WS-based substrate comprised various nutrient parameters, including OC, N, and metal elements. Moreover, the two different casing materials, i.e., loam soil and SS, showed significant variation ($p < 0.05$) in terms of physicochemical and metal element parameters such as pH, EC, OC, TKN, C/N ratio, Cu, Cr, Cd, Fe, Mn, and Zn (Table 1). Moreover, the metal elements and other major nutrient parameters (OC and TKN) were significantly ($p < 0.05$) increased in the casing material after SS mixing when compared to the control treatment (Table 2). The casing material had Fe as the most dominating metal element, followed by Mn, Zn, Cu, Cr, and Cd, respectively. The coefficient of variance (CV) values showed lesser uncertainty in parameter measurement (<2%). Thus, the mixing rate of SS greatly contributed to increasing the levels of various nutrients and metal elements in the casing material. Previously, Eid and coworkers [44] determined the physiochemical and metal element properties of SS and reported that SS had significant levels of various nutrients, which were further useful for increasing the soil fertility under *Abelmoschus esculentus* crop cultivation.

3.2. Effects of Sewage Sludge as Casing Material on Yield and Productivity of A. bisporus

The amendment of SS showed increased production of *A. bisporus* as compared to control treatment with no SS application (Table 3). In a total of three harvested flushes, the minimum yield (165.53 g) of *A. bisporus* was observed using control treatment, whereas the maximum yield (195.07 g) was achieved using a 200 g/kg SS amendment rate. As the SS amendment rate increased, the yield also increased significantly ($p < 0.05$). Biological efficiency corresponds to net mushroom yield from the unit dry weight of the substrate; it was the highest (65.02%) for the 200 g/kg SS rate. Mushroom yield and productivity are largely impacted by the type of substrate used for their cultivation along with the suitable environmental and irrigation conditions. Certain additives such as organic manure, chemical fertilizers, nano-fertilizers, and biofertilizers are widely used to increase production [45]. OC and TKN are main nutrient parameters that should be at optimum levels. The enzyme system of fungi helps in the breakdown of carbon-based compounds and utilizes them as their primary energy source. The polysaccharides and the polymer lignin compounds from dead organic wastes breakdowns into simpler carbon molecules, which are further taken by mycelia and help in mushroom growth. Thus, a significant amount of OC is utilized by mushrooms during their vegetative growth phases. Similarly, the optimum level of N is essential to be supplied to obtain an efficient yield and productivity of mushrooms [8]. Apart from the nutrient availability in the substrate, the growth of mushrooms also depends on the casing material on which they form a primordium and start forming fruiting bodies. Therefore, certain physiochemical and nutrient interactions take place between the mycelial network and the casing material [46,47]. The SS composition provided a better structure to the thicker mycelia necessary for the production of *A. bisporus* and did not inhibit its production.

Sewage sludge is a well-established soil amendment for various crops (*Abelmoschus esculentus*, *Brassica oleracea*, etc.) [8,48,49]. However, to date, no study is available on the use of SS as a casing material for mushroom cultivation [50]. Different types of casing materials (clay, sand, loam, Fargo silty, chalk, forest soils, vermiculite Rockwood, peat wool, sawdust, coir pith, wood charcoal, paper sludge waste) have shown promising impacts on mushroom growth and productivity [51]. Khakimov et al. [52] assessed the impact of sierozem, bio-humus, decomposed manure, sawdust, the different proportion of chalk on

A. bisporus production and found that a combination of sierozem + decomposed manure + bio-humus + chalk gave the highest mushroom yield due to the presence of optimum nutrient levels.

Table 1. Characteristics of the wheat straw-based mushroom substrate, loam soil, and sewage sludge used for *A. bisporus* cultivation.

Parameter	WS-Substrate		Loam Soil (LS)		Sewage Sludge (SS)	
	Value	CV (%)	Value	CV (%)	Value	CV (%)
pH	7.06 ± 0.03	0.37	7.31 ± 0.01	0.16	6.10 ± 0.01	0.16
Electrical conductivity (dS/m)	6.13 ± 0.04	0.60	5.31 ± 0.04	0.75	6.92 ± 0.02	0.30
Organic carbon (g/kg)	494.86 ± 2.86	0.57	10.10 ± 0.61	0.53	122.62 ± 0.42	1.00
TKN (g/kg)	17.26 ± 0.03	0.15	0.56 ± 0.04	7.14	6.44 ± 0.10	1.55
C/N ratio	28.67	-	18.03	-	18.63	-
Cu (mg/kg)	34.43 ± 0.57	1.67	5.07 ± 0.03	0.68	56.03 ± 0.07	0.12
Cr (mg/kg)	8.20 ± 0.12	1.54	2.75 ± 0.01	0.15	10.51 ± 0.03	0.26
Cd (mg/kg)	0.54 ± 0.01	0.56	0.15 ± 0.01	1.31	2.49 ± 0.01	0.24
Fe (mg/kg)	3091.92 ± 0.38	0.01	16.68 ± 0.09	0.52	51.87 ± 0.17	0.32
Mn (mg/kg)	310.73 ± 0.89	0.28	5.80 ± 0.01	0.14	12.06 ± 0.05	0.44
Zn (mg/kg)	385.89 ± 0.80	0.20	3.05 ± 0.01	0.34	117.98 ± 0.45	0.38

Values are mean ± SD of three analyses; CV: coefficient of variance (%); All parameters of LS and SS were significantly different from WS-substrate at $p < 0.05$.

Table 2. Physicochemical and metal element characteristics of casing material prepared using loam soil and sewage sludge for *A. bisporus* cultivation.

Characteristics	Sewage Sludge Treatment				
	Control (0 g/kg)	50 g/kg	100 g/kg	150 g/kg	200 g/kg
pH	7.31 ± 0.01	7.19 ± 0.01 a	7.05 ± 0.03 a	6.77 ± 0.02 a	6.50 ± 0.02 a
Electrical conductivity (dS/m)	5.31 ± 0.04	5.61 ± 0.01 a	6.03 ± 0.03 a	6.29 ± 0.03 a	6.51 ± 0.03 a
Organic carbon (g/kg)	10.10 ± 0.61	12.41 ± 0.73 a	14.76 ± 0.58 a	20.15 ± 1.03 a	23.44 ± 0.82 a
TKN (g/kg)	0.56 ± 0.04	0.70 ± 0.05 a	0.84 ± 0.09 a	0.94 ± 0.06 a	1.12 ± 0.10 a
Cu (mg/kg)	5.07 ± 0.03	6.53 ± 1.16 b	10.53 ± 0.16 a	13.40 ± 0.15 a	16.24 ± 0.08 a
Cr (mg/kg)	2.75 ± 0.01	3.24 ± 0.03 a	3.75 ± 0.05 a	4.31 ± 0.06 a	4.80 ± 0.04 a
Cd (mg/kg)	0.15 ± 0.01	0.26 ± 0.01 a	0.40 ± 0.01 a	0.52 ± 0.01 a	0.64 ± 0.01 a
Fe (mg/kg)	16.68 ± 0.09	19.25 ± 0.06 a	21.61 ± 0.26 a	24.32 ± 0.18 a	27.16 ± 0.14 a
Mn (mg/kg)	5.80 ± 0.01	6.40 ± 0.01 a	7.05 ± 0.04 a	7.60 ± 0.02 a	8.21 ± 0.01 a
Zn (mg/kg)	3.05 ± 0.01	8.98 ± 0.03 a	14.79 ± 0.14 a	20.88 ± 0.17 a	26.61 ± 0.12 a

Values are mean ± SD of three analyses; a and b: Significantly different and not different from the control group at $p < 0.05$ and $p > 0.5$.

Table 3. Mushroom yield and biological efficiency of *A. bisporus* cultivated on sewage sludge amended casing material.

Parameter	Flush	Sewage Sludge Treatment				
		Control (0 g/kg)	50 g/kg	100 g/kg	150 g/kg	200 g/kg
Mushroom yield (g/kg fresh substrate)	1st	64.32 ± 0.26	72.55 ± 0.28 a	75.75 ± 0.55 a	80.79 ± 0.40 a	76.74 ± 2.24 a
	2nd	57.77 ± 0.35	60.95 ± 0.49 a	65.70 ± 0.22 a	64.83 ± 0.65 a	65.95 ± 1.69 a
	3rd	43.45 ± 0.33	45.99 ± 0.39 a	48.24 ± 0.47 a	46.81 ± 1.47 a	52.39 ± 0.70 a
	Average	55.18 ± 10.67	59.83 ± 13.32 b	63.23 ± 13.92 b	64.14 ± 17.00 b	65.02 ± 12.20 b
	Total	165.53	179.49	189.69	192.43	195.07
Biological efficiency (%)	-	55.18	59.83	63.23	64.14	65.02

Values are mean ± SD of three replicates. a and b: Significantly different and not different from control group at $p < 0.05$ and $p > 0.5$.

3.3. Effects of Sewage Sludge as Casing Material on Metal Elements Accumulation by *A. bisporus*

Sewage sludge as an additive of casing material increased the bioaccumulation of various metal elements into the fruiting bodies of *A. bisporus*. In a total of three harvests, the accumulation of selected metal elements (Cu, Cr, Cd, Fe, Mn, and Zn) was significantly ($p < 0.05$) higher in all SS treatments as compared to control treatment (Table 4). However, the first mushroom flush showed slightly higher metal accumulation as compared to the second and third flush, which might be due to the high accumulation efficiency of mushroom mycelia at the early development stage. Moreover, the BAF of selected metal elements was less than 1, which indicated that their levels in the edible fruiting bodies might not have exceeded the toxic limits (Figure 2). The decreasing order of metal elements accumulated by *A. bisporus* was ranged as Fe > Zn > Cu > Mn > Cr > Cd. However, Cd is primarily considered a hazardous metal that can cause severe health hazards in mushroom consumers. As it is a Zn substitute, Cd might become accumulated in mushroom bodies as a competing element if it is available in substrates/casing materials. However, the levels of Cd did not reach high levels as compared with control samples. Other metal elements (Cu, Cr, Fe, Mn, Zn) are essentially taken by fungal hyphae, which play an important role in certain physiological, biochemical, and virulence mechanisms in mushrooms [8].

Table 4. Contents of metal elements accumulated by *A. bisporus* fruiting bodies under different sewage sludge treatments.

Metal Element (mg/kg dwt.)	**Flush**	**Sewage Sludge Treatment**				
		Control (0 g/kg)	**50 g/kg**	**100 g/kg**	**150 g/kg**	**200 g/kg**
Cu	1st	21.52 ± 0.02	24.97 ± 0.01	27.06 ± 0.01	31.12 ± 0.01	33.46 ± 0.01
	2nd	20.28 ± 0.01	24.28 ± 0.06	26.58 ± 0.01	31.01 ± 0.01	33.40 ± 0.01
	3rd	18.97 ± 0.01	23.86 ± 0.01	26.23 ± 0.01	30.94 ± 0.01	33.31 ± 0.01
	Average	20.25 ± 1.27	24.37 ± 0.55 a	26.62 ± 0.41 a	31.02 ± 0.09 a	33.39 ± 0.07 a
Cr	1st	1.07 ± 0.05	1.41 ± 0.01	1.69 ± 0.01	1.93 ± 0.01	2.28 ± 0.01
	2nd	1.00 ± 0.01	1.38 ± 0.01	1.63 ± 0.01	1.91 ± 0.01	2.25 ± 0.01
	3rd	0.98 ± 0.01	1.27 ± 0.01	1.59 ± 0.01	1.87 ± 0.01	2.45 ± 0.01
	Average	1.02 ± 0.01	1.35 ± 0.07 a	1.64 ± 0.05 a	1.90 ± 0.03 a	2.33 ± 0.11 a
Cd	1st	0.13 ± 0.01	0.15 ± 0.01	0.17 ± 0.01	0.20 ± 0.01	0.21 ± 0.01
	2nd	0.12 ± 0.01	0.14 ± 0.01	0.16 ± 0.01	0.19 ± 0.01	0.21 ± 0.01
	3rd	0.12 ± 0.01	0.13 ± 0.01	0.16 ± 0.01	0.19 ± 0.01	0.21 ± 0.01
	Average	0.12 ± 0.01	0.14 ± 0.01 a	0.17 ± 0.01 a	0.19 ± 0.01 a	0.21 ± 0.01 a
Fe	1st	48.77 ± 0.01	52.42 ± 0.33	58.13 ± 0.01	63.71 ± 0.01	70.45 ± 0.04
	2nd	44.28 ± 0.01	51.57 ± 0.22	57.57 ± 0.09	62.50 ± 0.01	69.70 ± 0.02
	3rd	43.06 ± 0.01	50.17 ± 0.06	57.23 ± 0.06	62.13 ± 0.01	68.37 ± 0.01
	Average	45.37 ± 3.01	51.39 ± 1.13 a	57.64 ± 0.45 a	62.78 ± 0.82 a	69.51 ± 1.05 a
Mn	1st	3.53 ± 0.01	4.76 ± 0.04	5.58 ± 0.01	6.14 ± 0.01	7.53 ± 0.25
	2nd	3.40 ± 0.01	4.64 ± 0.01	5.59 ± 0.05	6.13 ± 0.01	7.34 ± 0.01
	3rd	3.28 ± 0.01	4.41 ± 0.17	5.46 ± 0.11	6.12 ± 0.01	7.30 ± 0.01
	Average	3.40 ± 0.12	4.60 ± 0.20 a	5.54 ± 0.07 a	6.13 ± 0.01 a	7.39 ± 0.12 a
Zn	1st	30.28 ± 0.08	32.52 ± 0.03	36.39 ± 0.01	41.81 ± 0.01	45.15 ± 0.01
	2nd	27.29 ± 0.05	32.21 ± 0.01	36.11 ± 0.01	41.00 ± 0.01	44.68 ± 0.01
	3rd	26.13 ± 0.01	31.47 ± 0.02	35.86 ± 0.01	40.82 ± 0.01	44.11 ± 0.01
	Average	27.90 ± 2.14	32.07 ± 0.54 a	36.12 ± 0.26 a	41.21 ± 0.52 a	44.65 ± 0.52 a

Values are mean ± SD of three replicates; a: Significantly different from the average value of control group at $p < 0.05$.

Despite the nutritional benefits, SS is also known to bear many potential pathogens. Therefore, its proper sterilization should never be omitted. Even if no living organisms are present, the SS may contain mobile genetic elements, such as plasmids and viral particles that strongly bind with charged particles and resist degradation for prolonged periods. Those genetic elements often carry genes for antibiotic resistance and can be passed on to microorganisms present in the soil that was used as a casing material or as a substrate mixture, thus increasing antibiotic resistance in microorganisms that can be regarded as a hazard for human health. Additionally, some microorganisms sporulate and can survive poorly executed sterilization. Therefore, it is highly recommended that the sewage sludge should be fully inactivated before using it as a casing material and monitor the occurrence of certain toxic metals (Cd, As, Pb).

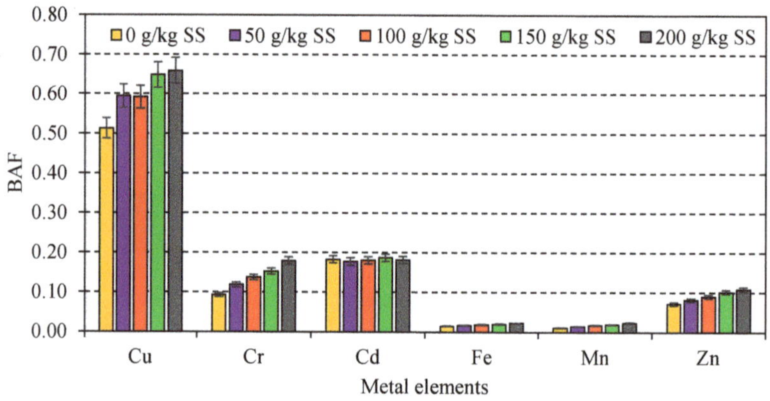

Figure 2. Bioaccumulation factor of metal elements uptake by *A. bisporus* cultivated on sewage sludge-based casing material.

Previous studies have shown that increased levels of different metal elements, including Fe in the casing soil, might be beneficial for higher mushroom production [53]. Similarly, Bhupathi et al. [54] emphasized the role of mineral properties of casing soil on the yield and productivity of milky mushrooms (*Calocyble indica*). They reported that the near-neutral (7.6) pH was more suitable for the rapid fruiting body formation and nutrient accumulation. Recently, the authors of [8] studied the impact of wastewater loading on the growth and productivity of *A. bisporus* mushroom cultivation on WH and sugar cane bagasse substrate and found that the metal elements in mushroom bodies increased as the wastewater concentration increased.

3.4. Prediction Models for Evaluating Metal Elements Accumulation by *A. bisporus*

Prediction modeling is an efficient method of determining the metal element enrichment by various organisms, including crop plants and fungi [44,48]. Multiple linear regression is one of the most efficient among various methods. The present investigation used a three-factor approach for determining the effective metal element uptake by *A. bisporus* grown on WH-substrate and SS amended casing material. The results revealed that the selected independent variables (initial pH, OC, and metal elements) were highly correlated with metal element uptake by *A. bisporus*. However, their affinity toward metal elements varied widely based on their uptake amount, as shown by the estimated regression coefficients (Table 5). More specifically, the pH of casing material was positively correlated for Cu, Fe, and Zn uptake while negatively for Cr, Cd, and Mn elements. On the other hand, the metal elements uptake was positively influenced by the OC of casing material for Cu, Cd, Fe, and Zn while negatively for Cr and Mn uptake, respectively. Similarly, initial metal element availability in both substrate and casing material also affected their uptake positively except for Cu, which showed a negative association. The metal uptake by mushroom mycelia depends on the adsorption and absorption of their specific ions, which are greatly impacted by both pH and OC, as discussed in previous studies [55,56]. Overall, the model equations given in Table 5 can be used for the precise prediction of metal element uptake by *A. bisporus* grown on SS amended casing materials. The models were characterized by a good coefficient of determination ($R^2 > 0.9900$) high model efficiency (ME > 0.98) and low root means square error (RMSE < 0.410), respectively. Therefore, the models showed acceptable prediction performance when compared to the actual observations as specified in Figure 3.

Table 5. Regression models and their validation results for metal elements uptake by *A. bisporus* cultivated on sewage sludge amended casing material.

Metal Element	Model Equation	R^2	ME	RMSE	F	p
Cu	$y = 20.217 + (0.067 \times pH_C) + (1.276 \times OC_C) - (0.360 \times ME_{S+C})$	0.9918	0.98	0.375	588.47	<0.001
Cr	$y = -2.814 - (0.335 \times pH_C) - (0.017 \times OC_C) + (0.593 \times ME_{S+C})$	0.9900	0.99	0.044	375.89	<0.001
Cd	$y = 0.079 - (0.003 \times pH_C) + (0.006 \times OC_C) + (0.004 \times ME_{S+C})$	0.9956	0.99	0.004	833.15	<0.001
Fe	$y = -6963.721 + (4.645 \times pH_C) + (0.313 \times OC_C) + (2.242 \times ME_{S+C})$	0.9976	0.98	0.410	1542.49	<0.001
Mn	$y = -906.571 - (0.096 \times pH_C) - (0.237 \times OC_C) + (2.885 \times ME_{S+C})$	0.9924	0.98	0.003	479.34	<0.001
Zn	$y = -63.503 + (0.058 \times pH_C) + (0.891 \times OC_C) + (0.208 \times ME_{S+C})$	0.9988	0.99	0.002	323.58	<0.001

pH_C: pH of casing material; OC_C: organic carbon of casing material; ME_{S+C}: metal element concentration in substrate + casing material; ME: model efficiency; RMSE: root mean square error.

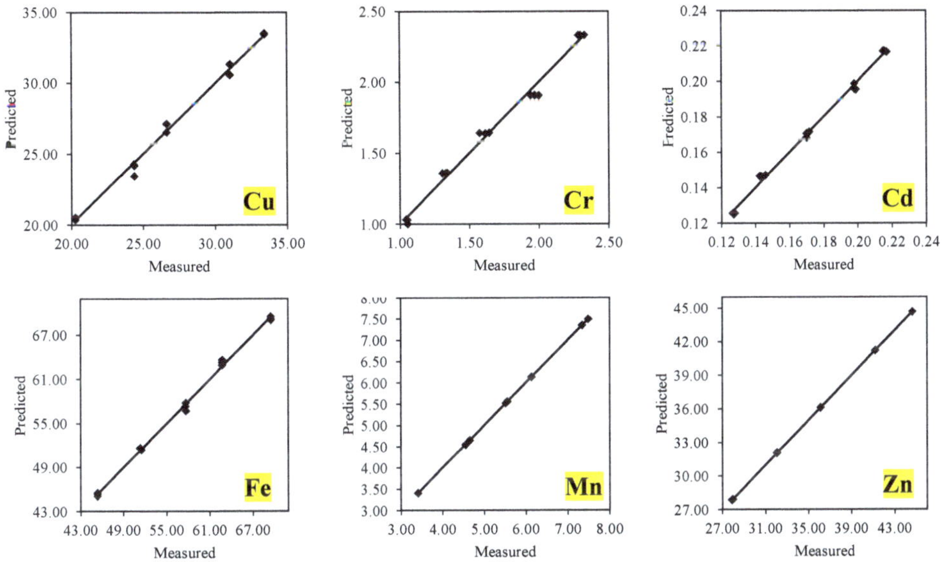

Figure 3. Measured vs. predicted metal elements uptake by *A. bisporus* cultivated on sewage sludge-based casing material.

There are not many studies conducted on the implementation of regression models for metal elements uptake by cultivated *A. bisporus* mushroom. Out of certain recent studies, Kumar et al. [8] constructed the regression and artificial neural network models for metal elements uptake by *A. bisporus* grown on the wheat straw substrate and reported that both methods were efficient in uptake prediction and health risk assessment. Similarly, Indolean et al. [57] also performed laboratory experiments for Cu uptake by *Lactarius piperatus* mushroom and evaluated the uptake process using artificial neural network models. They also reported that the absorption capacity of models was significantly correlated with the initial Cu dose, the weight of substrate used for *L. piperatus* cultivation.

4. Conclusions

This study concluded that the SS could be successfully used as a casing amendment for *A. bisporus* cultivation. The bioaccumulation of different metal elements (Cu, Cr, Cd, Fe, Mn, and Zn) increased with an increase in the SS rate from 0 to 200 g/kg. The ICP-

OES analysis revealed that the highest bioaccumulation (BAF $\leq$ 1)) of metal elements was observed using 200 g/kg SS application as casing material. The higher BAF may be alarming due to the high toxicity of some metal elements such as Cd. Thus, it needs to be properly monitored and regulated while using SS as a casing additive. Moreover, the multiple linear regression models of high efficiency were useful to monitor the metal element uptake by *A. bisporus*. This study demonstrated the sustainable use of SS for the formation of mushroom casing material, which might contribute to the emergence of a synergistic agro-economy creation and biosolid management practice. Further studies on large-scale production and biomonitoring of other metal elements uptake by *A. bisporus* are highly recommended.

Author Contributions: Conceptualization, P.K., V.K., M.G. and E.M.E.; Data curation, P.K., M.G. and B.A.; Formal analysis, P.K. and M.G.; Funding acquisition, P.K., E.M.E. and I.Š.; Investigation, P.K. and M.G.; Methodology, P.K. and V.K.; Project administration, V.K. and E.M.E.; Resources, V.K.; Software, P.K. and B.A.; Supervision, V.K. and E.M.E.; Validation, D.B., I.K., I.Š., S.A.M.A., S.A.A., A.K.A., A.B., K.S.C., F.O.A. and L.F.O.S.; Visualization, M.G.; Writing—original draft, P.K., M.G., B.A. and S.A.F.; Writing—Review and editing, D.B., I.K., I.Š., S.A.M.A., S.A.A., A.B., K.S.C., F.O.A. and L.F.O.S. All authors have read and agreed to the published version of the manuscript.

Funding: This research was funded by the Deanship of Scientific Research at King Khalid University (grant number RGP.1/7/43).

Institutional Review Board Statement: Not applicable.

Informed Consent Statement: Not applicable.

Data Availability Statement: Not applicable.

Acknowledgments: The authors are grateful to their host institutes for providing the necessary facilities to conduct this study. All individuals included in this section have consented to the acknowledgement.

Conflicts of Interest: The authors declare no conflict of interest.

References

1. Shaddel, S.; Bakhtiary-Davijany, H.; Kabbe, C.; Dadgar, F.; Østerhus, S.W. Sustainable sewage sludge management: From current practices to emerging nutrient recovery technologies. *Sustainability* **2019**, *11*, 3435. [CrossRef]
2. Wang, K.; Mao, H.; Li, X. Functional characteristics and influence factors of microbial community in sewage sludge composting with inorganic bulking agent. *Bioresour. Technol.* **2018**, *249*, 527–535. [CrossRef] [PubMed]
3. Zhang, Z.; Ju, R.; Zhou, H.; Chen, H. Migration characteristics of heavy metals during sludge pyrolysis. *Waste Manag.* **2021**, *120*, 25–32. [CrossRef] [PubMed]
4. Spinosa, L. *Wastewater Sludge: A Global Overview of the Current Status and Future Prospects*, 2nd ed.; IWA Publishing: London, UK, 2011; Volume 10, ISBN 9781843391425.
5. Aleisa, E.; Alsulaili, A.; Almuzaini, Y. Recirculating treated sewage sludge for agricultural use: Life cycle assessment for a circular economy. *Waste Manag.* **2021**, *135*, 79–89. [CrossRef]
6. CPCB Parivesh Sewage Pollution—News Letter. East Arjun Nagar: Delhi, India. 2005. Available online: https://cpcb.nic.in/openpdffile.php?id=TGF0ZXN0RmlsZS9MYXRlc3RfMTIzX1NVTU1BUllfQk9PS19GUy5wZGY (accessed on 29 December 2021).
7. Kacprzak, M.; Neczaj, E.; Fijałkowski, K.; Grobelak, A.; Grosser, A.; Worwag, M.; Rorat, A.; Brattebo, H.; Almås, Å.; Singh, B.R. Sewage sludge disposal strategies for sustainable development. *Environ. Res.* **2017**, *156*, 39–46. [CrossRef]
8. Kumar, P.; Kumar, V.; Goala, M.; Singh, J.; Kumar, P. Integrated use of treated dairy wastewater and agro-residue for *Agaricus bisporus* mushroom cultivation: Experimental and kinetics studies. *Biocatal. Agric. Biotechnol.* **2021**, *32*, 101940. [CrossRef]
9. Abou Fayssal, S.; Alsanad, M.A.; El Sebaaly, Z.; Ismail, A.I.H.; Sassine, Y.N. Valorization of Olive Pruning Residues through Bioconversion into Edible Mushroom *Pleurotus ostreatus* (Jacq. Ex Fr.) P. Kumm. (1871) of Improved Nutritional Value. *Scientifica* **2020**, *2020*, 3950357. [CrossRef]
10. Alsanad, M.A.; Sassine, Y.N.; El Sebaaly, Z.; Fayssal, S.A. Spent coffee grounds influence on *Pleurotus ostreatus* production, composition, fatty acid profile, and lignocellulose biodegradation capacity. *CYTA-J. Food* **2021**, *19*, 11–20. [CrossRef]
11. Naim, L.; Alsanad, M.A.; El Sebaaly, Z.; Shaban, N.; Fayssal, S.A.; Sassine, Y.N. Variation of *Pleurotus ostreatus* (Jacq. Ex Fr.) P. Kumm. (1871) performance subjected to differentdoses and timings of nano-urea. *Saudi. J. Biol. Sci.* **2020**, *27*, 1573–1579. [CrossRef]
12. Naim, L.; Alsanad, M.A.; Shaban, N.; El Sebaaly, Z.; Abou Fayssal, S.; Sassine, Y.N. Production and composition of *Pleurotus ostreatus* cultivated on Lithovit®-Amino25 supplemented spent substrate. *AMB Express* **2020**, *10*, 188. [CrossRef]

13. Naim, L.; El Sebaaly, Z.; Abou Fayssal, S.; Alsanad, M.A.; Najjar, R.; Yordanova, M.H.; Sassine, Y.N. The use of nitrogen-rich nano-supplements affects substrate temperature, delays the production cycle, and increases yield of *Pleurotus ostreatus*. *Acta Hortic.* **2021**, *1327*, 831–840. [CrossRef]
14. Abou Fayssal, S.; El Sebaaly, Z.; Alsanad, M.A.; Najjar, R.; Bohme, M.; Yordanova, M.H.; Sassine, Y.N. Combined effect of olive pruning residues and spent coffee grounds on *Pleurotus ostreatus* production, composition, and nutritional value. *PLoS ONE* **2021**, *16*, e0255794. [CrossRef] [PubMed]
15. Abou Fayssal, S.; Alsanad, M.A.; Yordanova, M.H.; El Sebaaly, Z.; Najjar, R.; Sassine, Y.N. Effect of olive pruning residues on substrate temperature and production of oyster mushroom (*Pleurotus ostreatus*). *Acta Hortic.* **2021**, *1327*, 245–252. [CrossRef]
16. Sebaaly, Z.; AlSanad, M.A.; Hayek, P.; Kfoury, L.; Shaban, N.; Sassine, Y.N. Using locally available chicken manure as a substitute to horse manure in compost formulas for growing *Agaricus bisporus* in Lebanon. *Acta Hortic.* **2020**, *1287*, 337–344. [CrossRef]
17. Sebaaly, Z.; AlSanad, M.A.; Seeman, H.; Rizkallah, J.; Shaban, N.; Sassine, Y.N. Investigating the potential use of composted grape marc in the production of *Agaricus bisporus*. *Acta Hortic.* **2020**, *1287*, 329–336. [CrossRef]
18. Kerketta, A.; Shukla, C.S.; Singh, H.K. Evaluation of different casing materials for growth and yield of button Musrhroom (*Agaricus bisporus* (L.) Sing.). *J. Pharmacogn. Phytochem.* **2019**, *8*, 207–209.
19. Sajyan, T.K.; Abou Fayssal, S.; Bejjani, R.; El Sebaaly, Z.; Sassine, Y.N. Casing and cropping. In *Mushrooms Agaricus bisporus*; Sassine, Y.N., Ed.; CABI: Oxfordshire, UK, 2021; pp. 240–278. ISBN 9781800620438.
20. Cunha Zied, D.; Pardo-Gonzalez, J.E.; Almeida Minhoni, M.T.; Arturo Pardo-Gimenez, A. A Reliable Quality Index for Mushroom Cultivation. *J. Agric. Sci.* **2011**, *3*, 50–61. [CrossRef]
21. Pardo-Giménez, A.; Zied, D.C.; Pardo-González, J.E. The use of peat as casing material in mushroom cultivation. In *Peat: Formation, Uses and Biological Effects*; Draguhn, C., Ciarimboli, N., Eds.; Nova Science Publishers Inc.: New York, NY, USA, 2012; pp. 81–100. ISBN 9781621001621.
22. De Juan, J.A.; Pardo, A.; Pardo, J.E. Effect of different casing materials on production and quality of the cultivated mushroom. *Adv. Hortic. Sci.* **2003**, *17*, 141–148. [CrossRef]
23. Pardo-Giménez, A.; Pardo-González, J.E. Evaluation of casing materials made from spent mushroom substrate and coconut fibre pith for use in production of *Agaricus bisporus* (Lange) Imbach. *Span. J. Agric. Res.* **2008**, *6*, 683–690. [CrossRef]
24. Sassine, Y.N.; Ghora, Y.; Kharrat, M.; Bohme, M.; Abdel-Mawgoud, A.M.R. Waste Paper as an Alternative for Casing Soil in Mushroom (*Agaricus bisporus*) Production. *J. Appl. Sci. Res.* **2005**, *1*, 277–284.
25. Sassine, Y.N.; Ghorra, Y.; Bohme, M.; Abdel-Mawgoud, A.M.R.; Kharrat, M. Effect of different mixtures with waste paper on the growth and production of mushroom (*Agaricus bisporus*). *Eur. J. Sci. Res.* **2006**, *13*, 281–293.
26. Mami, Y.; Peyvast, G.; Ghasemnezhad, M.; Ziaie, F. Supplementation at casing to improve yield and quality of white button mushroom. *Agric. Sci.* **2013**, *4*, 27–33. [CrossRef]
27. Beyer, D.M.; Beelman, R.B.; Kremser, J.J.; Rhodes, T.W. Casing additives and their influence on yield and fresh quality of *Agaricus bisporus*. In *Mushroom Biology and Mushroom Products*; Sánchez, J.E., Huerta, G., Montiel, E., Eds.; John Willey and Sons: Hoboken, NJ, USA, 2002; pp. 311–321.
28. Adibian, M.; Mami, Y. Mushroom supplement added to casing to improve postharvest quality of white button mushroom. *Eur. J. Hortic. Sci.* **2015**, *80*, 240–248. [CrossRef]
29. Block, S.S. Garbage Composting for Mushroom Production. *Appl. Microbiol.* **1965**, *13*, 5–9. [CrossRef] [PubMed]
30. Benbrahim, M.; Denaix, L.; Thomas, A.L.; Balet, J.; Carnus, J.M. Metal concentrations in edible mushrooms following municipal sludge application on forest land. *Environ. Pollut.* **2006**, *144*, 847–854. [CrossRef] [PubMed]
31. Kaposts, V.; Karins, Z.; Lazdins, A. Use of sewage sludge in forest cultivation. *Balt. For.* **2000**, *6*, 24–28.
32. Yamauchi, M.; Chishaki, N.; Yamada, M.; Katahira, T.; Harada, R.; Suemitsu, T.; Yagi, F.; Yamaguchi, T. Mass Production of Mushrooms using Sewage Sludge Compost and Utilization of Waste Beds for Crop Cultivation. *J. Jpn. Soc. Mater. Cycles Waste Manag.* **2020**, *31*, 88–97. [CrossRef]
33. Towers, W.; Home, P. Sewage sludge recycling to agricultural land: The environmental scientist's perspective. *J. Comm. Int. Water Environ. J.* **1997**, *11*, 126–132. [CrossRef]
34. Kominko, H.; Gorazda, K.; Wzorek, Z. Formulation and evaluation of organo mineral fertilizers based on sewage sludge optimized for maize and sunflower crops. *Waste Manag.* **2021**, *136*, 57–66. [CrossRef]
35. Cheng, J.; Ding, L.; Lin, R.; Yue, L.; Liu, J.; Zhou, J.; Cen, K. Fermentative biohydrogen and biomethane co-production from mixture of food waste and sewage sludge: Effects of physiochemical properties and mix ratios on fermentation performance. *Appl. Energy* **2016**, *184*, 1–8. [CrossRef]
36. Zhang, C.; Su, H.; Baeyens, J.; Tan, T. Reviewing the anaerobic digestion of food waste for biogas production. *Renew. Sustain. Energy Rev.* **2014**, *38*, 383–392. [CrossRef]
37. Malińska, K.; Zabochnicka-Świątek, M. Selection of bulking agents for composting of sewage sludge. *Environ. Prot. Eng.* **2013**, *39*, 91–103. [CrossRef]
38. Cibulka, J.; Sisak, L.; Pu Lkrab, K.; Miholova, D.; Szakova, J.; Fucikova, A.; Slamova, A.; Stehulova, I.; Arlakova, S. Cadmium, lead, mercury and caesium levels in wild mushrooms and forest berries from different localities of the Czech Republic. *Sci. Agric. Bohem.* **1996**, *27*, 113–129.
39. Sinden, J.W.; Mauser, E. The nature of the composting process and its relation to short composting. *Mushroom Sci.* **1953**, *2*, 123–130.

40. Sharma, V.P.; Kumar, S.; Sharma, S. *Technologies Developed by ICAR-Directorate of Mushroom Research for Commercial Use*; ICAR-DMR: Solan, India, 2020.
41. Abou Fayssal, S.; Hammoud, M.; El Sebaaly, Z.; Sassine, Y.N. Improvement of compost quality. In *Mushrooms Agaricus bisporus*; Sassine, Y.N., Ed.; CABI: Oxfordshire, UK, 2021; pp. 136–175. ISBN 9781800620438.
42. *AOAC Official Methods of Analysis*, 21st ed.; AOAC International: Rockville, MD, USA, 2019.
43. Walkley, A.; Black, I.A. An examination of the degtjareff method for determining soil organic matter, and a proposed modification of the chromic acid titration method. *Soil Sci.* **1934**, *37*, 29–38. [CrossRef]
44. Eid, E.M.; Shaltout, K.H.; Alamri, S.A.M.; Alrumman, S.A.; Sewelam, N. Prediction Models Founded on Soil Characteristics for the Estimated Uptake of Nine Metals by Okra Plant, *Abelmoschus esculentus* (L.) Moench., Cultivated in Agricultural Soils Modified with Varying Sewage Sludge Concentrations. *Sustainability* **2021**, *13*, 12356. [CrossRef]
45. Ghasemi, K.; Emadi, M.; Bagheri, A.; Mohammadi, M. Casing Material and Thickness Effects on the Yield and Nutrient Concentration of *Agaricus bisporus*. *Sarhad J. Agric.* **2020**, *36*, 734–1009. [CrossRef]
46. Masaphy, S.; Levanon, D.; Tchelet, R.; Henis, Y. Scanning Electron Microscope Studies of Interactions between *Agaricus bisporus* (Lang) Sing Hyphae and Bacteria in Casing Soil. *Appl. Environ. Microbiol.* **1987**, *53*, 1132–1137. [CrossRef]
47. Dias, E.S.; Zied, D.C.; Pardo-Gimenez, A. Revisiting the casing layer: Casing materials and management in Agaricus mushroom cultivation. *Ciênc. Agrotecnol.* **2021**, *45*, 1–11. [CrossRef]
48. Kumar, V.; Kumar, P.; Singh, J.; Kumar, P. Use of sugar mill wastewater for *Agaricus bisporus* cultivation: Prediction models for trace metal uptake and health risk assessment. *Environ. Sci. Pollut. Res.* **2021**, *28*, 26923–26934. [CrossRef]
49. Goala, M.; Yadav, K.K.; Alam, J.; Adelodun, B.; Choi, K.S.; Cabral-Pinto, M.M.S.; Hamid, A.A.; Alhoshan, M.; Ali, F.A.A.; Shukla, A.K. Phytoremediation of dairy wastewater using *Azolla pinnata*: Application of image processing technique for leaflet growth simulation. *J. Water Process Eng.* **2021**, *42*, 102152. [CrossRef]
50. Hu, Y.; Mortimer, P.E.; Hyde, K.D.; Kakumyan, P.; Thongklang, N. Mushroom cultivation for soil amendment and bioremediation. *Circ. Agric. Syst.* **2021**, *1*, 11. [CrossRef]
51. Pardo-Giménez, A.; Pardo González, J.E.; Zied, D.C. Casing Materials and Techniques in *Agaricus bisporus* Cultivation. In *Edible and Medicinal Mushrooms*; John Wiley & Sons, Ltd.: Chichester, UK, 2017; Volume 109, pp. 149–174. ISBN 9781119149415.
52. Khakimov, A.; Ismailov, A.; Murodullaeva, M. Casing soils selection for cultivation of common mushroom *Agaricus bisporus* (JE Lange) Imbach, 1946. *Bul. Sci. Pract.* **2018**, *7*, 64–71.
53. Rodrigues, M.M.; Viana, D.G.; Oliveira, F.C.; Alves, M.C.; Regitano, J.B. Sewage sludge as organic matrix in the manufacture of organomineral fertilizers: Physical forms, environmental risks, and nutrients recycling. *J. Clean. Prod.* **2021**, *313*, 127774. [CrossRef]
54. Maknali, F.; Kashi, A.; Salehi Mohammadi, R.; Khalighi, A. Enrichment of casing soil with Fe and soy-flour under Pseudomonas inoculation on yield and quality of button mushroom. *Rev. Agric. Neotrop.* **2021**, *8*, e5111. [CrossRef]
55. Bhupathi, P.; Subbaiah, K.A.; Balan, V. Influence of Soil Nutrition and Rhizobiome Bacteria on the Growth of *Calocybe Indica* (P & C). *Ann. Rom. Soc. Cell Biol.* **2021**, *25*, 4567–4575.
56. Eid, E.M.; Kumar, P.; Adelodun, B.; Choi, K.S.; Singh, J.; Kumari, S.; Kumar, V. Modeling of mineral elements uptake and localization in cabbage inflorescence (*Brassica oleracea var. capitata*) grown on sugar mill pressmud-amended soils. *Environ. Monit. Assess.* **2021**, *193*, 586. [CrossRef]
57. Indolean, C.; Măicăneanu, A.; Cristea, V.M. Prediction of Cu(II) biosorption performances on wild mushrooms *Lactarius piperatus* using Artificial Neural Networks (ANN) model. *Can. J. Chem. Eng.* **2017**, *95*, 615–622. [CrossRef]

MDPI
St. Alban-Anlage 66
4052 Basel
Switzerland
www.mdpi.com

Journal of Fungi Editorial Office
E-mail: jof@mdpi.com
www.mdpi.com/journal/jof

www.ingramcontent.com/pod-product-compliance
Lightning Source LLC
Chambersburg PA
CBHW061517210926
51850CB00008B/1953
9783036598130